PERGAMON INTERNATIONAL LIBRARY
of Science, Technology, Engineering and Social Studies

*The 1000-volume original paperback library in aid of education,
industrial training and the enjoyment of leisure*

Publisher: Robert Maxwell, M.C.

Atoms, Radiation, and Radiation Protection

NUCLEAR MEDICINE & RADIATION BIOLOGY

THE PERGAMON TEXTBOOK
INSPECTION COPY SERVICE

An inspection copy of any book published in the Pergamon International Library
will gladly be sent to academic staff without obligation for their consideration for
course adoption or recommendation. Copies may be retained for a period of 60 days
from receipt and returned if not suitable. When a particular title is adopted or
recommended for adoption for class use and the recommendation results in a sale
of 12 or more copies the inspection copy may be retained with our compliments.
The Publishers will be pleased to receive suggestions for revised editions and new
titles to be published in this important international Library.

Pergamon Titles of Related Interest

Bentel TREATMENT PLANNING AND DOSE CALCULATION IN
RADIATION ONCOLOGY, THIRD EDITION
Cember INTRODUCTION TO HEALTH PHYSICS, SECOND EDITION
Gollnick EXPERIMENTAL RADIOLOGICAL HEALTH PHYSICS
ICRP LIMITS FOR INTAKES OF RADIONUCLIDES BY WORKERS (#30)
Kase/Nelson CONCEPTS OF RADIATION DOSIMETRY
Kathren HEALTH PHYSICS: A BACKWARD GLANCE

Related Journals
(Free sample copies available upon request.)

ANNALS OF THE ICRP
HEALTH PHYSICS
INTERNATIONAL JOURNAL OF NUCLEAR MEDICINE AND BIOLOGY
INTERNATIONAL JOURNAL OF APPLIED RADIATION AND ISOTOPES
RADIATION PHYSICS AND CHEMISTRY

Atoms, Radiation, and Radiation Protection

James E. Turner

Oak Ridge National Laboratory

PERGAMON PRESS

New York Oxford Toronto Sydney Frankfurt

Pergamon Press Offices:

U.S.A.   Pergamon Press Inc., Maxwell House, Fairview Park,
                    Elmsford, New York 10523, U.S.A.

U.K.   Pergamon Press Ltd., Headington Hill Hall,
                    Oxford OX3 0BW, England

CANADA   Pergamon Press Canada Ltd., Suite 104, 150 Consumers Road,
                    Willowdale, Ontario M2J 1P9, Canada

AUSTRALIA   Pergamon Press (Aust.) Pty. Ltd., P.O. Box 544,
                    Potts Point, NSW 2011, Australia

FEDERAL REPUBLIC Pergamon Press GmbH, Hammerweg 6,
OF GERMANY   D-6242 Kronberg-Taunus, Federal Republic of Germany

BRAZIL   Pergamon Editora Ltda., Rua Eça de Queiros, 346,
                    CEP 04011, São Paulo, Brazil

JAPAN   Pergamon Press Ltd., 8th Floor, Matsuoka Central Building,
                    1-7-1 Nishishinjuku, Shinjuku, Tokyo 160, Japan

PEOPLE'S REPUBLIC Pergamon Press, Qianmen Hotel, Beijing,
OF CHINA   People's Republic of China

First printing 1986

Library of Congress Cataloging in Publication Data

Turner, J.E. (James Edward)
    Atoms, radiation, and radiation protection.

    1. Ionizing radiation. 2. Radiation--Safety
measures. I. Title. [DNLM: 1. Nuclear Energy.
2. Radiation Effects. 3. Radiation Protection.
4. Radiometry. WN 415 T948a]
QC795.T87   1986   539.2′028′9   85-28387
ISBN 0-08-031937-8
ISBN 0-08-031949-1 (pbk.)

Printed in Great Britain by A. Wheaton & Co. Ltd., Exeter

To Renate

CONTENTS

PREFACE

Atoms, Radiation, and Radiation Protection was written from material developed by the author over a number of years of teaching courses in the Oak Ridge Resident Graduate Program of the University of Tennessee's Evening School. The courses dealt with introductory health physics, preparation for the American Board of Health Physics certification examinations, and related specialized subjects such as microdosimetry and the application of Monte Carlo techniques to radiation protection. As the title of the book is meant to imply, atomic and nuclear physics and the interaction of ionizing radiation with matter are central themes. These subjects are presented in their own right at the level of basic physics, and the discussions are developed further into the areas of applied radiation protection. Radiation dosimetry, instrumentation, and external and internal radiation protection are extensively treated. The chemical and biological effects of radiation are not dealt with at length, but are presented in a summary chapter preceding the discussion of radiation-protection criteria and standards. Non-ionizing radiation is not included. The book is written at the senior or beginning graduate level as a text for a one-year course in a curriculum of physics, nuclear engineering, environmental engineering, or an allied discipline. A large number of examples are worked in the text. The traditional units of radiation dosimetry are used in much of the book; SI units are employed in discussing newer subjects, such as ICRP Publications 26 and 30. SI abbreviations are used throughout. With the inclusion of formulas, tables, and specific physical data, *Atoms, Radiation, and Radiation Protection* is also intended as a reference for professionals in radiation protection.

I have tried to include some important material not readily available in textbooks on radiation protection. For example, the description of the electronic structure of isolated atoms, fundamental to understanding so much of radiation physics, is further developed to explain the basic physics of "collective" electron behavior in semiconductors and their special properties as radiation detectors. In another area, under active research today, the details of charged-particle tracks in water are described from the time of the initial physical, energy-depositing events through the subsequent chemical changes that take place within a track. Such concepts are basic for relating the biological effects of radiation to particle-track structure.

I am indebted to my students and a number of colleagues and organizations, who contributed substantially to this book. Many individual contributions are acknowledged in figure captions. In addition, I would like to thank J. H. Corbin and W. N. Drewery of Martin Marietta Energy Systems, Inc.; Joseph D. Eddleman of Pulcir, Inc.; Michael D. Shepherd of Eberline; and Morgan Cox of Victoreen for their interest and help. I am especially indebted to my former teacher, Myron F. Fair, from whom I learned many of the things found in this book in countless discussions since we first met at Vanderbilt University in 1952.

It has been a pleasure to work with the professional staff of Pergamon Press, to whom I express my gratitude for their untiring patience and efforts throughout the production of this volume.

The last, but greatest, thanks are reserved for my wife, Renate, to whom this book is dedicated. She typed the entire manuscript and the correspondence that went with it. Her constant encouragement, support, and work made the book a reality.

James E. Turner
Oak Ridge, Tennessee
November 20, 1985

CHAPTER 1
ABOUT ATOMIC PHYSICS
AND RADIATION

1.1 CLASSICAL PHYSICS

As the 19th century drew to a close, man's physical understanding of the world appeared to rest on firm foundations. Newton's three laws accounted for the motion of objects as they exerted forces on one another, exchanging energy and momentum. The movements of the moon, planets, and other celestial bodies were explained by Newton's gravitation law. Classical mechanics was then over 200 years old, and experience showed that it worked well.

Early in the century Dalton's ideas revealed the atomic nature of matter, and in the 1860s Mendeleev proposed the periodic system of the chemical elements. The seemingly endless variety of matter in the world was reduced conceptually to the existence of a finite number of chemical elements, each consisting of identical smallest units, called atoms. Each element emitted and absorbed its own characteristic light, which could be analyzed in a spectrometer as a precise signature of the element.

Maxwell proposed a set of differential equations that explained known electric and magnetic phenomena and also predicted that an accelerated electric charge would radiate energy. In 1888 such radiated electromagnetic waves were generated and detected by Hertz, beautifully confirming Maxwell's theory.

In short, near the end of the 19th century man's insight into the nature of space, time, matter, and energy seemed to be fundamentally correct. While much exciting research in physics continued, the basic laws of the universe were considered to be known. Not many voices forecasted the complete upheaval in physics that would transform man's perception of the universe into something undreamed of as the 20th century began to unfold.

1.2 DISCOVERY OF X-RAYS

The totally unexpected discovery of X-rays by Roentgen on November 8, 1895 in Wuerzburg, Germany, is a convenient point to regard as marking the beginning of the story of ionizing radiation in modern physics. Roentgen was conducting experiments with a Crooke's tube — an evacuated glass enclosure, similar to a television picture tube, in which an electric current can be passed from one electrode to another through a high vacuum (Fig. 1.1). The current, which emanated from the cathode and was given the name cathode rays, was regarded by Crooke as a fourth state of matter. When the Crooke's tube was operated, fluorescence was excited in the residual gas inside and in the glass walls of the tube itself.

It was this fluorescence that Roentgen was studying when he made his discovery. By chance, he noticed in a darkened room that a small screen he was using fluoresced when the tube was turned on, even though it was some distance away. He soon recognized that

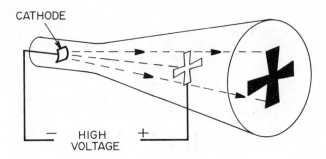

Figure 1.1. Schematic diagram of an early Crooke's, or cathode-ray, tube. A Maltese cross of mica placed in the path of the rays casts a shadow on the phosphorescent end of the tube.

he had discovered some previously unknown agent, to which he gave the name X-rays.† Within a few days of intense work, Roentgen had observed the basic properties of X-rays — their penetrating power in light materials such as paper and wood, their stronger absorption by aluminum and tin foil, and their differential absorption in equal thicknesses of glass that contained different amounts of lead. Figure 1.2 shows a picture that Roentgen made of a hand on December 22, 1895, contrasting the different degrees of absorption in soft tissue and bone. Roentgen demonstrated that, unlike cathode rays, X-rays are not deflected by a magnetic field. He also found that the rays affect photographic plates and cause a charged electroscope to lose its charge. Unexplained by Roentgen, the latter phenomenon is due to the ability of X-rays to ionize air molecules, leading to the neutralization of the electroscope's charge. He had discovered the first example of ionizing radiation.

1.3 SOME IMPORTANT DATES IN ATOMIC AND RADIATION PHYSICS

Events moved rapidly following Roentgen's communication of his discovery and subsequent findings to the Physical–Medical Society at Wuerzburg in December 1895. In France, Becquerel studied a number of fluorescent and phosphorescent materials to see whether they might give rise to Roentgen's radiation, but to no avail. Using photographic plates and examining salts of uranium among other substances, he found that a strong penetrating radiation was given off, independently of whether the salt phosphoresced. The source of the radiation was the uranium metal itself. The radiation was emitted spontaneously in apparently undiminishing intensity and, like X-rays, could also discharge an electroscope. Becquerel announced the discovery of radioactivity to the Academy of Sciences at Paris in February, 1896.

The following tabulation highlights some of the important historical markers in the development of modern atomic and radiation physics.

---

†That discovery favors the prepared mind is exemplified in the case of X-rays. Several persons who noticed the fading of photographic film in the vicinity of a Crooke's tube either considered it to be defective or sought other storage areas. An interesting account of the discovery and near-discoveries of X-rays as well as the early history of radiation is given in an article by R. L. Kathren, "Historical Development of Radiation Measurement and Protection," in *CRC Handbook of Radiation Measurement and Protection*, Section A, Vol. I, Allen B. Brodsky, Ed., CRC Press, Boca Raton, FL (1978).

1810 Dalton's atomic theory.
1859 Bunsen and Kirchhoff originate spectroscopy.
1869 Mendeleev's periodic system of the elements.
1873 Maxwell's theory of electromagnetic radiation.
1888 Hertz generates and detects electromagnetic waves.
1895 Lorentz theory of the electron.
1895 Roentgen discovers X-rays.
1896 Becquerel discovers radioactivity.
1897 Thomson measures charge-to-mass ratio of cathode rays (electrons).
1898 Curies isolate polonium and radium.
1899 Rutherford finds two kinds of radiation, which he names "alpha" and "beta," emitted from uranium.
1900 Villard discovers gamma rays, emitted from radium.
1900 Thomson's "plum pudding" model of the atom.
1900 Planck's constant, $h = 6.63 \times 10^{-27}$ erg sec.
1901 First Nobel prize in physics awarded to Roentgen.
1902 Curies obtain 0.1 g pure $RaCl_2$ from several tons of pitchblend.
1905 Einstein's special theory of relativity ($E = mc^2$).

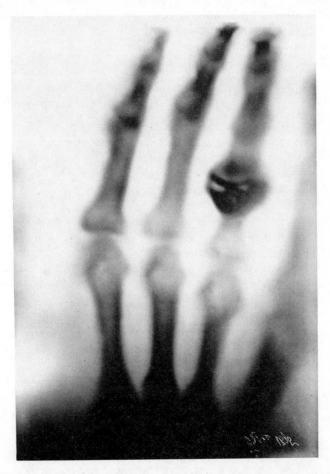

Figure 1.2. X-ray picture of the hand of Frau Roentgen made by Roentgen on December 22, 1895 and now on display at the Deutsches Museum. (Figure courtesy Deutsches Museum, Munich, West Germany)

1905 Einstein's explanation of photoelectric effect, introducing light quanta (photons of energy $E = h\nu$).

1909 Millikan's oil drop experiment, yielding precise value of electronic charge, $e = 4.80 \times 10^{-10}$ esu.

1910 Soddy establishes existence of isotopes.

1911 Rutherford discovers atomic nucleus.

1911 Wilson cloud chamber.

1912 von Laue demonstrates interference (wave nature) of X-rays.

1912 Hess discovers cosmic rays.

1913 Bohr's theory of the H atom.

1913 Coolidge X-ray tube.

1917 Rutherford produces first artificial nuclear transformation.

1922 Compton effect.

1924 de Broglie particle wavelength, $\lambda = h/$momentum.

1925 Uhlenbeck and Goudsmit ascribe electron with intrinsic spin $\hbar/2$.

1925 Pauli exclusion principle.

1925 Heisenberg's first paper on quantum mechanics.

1926 Schrodinger's wave mechanics.

1927 Heisenberg uncertainty principle.

1927 Mueller discovers that ionizing radiation produces genetic mutations.

1927 Birth of quantum electrodynamics, Dirac's paper on "The Quantum Theory of the Emission and Absorption of Radiation."

1928 Dirac's relativistic wave equation of the electron.

1930 Bethe quantum-mechanical stopping-power theory.

1930 Lawrence invents cyclotron.

1932 Anderson discovers positron.

1932 Chadwick discovers neutron.

1934 Joliot–Curie and Joliot produce artificial radioisotopes.

1935 Yukawa predicts the existence of mesons, responsible for short-range nuclear force.

1936 Gray's formalization of Bragg–Gray principle.

1937 Mesons found in cosmic radiation.

1938 Hahn and Strassmann observe nuclear fission.

1942 First nuclear chain reaction.

1945 First atomic bomb.

1948 Transistor invented by Shockley, Bardeen, and Brattain.

1952 Explosion of first fusion device (hydrogen bomb).

1956 Discovery of nonconservation of parity by Lee and Yang.

1958 Discovery of Van Allen radiation belts.

1960 First successful laser.

1972 First beam of 200 GeV protons at Fermilab.

1981 270 GeV proton–antiproton colliding-beam experiment at European Center for Nuclear Research (CERN); 540 GeV center-of-mass energy equivalent to laboratory energy of 150,000 GeV.

1983 Electron–positron colliders show continuing validity of radiation theory up to energy exchanges of 100 GeV and more.

1986 Plans for Superconducting Super Collider (SSC) under discussion by U.S. Department of Energy and Congress. SSC would have a circumference of 60 miles and accelerate protons to 40,000 GeV.

Figures 1.3 through 1.5 show how the complexity and size of particle accelerators have grown in 50 years. Lawrence's first cyclotron (1930) measured just 4 in. in diameter. With

Figure 1.3. E.O. Lawrence with his first cyclotron. (Photo by Watson Davis, Science Service; figure courtesy of American Institute of Physics Niels Bohr Library. Reprinted with permission from *Physics Today*, November 1981, p. 15. Copyright 1981 by the American Institute of Physics)

it he produced an 80 keV beam of protons. The Fermi National Accelerator Laboratory (Fermilab) is large enough to accommodate a herd of buffalo and other wildlife on its grounds. The LEP (large electron–positron) storage ring, under construction at the European Organization for Nuclear Research (CERN) on the border between Switzerland and France, near Geneva, will have a diameter of 8.6 km. The ring will allow electrons and positrons, circulating in opposite directions, to collide at very high energies in order to study elementary particles and forces in nature. The large size of the ring is needed to reduce the energy emitted as synchrotron radiation by the charged particles as they follow a curved path. The energy loss per turn is made up by an accelerator system in the ring structure. In Lawrence's day experimental equipment was usually put together by the individual researcher, possibly with the help of one or two associates. The huge machines of today require hundreds of technically trained persons to operate. Earlier radiation-protection practices were much less formalized than today, with little public involvement.

1.4 IMPORTANT DATES IN RADIATION PROTECTION

X-rays quickly came into widespread use following their discovery. Although it was not immediately clear that they presented a hazard to persons, mounting evidence during the first few years showed unequivocally that they did. Reports of skin burns, for example, became common. Recognition of the need for measures and devices to protect patients

Figure 1.4. Fermi National Accelerator Laboratory, Batavia, Illinois. Buffalo and other wildlife live on the 6800 acre site. The 1000 GeV proton synchrotron (Tevatron) will be operated in the late 1980s. (Figure courtesy of Fermi National Accelerator Laboratory. Reprinted with permission from *Physics Today*, November 1981, p. 23. Copyright 1981 by the American Institute of Physics)

Figure 1.5. Photograph showing location of underground LEP ring with its 27 km circumference. The SPS (super proton synchrotron) is comparable to Fermilab. Geneva airport is in foreground. [Figure courtesy of the European Organization for Nuclear Resarch (CERN)]

and operators from unnecessary exposure represented the beginning of radiation health protection.

Early criteria for limiting exposures both to X-rays and to radiation from radioactive sources were proposed by a number of individuals and groups. In time, organizations were founded to consider radiation problems and issue formal recommendations. Today, on the international scene, this role is fulfilled by the International Commission on Radiological Protection (ICRP) and, in the United States, by the National Council on Radiation Protection and Measurements (NCRP). The International Commission on Radiological Units and Measurements (ICRU) recommends radiation quantities and units, suitable measuring procedures, and numerical values for the physical data required. These organizations act as independent bodies comprised of specialists in a number of disciplines—physics, medicine, biology, dosimetry, instrumentation, administration, etc. They are not government affiliated and they have no legal authority to impose their recommendations. The NCRP today is a nonprofit corporation chartered by the United States Congress.

Some important dates and events in the history of radiation protection follow.

1895 Roentgen discovers ionizing radiation.
1900 American Roentgen Ray Society (ARRS) founded.
1915 British Roentgen Society adopts X-ray protection resolution; believed to be the first organized step toward radiation protection.

1920 ARRS establishes standing committee for radiation protection.

1921 British X-ray and Radium Protection Committee presents its first radiation protection rules.

1922 ARRS adopts British rules.

1922 American Registry of X-Ray Technicians founded.

1925 Mutscheller's "tolerance dose" for X-rays.

1925 First International Congress of Radiology, London, establishes ICRU.

1928 ICRP established under auspices of the Second International Congress of Radiology, Stockholm.

1928 ICRU adopts the roentgen as unit of exposure.

1929 Advisory Committee on X-Ray and Radium Protection (ACXRP) formed in United States (forerunner of NCRP).

1931 The roentgen adopted as unit of X-radiation.

1931 ACXRP publishes recommendations (*National Bureau of Standards Handbook 15*).

1934 ICRP recommends daily tolerance dose.

1941 ACXRP recommends first permissible body burden, for radium.

1942 Manhattan District begins to develop atomic bomb; beginning of health physics as a profession.

1946 U.S. Atomic Energy Commission created.

1946 NCRP formed as outgrowth of ACXRP.

1949 NCRP publishes recommendations and introduces risk/benefit concept.

1952 Radiation Research Society formed.

1953 ICRU introduces concept of absorbed dose.

1955 Health Physics Society formed.

1955 United Nations Scientific Committee on the Effects of Atomic Radiation (UNSCEAR) established.

1956 ICRP lowers basic permissible occupational dose to present level.

1957 NCRP introduces age proration for occupational doses and recommends nonoccupational exposure limits.

1957 U.S. Congressional Joint Committee on Atomic Energy begins series of hearings on radiation hazards, beginning with "The Nature of Radioactive Fallout and Its Effects on Man."

1958 United Nations Scientific Committee on the Effects of Atomic Radiation publishes study of exposure sources and biological hazards (first UNSCEAR Report).

1958 Society of Nuclear Medicine formed.

1959 ICRP recommends limitation of genetically significant dose to population.

1960 U.S. Congressional Joint Committee on Atomic Energy holds hearings on "Radiation Protection Criteria and Standards: Their Basis and Use."

1960 American Association of Physicists in Medicine formed.

1960 American Board of Health Physics begins certification of health physicists.

1964 International Radiation Protection Association (IRPA) formed.

1964 Act of Congress incorporates NCRP.

1974 Adoption by ICRP of Publication 23, "Report of Task Group on Reference Man."

1977 Adoption by ICRP of Publication 26, current limits based on risk concepts.

1978 Adoption by ICRP of Publication 30, "Limits for Intakes of Radionuclides by Workers."

1980 Publication of BEIR III Report, "The Effects on Populations of Exposure to Low Levels of Ionizing Radiation: 1980," by the National Academy of Sciences.

1.5 SOURCES OF RADIATION EXPOSURE

Everyone is continually exposed to ionizing radiation, which has always been present in the environment. Man has lived and evolved in a background of radiation from naturally occurring sources, such as radium in drinking water, ^{40}K in living tissues, the heavy radioactive elements in rock and stone, and cosmic rays. After 1895, man-made sources were added to those that occur naturally.

Table 1.1 gives a summary of various contributions to the average annual dose equivalent (Section 10.2) expressed in millirem per year (mrem/yr). As can be seen, an average person living in the United States is apt to receive somewhat more radiation from man-made sources than from natural background. Background can vary with location because of different amounts of radioactive minerals in soil, water, and rocks and because of the increase of cosmic radiation with altitude. Medical uses of radiation, particularly diagnostic X-rays, contribute the largest dose from man-made sources. The average levels of medical exposure also vary throughout the world because of differences in medical practice.

Table 1.1 Average Dose Equivalent to Persons in U.S. from Various Radiation Sources†

	Dose equivalent (mrem/yr)	Totals (mrem/yr)
Natural background		
cosmic rays	28	
terrestrial radiation	26	
internal sources	26	
	80	80
Medical		
medical diagnosis	77	
dental diagnosis	1.4	
radiopharmaceuticals	13.6	
	92.0	92.0
Atmospheric weapons tests		4.5
Nuclear industry		0.3
Consumer products		
building materials	3.5	
television receivers	0.5	
	4.0	4.0
Airline travel		0.5
Grand total (rounded off)		180

†Based on data obtained from *The Effects on Populations of Exposure to Low Levels of Ionizing Radiation: 1980*, National Academy of Sciences, Washington, D.C. (1980).

CHAPTER 2
ATOMIC STRUCTURE AND
ATOMIC RADIATION

2.1 THE ATOMIC NATURE OF MATTER (ca.1900)

The work of John Dalton in the early 19th century laid the foundation for modern analytic chemistry. Dalton formulated and interpreted the laws of definite, multiple, and equivalent proportions, based on the existence of identical atoms as the smallest indivisible unit of a chemical element. The law of definite proportions states that in every sample of a chemical compound, the proportion by mass or weight of the constituent elements is always the same. When two elements combine to form more than one compound, the law of multiple proportions says that the proportions by mass of the different elements are always in simple ratios to one another. When two elements react completely with a third, then the ratio of the masses of the two is the same, regardless of what the third element is, a fact expressed by the law of equivalent proportions. Dalton also assumed a rule of greatest simplicity—that elements forming only a single compound do so by means of a simple one-to-one combination of atoms. This rule does not always hold.

These ideas were supported by the work of Dalton's contemporary, Gay-Lussac, on the law of combining volumes of gases. This law states that the volumes of gases that enter into chemical combination with one another are in the ratio of simple whole numbers when all volumes are measured under the same conditions of pressure and temperature. Avogadro hypothesized that equal volumes of any gases at the same pressure and temperature contain the same number of molecules. Avogadro also suggested that the molecules of some gaseous elements could be comprised of two or more atoms of that element.

Today we recognize that a gram atomic weight of any element contains Avogadro's number, $N_0 = 6.023 \times 10^{23}$, of atoms. Furthermore, a gram molecular weight of any gas also contains N_0 molecules and occupies a volume of 22.4136 L (liters) at standard temperature and pressure (STP, 0°C (= 273 K on the absolute temperature scale) and 760 torr (1 torr = 1 mm Hg). The modern scale of atomic and molecular weights is set by stipulating that the gram atomic weight of the carbon isotope, ^{12}C, is exactly 12.000... g. A periodic chart, showing atomic numbers, atomic weights, densities, and other information about the chemical elements, is shown on the inside back cover of this book.

Example

How many grams of oxygen combine with 2.3 g of carbon in the reaction $C + O_2 \rightarrow CO_2$? How many molecules of CO_2 are thus formed? How many liters of CO_2 are formed at 20°C and 752 torr?

Solution

In the given reaction, 1 atom of carbon combines with one molecule (2 atoms) of oxygen. From the atomic weights given on the periodic chart on the inside back cover, it follows that 12.011 g of carbon reacts with $2 \times 15.9994 = 31.9988$ g of oxygen. Rounding off to three significant figures, letting y represent the number of grams of oxygen asked for, and taking simple proportions, we have $y = (2.3/12.0) \times 32.0 = 6.13$ g. The number N of molecules of CO_2 formed is equal to the number of atoms in 2.3 g of C, which is 2.3/12.0 times Avogadro's number: $N = (2.3/12.0) \times 6.02 \times 10^{23} = 1.15 \times 10^{23}$. Since Avogadro's number of molecules occupies 22.4 L at STP, the volume of CO_2 at STP is $(1.15 \times 10^{23}/6.02 \times 10^{23}) \times 22.4 = 4.28$ L. At the given higher temperature of 20°C = 293 K, the volume is larger by the ratio of the absolute temperatures, 293/273; the volume is also increased by the ratio of the pressures, 760/752. Therefore, the volume of CO_2 made from 2.3 g of C at 20°C and 752 torr is 4.28 (293/273) (760/752) = 4.64 L. This would also be the volume of oxygen consumed in the reaction under the same conditions of temperature and pressure, since 1 molecule of oxygen is used to form 1 molecule of carbon dioxide.

As mentioned in Chapter 1, mid-19th century scientists could analyze light to identify the elements present in its source. Light entering an optical spectrometer is collimated by a lens and slit system, through which it is then directed toward an analyzer (e.g., a diffraction grating or prism). The analyzer disperses the light, changing its direction by an amount that depends on its wavelength. White light, for example, is spread out into the familiar rainbow of colors. Light that is dispersed at various angles with respect to the incident direction can be seen with the eye, photographed, or recorded electronically. Light from a single chemical element is observed as a series of discrete line images of the entrance slit that emerge at various angles from the analyzer. The spectrometer can be calibrated so that the angles at which the lines occur give the wavelengths of the light that appears there. Each chemical element produces its own unique, characteristic series of lines which identify it. The series is referred to as the optical, or line, spectrum of the element, or simply as the spectrum. When a number of elements are present in a light source, their spectra appear superimposed in the spectrometer, and the individual elemental spectra can be sorted out. Elements absorb light of the same wavelengths they emit.

Figure 2.1 shows the lines in the visible and near-ultraviolet spectrum of atomic hydrogen. [The wavelength of visible light is between about 4000 Å (violet) and 7500 Å (red).] In 1885 Balmer published an empirical formula that gives these observed wavelengths, λ, in the hydrogen spectrum. His formula is equivalent to the following:

$$\frac{1}{\lambda} = R_H \left(\frac{1}{2^2} - \frac{1}{n^2} \right), \tag{2.1}$$

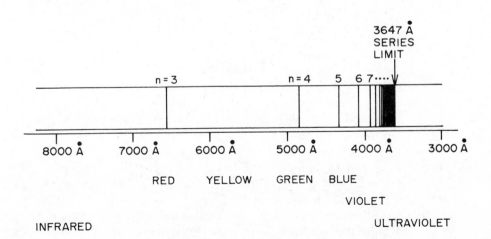

Figure 2.1. Balmer series of lines in the spectrum of atomic hydrogen.

where $R_H = 109,678$ cm^{-1} is called the Rydberg constant for hydrogen and $n = 3,4,5 \ldots$ represents any integer greater than 2. When $n = 3$, the formula gives $\lambda = 6562$ Å; when $n = 4$, $\lambda = 4861$ Å; and so on. The series of lines, which continue to get closer together as n increases, converges to the limit $\lambda = 3647$ Å in the ultraviolet as $n \to \infty$. Balmer correctly speculated that other series might exist for hydrogen, which could be described by replacing the 2^2 in Eq. (2.1) by the square of other integers. These other series, however, lie entirely in the ultraviolet or infrared portions of the electromagnetic spectrum. We shall see in Section 2.3 how the Balmer formula (2.1) was derived theoretically by Bohr in 1913.

As mentioned in Section 1.3, J. J. Thomson in 1897 measured the charge-to-mass ratio of cathode rays, which marked the experimental "discovery" of the electron as a particle of matter. The value he found for the ratio was about 1700 times that associated with the hydrogen atom in electrolysis. One concluded that the electron was less massive than the hydrogen atom by this factor. Thomson pictured atoms as containing a large number of the negatively charged electrons in a positively charged matrix filling the volume of the electrically neutral atom. When a gas was ionized by radiation, some electrons were knocked out of the atoms in the gas molecules, leaving behind positive ions of much greater mass. Thomson's concept of the structure of the atom is sometimes referred to as the "plum pudding" model.

2.2 THE RUTHERFORD NUCLEAR ATOM

The existence of alpha, beta, and gamma rays was known by 1900. With the discovery of these different kinds of radiation came their use as probes to study the structure of matter itself.

Rutherford and his students, Geiger and Marsden, investigated the penetration of alpha particles through matter. Because the range of these particles is small, an energetic source and thin layers of material were employed. In one set of experiments, 7.68 MeV collimated alpha particles from $^{214}_{84}$Po (RaC') were directed at a 6×10^{-5} cm thick gold foil. The relative number of particles leaving the foil at various angles with respect to the incident beam could be observed through a microscope on a scintillation screen. While most of the alpha particles passed through the foil with only slight deviation from their original direction, an occasional particle was scattered through a large angle, even backwards from the foil. About 1 in 8000 was deflected more than 90°. An enormously strong electric or magnetic field would be required to reverse the direction of the fast and relatively massive alpha particle. (In 1909 Rutherford conclusively established that alpha particles are doubly charged helium ions.) "It was about as credible as if you had fired a 15-in. shell at a piece of tissue paper and it came back and hit you," said Rutherford of this surprising discovery. He reasoned that the large-angle deflection of some alpha particles was evidence for the existence of a very small and massive nucleus, which was also the seat of the positive charge of an atom. The rare scattering of an alpha particle through a large angle could then be explained by the large repulsive force it experienced when it approached the tiny nucleus of a single atom almost head-on. Furthermore, the light electrons in an atom must move rapidly about the nucleus, filling the volume occupied by the atom. Indeed, atoms must be mostly empty space, allowing the majority of alpha particles to pass right through a foil with little or no scattering. Following these ideas, Rutherford calculated the distribution of scattering angles for the alpha particles and obtained quantitative agreement with the experimental data. In contrast to the plum pudding model, Rutherford's atom is sometimes called a planetary model, in analogy with the solar system.

Today we know that the radius of the nucleus of an atom of atomic mass number A is given approximately by the formula

$$R \cong 1.3A^{1/3} \times 10^{-13} \text{ cm.} \qquad (2.2)$$

The radius of the gold nucleus is $1.3(197)^{1/3} \times 10^{-13} = 7.56 \times 10^{-13}$ cm. The atomic radius of gold is 1.79×10^{-8} cm. The ratio of the two radii is $(7.56 \times 10^{-13}/1.79 \times 10^{-8}) = 4.22 \times 10^{-5}$. In physical extent, the massive nucleus is only a tiny speck at the center of the atom.

Nuclear size increases with atomic mass number A. Equation (2.2) indicates that the nuclear volume is proportional to A. The so-called strong, or nuclear, forces† that hold nucleons (protons and neutrons) together in the nucleus have short ranges ($\sim 10^{-13}$ cm). Nuclear forces saturate, i.e., a given nucleon interacts with only a few others. As a result, nuclear size is increased in proportion as more and more nucleons are merged to form heavier atoms. The size of all atoms, in contrast, is more or less the same. All electrons in an atom, no matter how many, are attracted to the nucleus and repelled by each other. Electric forces do not saturate—all pairs of charges interact with one another.

2.3 BOHR'S THEORY OF THE HYDROGEN ATOM

An object that does not move uniformly in a straight line is accelerated, and an accelerated charge emits electromagnetic radiation. In view of these laws of classical physics, it was not understood how Rutherford's planetary atom could be stable. Electrons orbiting about the nucleus should lose energy by radiation and spiral into the nucleus.

In 1913 Bohr put forward a bold new hypothesis, at variance with classical laws, to explain atomic structure. His theory gave correct predictions for the observed spectra of the H atom and single-electron atomic ions, such as He$^+$, but gave wrong answers for other systems, such as He and H$_2^+$. The discovery of quantum mechanics in 1925 and its subsequent development has led to the modern mathematical theory of atomic and molecular structure. Although it proved to be inadequate, Bohr's theory gives useful insight into the quantum nature of matter. We shall see that a number of properties of atoms and radiation can be understood from its basic concepts and their logical extensions.

Bohr assumed that an atomic electron moves without radiating only in certain discrete orbits about the nucleus. He further assumed that the transition of the electron from one orbit to another must be accompanied by the emission or absorption of a photon of light, the photon energy being equal to the orbital energy lost or gained by the electron. In principle, Bohr's ideas thus account for the existence of discrete optical spectra that characterize an atom and for the fact that an element emits and absorbs photons of the same wavelengths.

Bohr discovered that the proper electronic energy levels, yielding the observed spectra, were obtained by requiring that the angular momentum of the electron about the nucleus be an integral multiple of Planck's constant h divided by 2π ($\hbar = h/2\pi$). (Classically, any value of angular momentum is permissible.) For an electron of mass m moving uniformly with speed v in a circular orbit of radius r (Fig. 2.2), we thus write

$$mvr = n\hbar, \qquad (2.3)$$

---

†The four kinds of forces in nature are (1) gravitational, (2) electromagnetic, (3) strong (nuclear), and (4) weak (responsible for beta decay). The attractive nuclear force is strong enough to overcome the mutual Coulomb repulsion of protons in the nucleus (Section 3.1).

where n is a positive integer, called a quantum number ($n = 1,2,3\dots$). [Angular momentum, mvr, is defined in Appendix C; in cgs units, $\hbar = 1.0545 \times 10^{-27}$ erg sec (Appendix A)]. If the electron changes from an initial orbit in which its energy is E_i to a final orbit of lower energy E_f, then a photon of energy

$$h\nu = E_i - E_f \qquad (2.4)$$

is emitted, where ν is the frequency of the photon. ($E_f > E_i$ if a photon is absorbed.) Equations (2.3) and (2.4) are two succinct statements that embody Bohr's ideas quantitatively. We now use them to derive the properties of single-electron atomic systems.

When an object moves with constant speed v in a circle of radius r, it experiences an acceleration v^2/r, directed toward the center of the circle. By Newton's second law, the force on the object is mv^2/r, also directed toward the center. The force on the electron in Fig. 2.2 is supplied by the Coulomb attraction between the electronic and nuclear charges, $-e$ and $+Ze$. Therefore, we write for the equation of motion of the electron,

$$\frac{mv^2}{r} = \frac{Ze^2}{r^2}, \qquad (2.5)$$

giving

$$r = \frac{Ze^2}{mv^2}. \qquad (2.6)$$

Solving Eq. (2.3) for v and substituting into (2.6), we find for the radii r_n of the allowed orbits

$$r_n = \frac{n^2\hbar^2}{Ze^2m}. \qquad (2.7)$$

Substituting values of the constants from Appendix A, we obtain, in cgs units,

$$r_n = \frac{n^2(1.0545 \times 10^{-27})^2}{Z(4.80298 \times 10^{-10})^2(9.1091 \times 10^{-28})} = 5.29 \times 10^{-9}\frac{n^2}{Z} \text{ cm.} \qquad (2.8)$$

The innermost orbit ($n = 1$) in the hydrogen atom ($Z = 1$) thus has a radius of 5.29×10^{-9} cm $= 0.529$ Å, often referred to as the Bohr radius.

In similar fashion, eliminating r between Eqs. (2.3) and (2.6) yields the orbital velocities

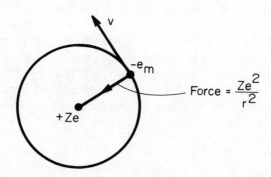

Figure 2.2. Schematic representation of electron (mass m, charge $-e$) in uniform circular motion (speed v, orbital radius r) about nucleus of charge $+Ze$.

$$v_n = \frac{Ze^2}{n\hbar} = 2.19 \times 10^8 \text{ cm/sec.} \tag{2.9}$$

The velocity of the electron in the first Bohr orbit ($n = 1$) of hydrogen ($Z = 1$) is 2.19×10^8 cm/sec. In terms of the speed of light c, the quantity $v_1/c = e^2/\hbar c \cong 1/137$ is called the fine-structure constant. Usually denoted by α, it determines the relativistic corrections to the Bohr energy levels, which give rise to a fine structure in the spectrum of hydrogen.

It follows that the kinetic and potential energies of the electron in the nth orbit are

$$\text{KE}_n = \frac{1}{2}mv_n^2 = \frac{Z^2e^4m}{2n^2\hbar^2} \tag{2.10}$$

and

$$\text{PE}_n = -\frac{Ze^2}{r_n} = -\frac{Z^2e^4m}{n^2\hbar^2}, \tag{2.11}$$

showing that the potential energy is twice as large in magnitude as the kinetic energy (virial theorem). The total energy of the electron in the nth orbit is therefore

$$E_n = \text{KE}_n + \text{PE}_n = -\frac{Z^2e^4m}{2n^2\hbar^2} = -\frac{13.6Z^2}{n^2} \text{ eV.} \tag{2.12}$$

[The energy unit, electron volt (eV), given in Appendix B, is defined as the energy acquired by an electron in moving freely through a potential difference of 1 V: 1 eV = 1.60×10^{-12} erg.] The lowest energy occurs when $n = 1$. For the H atom, this normal, or ground-state, energy is -13.6 eV; for He$^+$ ($Z = 2$) it is $-13.6 \times 4 = -54.4$ eV. The energy required to remove the electron from the ground state is called the ionization potential, which therefore is 13.6 eV for the H atom and 54.4 eV for the He$^+$ ion.

It remains to calculate the optical spectra for the single-electron systems based on Bohr's theory. Balmer's empirical formula (2.1) gives the wavelengths found in the visible spectrum of hydrogen. According to postulate (2.4), the energies of photons that can be emitted or absorbed are equal to the differences in the energy values given by Eq. (2.12). When the electron makes a transition from an initial orbit with quantum number n_i to a final orbit of lower energy with quantum number n_f (i.e., $n_i > n_f$), then from Eqs. (2.4) and (2.12) the energy of the emitted photon is

$$h\nu = \frac{hc}{\lambda} = \frac{Z^2e^4m}{2\hbar^2}\left(-\frac{1}{n_i^2} + \frac{1}{n_f^2}\right), \tag{2.13}$$

where λ is the wavelength of the photon and c is the speed of light. Substituting the numerical values† of the physical constants in cgs units, one finds from Eq. (2.13) that

$$\frac{1}{\lambda} = 109678Z^2\left(\frac{1}{n_f^2} - \frac{1}{n_i^2}\right) \text{ cm}^{-1}. \tag{2.14}$$

---

†For high accuracy, the *reduced* mass of the electron must be used. See last paragraph in this section.

When $Z = 1$, the constant in front of the parentheses is equal to the Rydberg constant R_H for hydrogen in Balmer's empirical formula (2.1). The integer 2 in the Balmer formula is interpretable from Bohr's theory as the quantum number of the orbit into which the electron falls when it emits the photon. Derivation of the Balmer formula and calculation of the Rydberg constant from the known values of e, m, h, and c provided undeniable evidence for the validity of Bohr's postulates for single-electron atomic systems, although the postulates were totally foreign to classical physics.

Figure 2.3 shows a diagram of the energy levels of the hydrogen atom, calculated from Eq. (2.12), together with vertical lines that indicate the electron transitions that result in the emission of photons with the wavelengths shown. There are infinitely many orbits in which the electron has negative energy (bound states of the H atom). The orbital energies

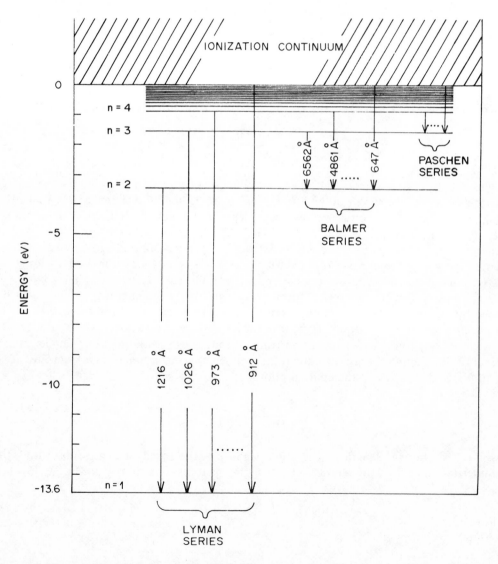

Figure 2.3. Energy levels of the hydrogen atom. Vertical lines represent transitions that electron can make between various levels with the associated emitted photon wavelengths shown.

get closer together near the ionization threshold, 13.6 eV above the ground state. When an H atom becomes ionized, the electron is not bound and can have any positive energy. In addition to the Balmer series, Bohr's theory predicts other series, each corresponding to a different final-orbit quantum number n_f and having an infinite number of lines. The set that results from transitions of electrons to the innermost orbit ($n_f = 1$, $n_i = 2,3,4,\ldots$) is called the Lyman series. The least energetic photon in this series has an energy

$$E = -13.6 \left(\frac{1}{2^2} - \frac{1}{1^2}\right) = -13.6\left(-\frac{3}{4}\right) = 10.2 \text{ eV,} \tag{2.15}$$

as follows from Eqs. (2.4) and (2.12) with $Z = 1$. Its wavelength is 1216 Å. As n_i increases, the Lyman lines get ever closer together, like those in the Balmer series, converging to the energy limit of 13.6 eV, the ionization potential of H. The photon wavelength at the Lyman series limit ($n_f = 1$, $n_i \to \infty$) is obtained from Eq. (2.14):

$$\frac{1}{\lambda} = 109678 \left(\frac{1}{1^2} - \frac{1}{\infty^2}\right) = 109678 \text{ cm}^{-1}, \tag{2.16}$$

or $\lambda = 912$ Å. The Lyman series lies entirely in the ultraviolet region of the electromagnetic spectrum. The series with $n_f \geq 3$ lies in the infrared. The shortest wavelength in the Paschen series ($n_f = 3$) is given by $1/\lambda = 109678/9$ cm^{-1}, or $\lambda = 8.21 \times 10^{-5}$ cm = 8210 Å.

Example
Calculate the wavelength of the third line in the Balmer series in Fig. 2.1. What is the photon energy in eV?

Solution
We use Eq. (2.14) with $Z = 1$, $n_f = 2$, and $n_i = 5$:

$$\frac{1}{\lambda} = 109678 \left(\frac{1}{4} - \frac{1}{25}\right) = 23032 \text{ cm}^{-1}. \tag{2.17}$$

Thus $\lambda = 4.34 \times 10^{-5}$ cm = 4340 Å. A photon of this wavelength has an energy (cgs units)

$$E = \frac{hc}{\lambda} = \frac{6.63 \times 10^{-27} \times 3 \times 10^{10}}{4.34 \times 10^{-5}} = 4.58 \times 10^{-12} \text{ erg,} \tag{2.18}$$

or 2.86 eV. Alternatively, we can obtain the photon energy from Eq. (2.12). The energy levels involved in the electronic transition are $13.6/4 = 3.40$ eV and $13.6/25 = 0.544$ eV; their difference is 2.86 eV.

Example.
What is the largest quantum number of a state of the Li^{2+} ion with an orbital radius less than 50 Å?

Solution
The radii of the orbits are described by Eq. (2.8) with $Z = 3$. Setting $r_n = 50$ Å $= 5 \times 10^{-7}$ cm and solving for n, we find that

$$n = \sqrt{\frac{r_n Z}{5.29 \times 10^{-9}}} = \sqrt{\frac{5 \times 10^{-7} \times 3}{5.29 \times 10^{-9}}} = 16.8. \tag{2.19}$$

A nonintegral quantum number is not defined in the Bohr theory. Equation (2.19) tells us, though, that $r_n > 50$ Å when $n = 17$ and $r_n < 50$ Å when $n = 16$. Therefore, $n = 16$ is the desired answer.

Example
Calculate the angular velocity of the electron in the ground state of He^+.

Solution
With quantum number n, the angular velocity ω_n in radians/sec is equal to $2\pi f_n$, where f_n is the frequency, or number of orbital revolutions of the electron about the nucleus per second. In general, $f_n = v_n/(2\pi r_n)$; and so $\omega_n = v_n/r_n$. With $n = 1$ and $Z = 2$, Eqs. (2.8) and (2.9) give $\omega_1 = v_1/r_1 = 1.66 \times 10^{17}$ sec^{-1}, where the dimensionless angular unit, radian, is understood.

In deriving Eq. (2.14) it was tacitly assumed that an electron of mass m orbits about a stationary nucleus. In reality, the electron and nucleus (mass M) orbit about their common center of mass. The energy levels are determined by the relative motion of the two, in which the effective mass is the reduced mass of the system (electron plus nucleus), given by

$$m_r = \frac{mM}{m + M}. \tag{2.20}$$

For the hydrogen atom, $M = 1836m$, and so the reduced mass $m_r = 1836m/1837 = 0.9995m$ is nearly the same as the electron mass. The heavier the nucleus, the closer the reduced mass is to the electron mass. Sometimes R_∞ is used to denote the Rydberg constant for a stationary (infinitely heavy) nucleus, with $m_r = m$. Then the Rydberg constant for ions with different nuclear masses M is given by

$$R_M = \frac{R_\infty}{1 + m/M}. \tag{2.21}$$

Problem 24 shows an example in which the reduced mass plays a significant role.

2.4 SEMICLASSICAL MECHANICS, 1913–1925

The success of Bohr's theory for hydrogen and single-electron ions showed that atoms are "quantized" systems. They radiate photons with the properties described earlier by Planck and by Einstein. At the same time, the failure of the Bohr theory to give correct predictions for other systems led investigators to search for a more fundamental expression of the quantum nature of atoms and radiation.

Between Bohr's 1913 theory and Heisenberg's 1925 discovery of quantum mechanics, methods of semiclassical mechanics were explored in physics. A general quantization procedure was sought that would incorporate Bohr's rules for single-electron systems and would also be applicable to many-electron atoms and to molecules. Basically, as we did above with Eq. (2.5), one used classical equations of motion to describe an atomic system and then superimposed a quantum condition, such as Eq. (2.3).

A principle of "adiabatic invariance" was used to determine which variables of a system should be quantized. It was recognized that quantum transitions occur as a result of *sudden* perturbations on an atomic system, not as a result of gradual changes. For example, the rapidly varying electric field of a passing photon can result in an electronic transition with photon absorption by a hydrogen atom. On the other hand, the electron is unlikely to make a transition if the atom is simply placed in an external electric field that is slowly increased in strength. The principle thus asserted that those variables in a system that were invariant under slow, "adiabatic" changes were the ones that should be quantized.

A generalization of Bohr's original quantum rule (2.3) was also worked out (by Wilson and Sommerfeld, independently) that could be applied to pairs of variables, such as momentum and position. So-called phase integrals were used to quantize systems after the classical laws of motion were applied.

These semiclassical procedures had some successes. For example, elliptical orbits were introduced into Bohr's picture and relativistic equations were used in place of the non-relativistic Eq. (2.5). The relativistic theory predicted a split in some atomic energy levels with the same quantum number, the magnitude of the energy difference depending on the fine-structure constant. The existence of the split gives rise to a fine structure in the spectrum of most elements in which some "lines" are observed under high resolution to be two closely separated lines. The well-known doublet in the sodium spectrum, consisting of two yellow lines at 5890 Å and 5896 Å, is due to transitions to two closely spaced energy levels, degenerate in nonrelativistic theory. In spite of its successes in some areas, semiclassical atomic theory did not work for many-electron atoms and for such simple systems as some diatomic molecules, for which it gave unambiguous but incorrect spectra.

As a guide for discovering quantum laws, Bohr in 1923 introduced his correspondence principle. This principle states that the predictions of quantum physics must be the same as those of classical physics in the limit of very large quantum number n. In addition, any relationships between states that are needed to obtain the classical results for large n also hold for all n. The diagram of energy levels in Fig. 2.3 illustrates the approach of a quantum system to a classical one when the quantum numbers become very large. Classically, the electron in a bound state has continuous, rather than discrete, values of the energy. As $n \to \infty$, the bound-state energies of the H atom get arbitrarily close together.

Advances toward the discovery of quantum mechanics were also being made along other lines. The classical Maxwellian wave theory of electromagnetic radiation seemed to be at odds with the existence of Einstein's corpuscular photons of light. How could light act like waves in some experiments and like particles in others? The diffraction and interference of X-rays was demonstrated in 1912 by von Laue, thus establishing their wave nature. The Braggs used X-ray diffraction from crystal layers of known separation to measure the wavelength of X-rays. In 1922, discovery of the Compton effect (Section 7.4)—the scattering of X-ray photons from atoms with a decrease in photon energy—demonstrated their nonwave, or corpuscular, nature in still another way. The experimental results were explained by assuming that a photon of energy E has a momentum $p = E/c = h\nu/c$, where ν is the photon frequency and c is the speed of light. In 1924, de Broglie proposed that the wave/particle dualism recognized for photons was a characteristic of all fundamental particles of nature. An electron, for example, hitherto regarded as a particle, also might have wave properties associated with it. The universal formula that links the property of wavelength, λ, with the particle property of momentum, p, is that which applies to photons: $p = h\nu/c = h/\lambda$. Therefore, de Broglie proposed that the wavelength associated with a particle be given by the relation

$$\lambda = \frac{h}{p} = \frac{h}{\gamma m v},\qquad (2.22)$$

where m and v are the rest mass and speed of the particle and γ is the relativistic factor defined in Appendix C.

Davisson and Germer in 1927 published the results of their experiments, which demonstrated that a beam of electrons incident on a single crystal of nickel is diffracted by the regularly spaced crystal layers of atoms. Just as the Braggs measured the wavelength of X-rays from crystal diffraction, Davisson and Germer measured the wavelength for electrons. They found excellent agreement with Eq. (2.22). The year before this experimental confirmation of the existence of electron waves, Schroedinger had extended de Broglie's ideas and developed his wave equation for the new quantum mechanics, as described in the next section.

A convenient formula can be used to obtain the wavelength λ of a nonrelativistic electron in terms of its kinetic energy T. (An electron is nonrelativistic as long as T is small compared with its rest energy, $mc^2 = 0.511$ MeV.) The nonrelativistic formula relating momentum p and kinetic energy T is $p = \sqrt{2mT}$ (Appendix C). It follows from Eq. (2.22) that

$$\lambda = \frac{h}{\sqrt{2mT}}. \tag{2.23}$$

It is often convenient to express the wavelength in Å and the energy in eV. Using these units for λ and T and cgs units elsewhere in Eq. (2.23), we write

$$\lambda_{\text{Å}} \times 10^{-8} = \frac{6.6256 \times 10^{-27}}{\sqrt{2 \times 9.1091 \times 10^{-28} \times T_{\text{eV}} \times 1.6021 \times 10^{-12}}}, \tag{2.24}$$

or

$$\lambda_{\text{Å}} = \frac{12.264}{\sqrt{T_{\text{eV}}}} = \frac{12.3}{\sqrt{T_{\text{eV}}}}. \tag{2.25}$$

The subscripts indicate the units for λ and T when this formula is used.

An analogous expression can be derived for photons. Since the photon energy is given by $E = h\nu$, the wavelength is $\lambda = c/\nu = ch/E$. Analogously to Eq. (2.25), we find

$$\lambda_{\text{Å}} = \frac{12398}{E_{\text{eV}}} = \frac{12400}{E_{\text{eV}}}. \tag{2.26}$$

Example
In some of their experiments, Davisson and Germer used electrons accelerated through a potential difference of 54 V. What is the de Broglie wavelength of these electrons?

Solution
The nonrelativistic formula (2.25) gives, with $T_{\text{eV}} = 54$ eV, $\lambda_{\text{Å}} = 12.3/\sqrt{54} = 1.67$ Å. Electron wavelengths much smaller than optical ones are readily obtainable. This is the basis for the vastly greater resolving power that electron microscopes have over optical microscopes (wavelengths $\gtrsim$ 4000 Å).

Example
Calculate the de Broglie wavelength of a 10 MeV electron.

Solution
We must treat the problem relativistically. We thus use Eq. (2.22) after determining γ and v. From Appendix C, with $T = 10$ MeV and $mc^2 = 0.511$ MeV, we have

$$10 = 0.511 \, (\gamma - 1), \tag{2.27}$$

giving $\gamma = 20.6$. We can compute v directly from γ. In this example, however, we know that v is very nearly equal to c. Using $v = c$ in Eq. (2.22), we therefore write

$$\lambda = \frac{h}{\gamma mc} = \frac{6.63 \times 10^{-27}}{20.6 \times 9.11 \times 10^{-28} \times 3 \times 10^{10}} = 1.18 \times 10^{-11} \text{ cm}. \tag{2.28}$$

Also, in the period just before the discovery of quantum mechanics, Pauli formulated his famous exclusion principle. This rule can be expressed by stating that no two electrons in an atom can have the same set of four quantum numbers. We shall discuss the Pauli principle in connection with the periodic system of the elements in Section 2.6.

2.5 QUANTUM MECHANICS

Quantum mechanics was discovered by Heisenberg in 1925 and, from a completely different point of view, independently by Schroedinger at about the same time. Heisenberg's formulation is termed matrix mechanics and Schroedinger's is called wave mechanics. Although they are entirely different in their mathematical formulation, Schroedinger showed in 1926 that the two systems are completely equivalent and lead to the same results. We shall discuss each in turn.

Heisenberg associated the failure of the Bohr theory with the fact that it was based on quantities that are not directly observable, like the classical position and speed of an electron in orbit about the nucleus. He proposed a system of mechanics based on observable quantities, notably the frequencies and intensities of the lines in the emission spectrum of atoms and molecules. He then represented dynamical variables (e.g., the position x of an electron) in terms of observables and worked out rules for representing x^2 when the representation for x is given. In so doing, Heisenberg found that certain pairs of variables did not commute multiplicatively (i.e., $xp \neq px$ when x and p represent position and momentum in the direction of x), a mathematical property of matrices recognized by others after Heisenberg's original formulation. Heisenberg's matrix mechanics was applied to various systems and gave results that agreed with those predicted by Bohr's theory where the latter was consistent with experiment. In other instances it gave new theoretical predictions that also agreed with observations. For example, Heisenberg explained the pattern of alternating strong and weak lines in the spectra of diatomic molecules, a problem in which Bohr's theory had failed. He showed that two forms of molecular hydrogen should exist, depending on the relative directions of the proton spin, and that the form with spins aligned (orthohydrogen) should be three times as abundant as the other with spins opposed (parahydrogen). This discovery was cited in the award of the 1932 Nobel Prize in physics to Heisenberg.

The concept of building an atomic theory on observables and its astounding successes led to a revolution in physics. In classical physics, objects move with certain endowed properties, such as a position and velocity at every moment in time. If one knows these two quantities at any one instant and also the total force that an object experiences, then its motion is determined completely for all times by Newton's second law. Such concepts are applied in celestial mechanics, where the positions of the planets can be computed backwards and forwards in time for centuries. The same determinism holds for the motion of familiar objects in everyday life. However, on the atomic scale, things are inherently different. In an experiment that would measure the position and velocity of an electron in orbit about a nucleus, the act of measurement itself introduces uncontrollable perturbations that prevent one's obtaining all the data precisely. For example, photons of very short wavelength would be required to localize the position of an electron within an atomic dimension. Such photons impart high momentum in scattering from an electron, thereby making simultaneous knowledge of the electron's position and momentum imprecise.

In 1927 Heisenberg enunciated the uncertainty principle, which sets the limits within which certain pairs of quantities can be known simultaneously. For momentum p and position x (in the direction of the momentum) the uncertainty relation states that

$$\Delta p \Delta x \geq \hbar. \tag{2.29}$$

Here Δp and Δx are the uncertainties (probable errors) in these quantities, determined simultaneously; the product of the two can never be smaller than $\hbar$, which it can approach under optimum conditions. Another pair of variables consists of energy E and time t, for which

$$\Delta E \Delta t \geq \hbar. \tag{2.30}$$

The energy of a system cannot be measured with arbitrary precision in a very short time interval. These uncertainties are not due to any shortcomings in our measuring ability. They are a result of the recognition that only observable quantities have an objective meaning in physics and that there are limits to making measurements on an atomic scale. The question of whether an electron "really" has a position and velocity simultaneously — whether or not we try to look — is metaphysical. Schools of philosophy differ on the fundamental nature of our universe and the role of the observer.

Whereas observation is immaterial to the future course of a system in classical physics, the observer's role is a basic feature of quantum mechanics, a formalism based on observables. The uncertainty relations rule out classical determinism for atomic systems. Knowledge obtained from one measurement, say, of an electron's orbital position, will not enable one to predict with certainty the result of a second measurement of the orbital position. Instead of this determinism, quantum mechanics enables one to predict only the *probabilities* of finding the electron in various positions when the second measurement is made. Operationally, such a probability distribution can be measured by performing an experiment a large number of times under identical conditions and compiling the frequency distribution of the different results. The laws of quantum mechanics are definite, but they are statistical, rather than deterministic, in nature. As an example of this distinction, consider a sample of 10^{16} atoms of a radioactive isotope that is decaying at an average rate of 10^4 atoms per second. We cannot predict which particular atoms will decay during any given second nor can we say exactly how many will do so. However, we can predict with assurance the probability of obtaining any given number of counts (e.g., 10,132) in a given second, as can be checked by observation.

Example
What is the minimum uncertainty in the momentum of an electron that is localized within a distance $\Delta x = 1 \text{Å}$, approximately the diameter of the hydrogen atom? How large can the kinetic energy of the electron be, consistent with this uncertainty?

Solution
The relation (2.29) requires that the uncertainty in the momentum be at least as large as the amount

$$\Delta p \cong \frac{\hbar}{\Delta x} = \frac{1.06 \times 10^{-27} \text{ erg sec}}{10^{-8} \text{ cm}} \sim 10^{-19} \frac{\text{g cm}}{\text{sec}}. \tag{2.31}$$

To estimate how large the kinetic energy of the electron can be, we note that its momentum p can be as large as Δp. With $p \sim \Delta p$, the kinetic energy T of the electron (mass m) is

$$T = \frac{p^2}{2m} \sim \frac{(10^{-19})^2}{2 \times 9.11 \times 10^{-28}} \sim 5 \times 10^{-12} \text{ erg}, \tag{2.32}$$

or about 3 eV. This analysis indicates that an electron confined within a distance $\Delta x \sim 1$ Å will have a kinetic energy in the eV range. In the case of the H atom we saw that the electron's kinetic energy is 13.6 eV. The uncertainty principle implies that electrons confined to even smaller regions become more energetic, as the next example illustrates.

Example

If an electron is localized to within the dimensions of an atomic nucleus, $\Delta x \sim 10^{-13}$ cm, estimate its kinetic energy.

Solution

In this case, we have $p \sim \Delta p \sim \hbar/\Delta x \sim 10^{-27}/10^{-13} \sim 10^{-14}$ g cm/sec. Comparison with the last example ($p \sim 10^{-19}$ g cm/sec) indicates that we must use the relativistic formula to find the kinetic energy from the momentum in this problem (Appendix C). The total energy E_T of the electron is given by

$$E_T^2 = p^2c^2 + m^2c^4 \tag{2.33}$$

$$= (10^{-14})^2 \, (3 \times 10^{10})^2 + (9 \times 10^{-28})^2 \, (3 \times 10^{10})^2 \tag{2.34}$$

$$= 9 \times 10^{-8} \text{ erg}^2, \tag{2.35}$$

from which we obtain $E_T = 200$ MeV. It follows that electrons in atomic nuclei would have energies of hundreds of MeV. Before the discovery of the neutron in 1932 it was speculated that a nucleus of atomic number Z and atomic mass number A consists of A protons and $A - Z$ electrons. The uncertainty principle argues against such a picture, since maximum beta-particle energies of only a few MeV are found. (In addition, some nuclear spins would be different from those observed if electrons existed in the nucleus). The beta particle is created in the nucleus at the time of decay.

We turn now to Schroedinger's wave mechanics. Schroedinger began with de Broglie's hypothesis [Eq. (2.22)] relating the momentum and wavelength of a particle. He introduced an associated oscillating quantity, ψ, and constructed a differential equation for it to satisfy. The coefficients in the equation involve the constants h and the mass and charge of the particle. Equations describing waves are well known in physics. Schroedinger's wave equation is a linear differential equation, second order in the spatial coordinates and first order in time. It is linear, so that the sum of two or more solutions is also a solution. Linearity thus permits the superposition of solutions to produce interference effects and the construction of wave packets to represent particles. The wave-function solution ψ must satisfy certain boundary conditions, which lead to discrete values called eigenvalues, for the energies of bound atomic states. Applied to the hydrogen atom, Schroedinger's wave equation gave exactly the Bohr energy levels. It also gave correct results for the other systems to which it was applied. Today it is widely used to calculate the properties of many-electron atomic and molecular systems, usually by numerical solution on a computer.

A rough idea can be given of how wave mechanics replaces Bohr's picture of the H atom. Instead of the concept of the electron moving in discrete orbits about the nucleus, we envision the electron as being represented by an oscillating cloud. Furthermore, the electron cloud oscillates in such a way that it sets up a standing wave about the nucleus. A familiar example of standing waves is provided by a vibrating string of length L stretched between two fixed points P_1 and P_2, as illustrated in Fig. 2.4. Standing waves are possible only with wavelengths λ given by

$$L = n \frac{\lambda}{2} \, , \quad n = 1,2,3,\dots. \tag{2.36}$$

This relation describes a discrete set of wavelengths λ. In an analogous way, an electron standing-wave cloud in the H atom can be envisioned by requiring that an integral number of wavelengths $n\lambda$ fit exactly into a circumferential distance $2\pi r$ about the nucleus: $2\pi r = n\lambda$. Using the de Broglie relation (2.22) then implies nonrelativistically ($\gamma = 1$)

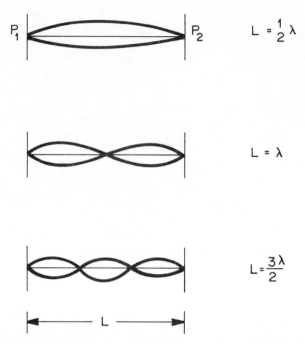

Figure 2.4. Examples of standing waves in string of length L stretched between two fixed points P_1 and P_2. Such waves exist only with discrete wavelengths given by $\lambda = 2L/n$, where $n = 1, 2, 3, \ldots$.

that $2\pi r = nh/mv$, or $mvr = n\hbar$. One thus arrives at Bohr's original quantization law, Eq. (2.3).

Schroedinger's wave equation is nonrelativistic, and he proposed a modification of it in 1926 to meet the relativistic requirement for symmetry between space and time. As mentioned above, the Schroedinger differential equation is second order in space and first order in time variables. His relativistic equation, which contained the second derivative with respect to time, led to a fine structure in the hydrogen spectrum, but the detailed results were wrong. Taking a novel approach, Dirac proposed a wave equation that was first order in both the space and time variables. In 1928 Dirac showed that the new equation automatically contained the property of intrinsic angular momentum for the electron, rotating about its own axis. The predicted value of the electron's spin angular momentum was $\hbar/2$, the value ascribed experimentally in 1925 by Uhlenbeck and Goudsmit to account for the structure of the spectra of the alkali metals. Furthermore, the fine structure for the hydrogen-atom spectrum came out correctly from the Dirac equation. Dirac's equation also implied the existence of a positive electron, found later by Anderson, who discovered the positron in cosmic radiation in 1932. In 1927 Dirac also laid the foundation for quantum electrodynamics—the modern theory of the emission and absorption of electromagnetic radiation by atoms. The reader is referred to the historical outline in Section 1.3 for a chronology of events that occurred with the discovery of quantum mechanics.

2.6 THE PAULI EXCLUSION PRINCIPLE

Originally based on the older, semiclassical quantum theory, the Pauli exclusion principle plays a vital role in modern quantum mechanics. The principle was developed for electrons in orbital states in atoms. It holds that no two electrons in an atom can be in the same state, characterized by four quantum numbers which we now define. The Pauli principle enables one to use atomic theory to account for the periodic system of the chemical elements.

In the Bohr theory with circular orbits, described in Section 2.3, only a single quantum number, n, was used. This is the first of the four quantum numbers and we designate it as the principal quantum number. In Bohr's theory each value of n gives an orbit at a given distance from the nucleus. For an atom with many electrons, we say that different values of n correspond to different electron shells. Each shell can accommodate only a limited number of electrons. In developing the periodic system, we postulate that, as more and more electrons are present in atoms of increasing atomic number, they fill the innermost shells. The outer-shell electrons determine the gross chemical properties of an element; these properties are thus repeated successively after each shell is filled. When $n = 1$, the shell is called the K shell; $n = 2$ denotes the L shell; $n = 3$, the M shell; and so on.

The second quantum number arises in the following way. As mentioned in Section 2.4, elliptical orbits and relativistic mechanics were also considered in the older quantum theory of the hydrogen atom. In nonrelativistic mechanics, the mean energy of an electron is the same for all elliptical orbits having the same major axis. Furthermore, the mean energy is the same as that for a circular orbit with a diameter equal to the major axis. (The circle is the limiting case of an ellipse with equal major and minor axes.) Relativistically, the situation is different because of the increase in velocity and hence mass that an electron experiences in an elliptical orbit when it comes closest to the nucleus. In 1916 Sommerfeld extended Bohr's theory to include elliptical orbits with the nucleus at one focus. The formerly degenerate energies of different ellipses with the same major axis are slightly different relativistically, giving rise to the fine structure in the spectra of elements. The observed fine structure in the hydrogen spectrum was obtained by quantizing the ratio of the major and minor axes of the elliptical orbits, thus providing a second quantum number, called the azimuthal quantum number. In modern theory it amounts to the same thing as the orbital angular-momentum quantum number, l, with values $l = 0, 1, 2, \ldots, n - 1$. Thus, for a given shell, the second quantum number can be any nonnegative integer smaller than the principal quantum number. When $n = 1$, $l = 0$ is the only possible azimuthal quantum number; when $n = 2$, $l = 0$ and $l = 1$ are both possible.

The magnetic quantum number m is the third. It was introduced to account for the splitting of spectral lines in a magnetic field (Zeeman effect). An electron orbiting a nucleus constitutes an electric current, which produces a magnetic field. When an atom is placed in an external magnetic field, its own orbital magnetic field lines up only in certain discrete directions with respect to the external field. The magnetic quantum number gives the component of the orbital angular momentum in the direction of the external field. Accordingly, m can have any integral value between $+l$ and $-l$; viz., $m = 0, \pm1, \pm2, \ldots, \pm l$. With $l = 1$, for example $m = -1, 0, 1$.

Although the Bohr–Sommerfeld theory explained a number of features of atomic spectra, problems still persisted. Unexplained was the fact that the alkali-metal spectra (e.g., Na) show a doublet structure even though these atoms have only a single valence electron in their outer shell (as we show in the next section). In addition, spectral lines do not split into a normal pattern in a weak magnetic field (anomalous Zeeman effect). These problems were cleared up when Pauli introduced a fourth quantum number of "two-valuedness", having no classical analogue. Then, in 1925, Uhlenbeck and Goudsmit proposed that the electron has an intrinsic angular momentum $\frac{1}{2}\hbar$ due to rotation about its own axis; thus the physical significance of Pauli's fourth quantum number was evident. The electron's intrinsic spin endows it with magnetic properties. The spin quantum number, s, has two values, $s = \pm\frac{1}{2}$. In an external magnetic field, the electron aligns itself either with "spin up" or "spin down" with respect to the field direction.

The Pauli exclusion principle states that no two electrons in an atom can occupy a state with the same set of four quantum numbers n, l, m, and s. The principle can also be expressed equivalently, but more generally, by saying that no two electrons in a system can have the same complete set of quantum numbers. Beyond atomic physics, the Pauli exclusion principle applies to all types of identical particles of half-integral spin (called fermions and having intrinsic angular momentum $\frac{1}{2}\hbar$, $\frac{3}{2}\hbar$, etc.). Such particles include positrons, protons, neutrons, muons, and others. Integral-spin particles (called bosons) do not obey the exclusion principle. These include photons, alpha particles, pions, and others.

We next apply the Pauli principle as a basis for understanding the periodic system of the elements.

2.7 ATOMIC THEORY OF THE PERIODIC SYSTEM

The K shell, with $n = 1$, can contain at most two electrons, since $l = 0$, $m = 0$, and $s = \pm\frac{1}{2}$ are the only possible values of the other three quantum numbers. The two electrons in the K shell differ only in their spin directions. The element with atomic number $Z = 2$ is the noble gas helium. Like the other noble-gas atoms it has a completed outer shell and is chemically inert. The electron configurations of H and He are designated, respectively, as $1s^1$ and $1s^2$. The symbols in the configurations give the principal quantum number, a letter designating the azimuthal quantum number (s denotes $l = 0$; p denotes $l = 1$; d, $l = 2$; and f, $l = 3$), and a superscript giving the total number of electrons in the states with the given values of n and l. The electron configurations of each element are shown in the periodic table on the inside back cover of this book, to which the reader is referred in this discussion. The first period contains only hydrogen and helium.

The next element, Li, has three electrons. Two occupy the full K shell and the third occupies a state in the L shell ($n = 2$). Electrons in this shell can have $l = 0$ (s states) or $l = 1$ (p states). The 2s state has lower energy than the 2p, and so the electron configuration of Li is $1s^2 2s^1$. With $Z = 4$ (Be), the other 2s state is occupied, and the configuration is $1s^2 2s^2$. No additional s electrons ($l = 0$) can be added in these two shells. However, the L shell can now accommodate electrons with $l = 1$ (p electrons) and with three values of m: -1, 0, $+1$. Since two electrons with opposite spins (spin quantum numbers $\pm\frac{1}{2}$ can occupy each state of given n, l, and m, there can be a total of six electrons in the 2p states. The configurations for the next six elements involve the successive filling of these states, from $Z = 5$ (B), $1s^2 2s^2 p^1$, to $Z = 10$ (Ne), $1s^2 2s^2 p^6$. The noble gas neon has the completed L shell. To save repeating the writing of the identical inner-shell configurations for other elements, one denotes the neon configuration by [Ne]. The second period of the table begins with Li and ends with Ne.

With the next element, sodium, the filling of the M shell begins. Sodium has a 3s electron and its configuration is $[Ne]3s^1$. Its single outer-shell electron gives it properties akin to those of lithium. One sees that the other alkali metals in the group IA of the periodic table are all characterized by having a single s electron in their outer shell. The third period ends with the filling of the $3s^2 p^6$ levels in the noble gas, Ar ($Z = 18$). The chemical and physical properties of the eight elements in the third period are similar to those of the eight elements in the second period with the same outer-shell electron configurations.

The configuration of Ar ($Z = 18$), which is $[Ne]3s^2 3p^6$, is also designated as [Ar]. All of the states with $n = 1$, $l = 0$; $n = 2$, $l = 0$, 1; and $n = 3$, $l = 0$, 1 are occupied in Ar. However, the M shell is not yet filled, because d states ($l = 2$) are possible when $n = 3$.

Because there are five values of m when $l = 2$, there are five d states, which can accommodate a total of ten electrons (five pairs with opposite spin), which is the number needed to complete the M shell. It turns out that the 4s energy levels are lower than the 3d. Therefore, the next two elements, K and Ca, that follow Ar have the configurations $[Ar]4s^1$ and $[Ar]4s^2$. The next ten elements, from Sc ($Z = 21$) through Zn ($Z = 30$), are known as the transition metals. This series fills the 3d levels, sometimes in combination with $4s^1$ and sometimes with $4s^2$ electrons. The configuration of Zn is $[Ar]3d^{10}4s^2$, at which point the M shell ($n = 3$) is complete. The next six elements after Zn fill the six 4p states, ending with the noble gas, Kr, having the configuration $[Ar]3d^{10}4s^2p^6$.

After Ar, the shells with a given principal quantum number do not get filled in order. Nevertheless, one can speak of the filling of certain subshells in order, such as the 4s subshell and then the 3d in the transition metals. The lanthanide series of rare-earth elements, from $Z = 58$ (Ce) to $Z = 71$ (Lu), occurs when the 4f subshell is being filled. For these states $l = 3$, and since $-3 \leq m \leq 3$, a total of $7 \times 2 = 14$ elements comprise the series. Since it is an inner subshell that is being filled, these elements all have very nearly the same chemical properties. The situation is repeated with the actinide elements from $Z = 90$ (Th) to $Z = 103$ (Lr), in which the 5f subshell is being filled.

The picture given here is that of an independent-electron model of the atom, in which each electron independently occupies a given state. In reality, the atomic electrons are indistinguishable from one another and an atomic wave function is one in which any electron can occupy any state with the same probability as any other electron. Moreover, hybrid atomic states of mixed configurations are used to explain still other phenomena (e.g., the tetrahedral bonds in CH_4).

2.8 MOLECULES

Quantum mechanics has also been very successful in areas other than atomic structure and spectroscopy. It has also explained the physics of molecules and condensed matter (liquids and solids). Indeed, the nature of the chemical bond between two atoms, of either the same or different elements, is itself quantum mechanical in nature, as we now describe.

Consider the formation of the H_2 molecule from two H atoms. Experimentally, it is known from the vibrational spectrum of H_2 that the two protons' separation oscillates about an equilibrium distance of 0.74 Å and the dissociation energy of the molecule is 4.7 eV. The two electrons move very rapidly about the two nuclei, which, by comparison, move slowly back and forth along the direction between their centers. When the nuclei approach each other, their coulomb repulsion causes them to reverse their directions and move apart. The electrons more than keep pace and move so that the separating nuclei again reverse directions and approach one another. Since the electrons move so quickly, they make many passes about the nuclei during any time in which the latter move appreciably. Therefore, one can gain considerable insight into the structure of H_2 and other molecules by considering the electronic motion at different fixed separations of the nuclei (Born–Oppenheimer approximation).

To analyze H_2, we begin with the two protons separated by a large distance R, as indicated in Fig. 2.5(a). The lowest energy of the system will then occur when each electron is bound to one of the protons. Thus the ground state of the H_2 system at large nuclear separations is that in which the two hydrogen atoms, H_A and H_B, are present in their ground states. We denote this structure by writing (H_A1, H_B2), indicating that electron number one is bound in the hydrogen atom H_A and electron number two in H_B. Another stable structure at large R is an ionic one, (H_A12^-, H_B^+), in which both electrons orbit

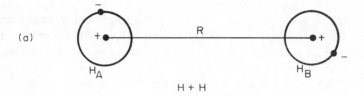

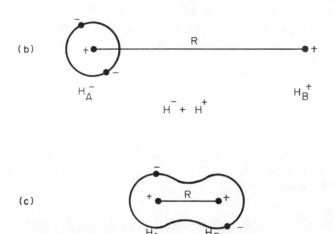

Figure 2.5. At large internuclear separation R, the structure of the H_2 molecule can approach (a) that of two neutral H atoms, H + H, or (b) that of two ions, H^- and H^+. These structures merge at close separation in (c). The indistinguishability of the two electrons gives stability to the bond formed through the quantum-mechanical phenomenon of resonance.

one of the protons. This structure is shown in Fig. 2.5(b). Since 13.6 eV is required to remove an electron from H and its binding energy in the H^- ion is only 0.80 eV, the ionic structure in Fig. 2.5(b) has less binding energy than the neutral one in (a). In addition to the states shown in the figure, one can consider the same two structures in which the two electrons are interchanged, (H_A2, H_B1) and (H_A21^-, H_B^+).

Next, we consider what happens when R becomes smaller. When the nuclei move close together, as in Fig. 2.5(c), the electron wave functions associated with each nucleus overlap. Detailed calculations show that neither of the structures shown in Fig. 2.5(a) or (b) nor a combination of the two leads to the formation of a stable molecule. Instead, stability arises from the indistinguishable participation of both electrons. The neutral structure alone will bind the two atoms when, in place of either (H_A1, H_B2) or (H_A2, H_B1) alone, one uses the superposed structure (H_A1, H_B2) + (H_A2, H_B1). The need for the superposed structure is a purely quantum-mechanical concept, and is due to the fact that the two electrons are indistinguishable and their roles must be exchangeable without affecting observable quantities. The energy contributed to the molecular binding by the electron exchange is called the resonance energy, and its existence with the neutral structure in Fig. 2.5(c) accounts for ~80% of the binding energy of H_2. The type of electron-pair bond that is thus formed by the exchange is called covalent. Resonance also occurs between the ionic structures (H_A12^-, H_B^+) and (H_A21^-, H_B^+) from Fig. 2.5(b) and contributes ~5% of the binding energy, giving the H_2 bond a small ionic character. The remaining 15% of the binding energy comes from other effects, such as deformation of the electron wave functions from the simple structures discussed here and from partial shielding of the nuclear charges by each electron from the other. In general, for covalent bonding to occur, the two atoms involved must have the same number of unpaired electrons, as is the case with hydrogen. However, the atoms need not be identical.

The character of the bond in HF and HCl, for example, is more ionic than in the homonuclear H_2 or N_2. The charge distribution in a heteronuclear diatomic molecule is not symmetric, and so the molecule has a permanent electric dipole moment. (The two types of bonds are called homopolar and heteropolar.)

Figure 2.6 shows the total energy of the H_2 molecule as a function of the internuclear separation R. (The total energy of the two H atoms at large R is taken as the reference level of zero energy.) The bound state has a minimum energy of -4.7 eV at the equilibrium separation of 0.74 Å, in agreement with the data given earlier in this section. In this state the spins of the bonding electron pair are antiparallel. A second, nonbonding state is formed with parallel spins. These two energy levels are degenerate at large R, where the wave functions of the two atomic electrons do not overlap appreciably.

Molecular spectra are very complicated. Changes in the rotational motion of molecules

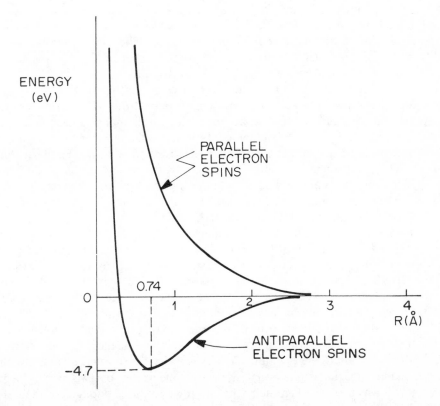

Figure 2.6. Total energy of the H_2 molecule as function of internuclear separation R. A stable molecule is formed when the spins of the two electrons are antiparallel. A nonbonding energy level is formed when the spins are parallel. The two energies coincide at large R.

accompanies the emission or absorption of photons in the far infrared. Vibrational changes together with rotational ones usually produce spectra in the near infrared. Electronic transitions are associated with the visible and ultraviolet part of molecular spectra. Electronic molecular spectra have a fine structure due to the vibrational and rotational motions of the molecule. Molecular spectra also show isotopic structure. The presence of the naturally occurring ^{35}Cl and ^{37}Cl isotopes in chlorine, for example, gives rise to two sets of vibrational and rotational energy-level differences in the spectrum of HCl.

2.9 SOLIDS AND ENERGY BANDS

We briefly discuss the properties of solids and the origin of energy bands, which are essential for understanding how semiconductor materials can be used as radiation detectors (Chapter 9).

Solids can be crystalline or noncrystalline (e.g., plastics). Crystalline solids, of which semiconductors are an example, can be put into four groups according to the type of binding that exists between atoms. Crystals are characterized by regular, repeated atomic arrangements in a lattice.

In a *molecular* solid the bonds between molecules are formed by the weak, attractive van der Waals forces. Examples are the noble gases, H_2, N_2, and O_2, which are solids only at very low temperatures and can be easily deformed and compressed. All electrons are paired and hence molecular solids are poor electrical conductors.

In an *ionic* solid all electrons are also paired. A crystal of NaCl, for example, exists as alternating charged ions, Na^+ and Cl^-, in which all atomic shells are filled. These solids are also poor conductors. The electrostatic forces between the ions are very strong and hence ionic solids are hard and have high melting points. They are generally transparent to visible light, because their electronic absorption frequencies are in the ultraviolet region and lattice vibration frequencies are in the infrared.

A *covalent* solid is one in which adjacent atoms are covalently bound by shared valence electrons. Such bonding is possible only with elements in Group IVB of the periodic system; diamond, silicon, and germanium are examples. In diamond, a carbon atom (electronic configuration $1s^2 2s^2 p^2$) shares one of its four L-shell electrons with each of four neighbors, which, in turn, donates one of its L electrons for sharing. Each carbon atom thus has its full complement of eight L-shell electrons through tight binding with its neighbors. Covalent solids are very hard and have high melting points. They have no free electrons, and are therefore poor conductors. Whereas diamond is an insulator, Si and Ge are semiconductors, as will be discussed in Chapter 9.

In a *metallic* solid (e.g., Cu, Au) the valence electrons in the outermost shells are weakly bound and shared by all of the atoms in the crystal. Vacancies in these shells permit electrons to move with ease through the crystal in response to the presence of an electric field. Metallic solids are good conductors of electricity and heat. Many of their properties can be understood by regarding some of the electrons in the solid as forming an "electron gas" moving about in a stationary lattice of positively charged ions. The electrons satisfy the Pauli exclusion principle and occupy a range of energies consistent with their temperature. This continuous range of energies is called a conduction band.

To describe the origin of energy bands in a solid, we refer to Fig. 2.6. We saw that the twofold exchange degeneracy between the electronic states of two widely separated hydrogen atoms was broken when their separation was reduced enough for their electron wave functions to overlap appreciably. The same twofold splitting occurs whenever any two identical atoms bind together. Moreover, the excited energy levels of isolated atoms also undergo a similar twofold splitting when the two atoms unite. Such splitting of exchange-degenerate energy levels is a general quantum-mechanical phenomenon. If three identical atoms are present, then the energy levels at large separations are triply degenerate and split into three different levels when the atoms are brought close together. In this case, the three levels all lie in about the same energy range as the first two if the interatomic distances are comparable. If N atoms are brought together in a regular arrangement, such as a crystal solid, then there are N levels in the energy interval.

Figure 2.7 illustrates the splitting of electronic levels for $N = 2$, 4, and 8 and the onset of band formation. Here the bound-state energies E are plotted schematically as functions of the atomic separations R, with R_0 being the normal atomic spacing in the solid. When N becomes very large, as in a crystal, the separate levels are "compressed together" into a band, within which an electron can have any energy. At a given separation, the band structure is most pronounced in the weakly bound states, in which the electron cloud extends over large distances. The low-lying levels, with tightly bound electrons, remain discrete and unperturbed by the presence of neighboring atoms. Just as for the discrete levels, electrons cannot exist in the solid with energies between the allowed bands. (The existence of energy bands also arises directly out of the quantum-mechanical treatment of the motion of electrons in a periodic lattice.)

While the above properties are those of "ideal" crystalline solids, the presence of impurities — even in trace amounts — often changes the properties markedly. We shall see in Chapter 9 how doping alters the behavior of intrinsic semiconductors and the scintillation characteristics of crystals.

2.10 CONTINUOUS AND CHARACTERISTIC X-RAYS

Roentgen discovered that X-rays are produced when a beam of electrons strikes a target. The electrons lose most of their energy in collisions with atomic electrons in the target, causing the ionization and excitation of atoms. In addition, they can be sharply deflected in the vicinity of the atomic nuclei, thereby losing energy by irradiating X-ray photons. Heavy nuclei are much more efficient than light nuclei in producing the radiation because the deflections are stronger. A single electron can emit an X-ray photon having any energy up to its own kinetic energy. As a result, a monoenergetic beam of electrons produces a continuous spectrum of X-rays with photon energies up to the value of the beam energy. The continuous X-rays are also called bremsstrahlung, or "braking radiation."

A schematic diagram, showing the basic elements of a modern X-ray tube, is shown in Fig. 2.8. The tube has a cathode and anode sealed inside under high vacuum. The cathode assembly consists of a heated tungsten filament contained in a focusing cup. When the tube operates, the filament, heated white hot, "boils off" electrons, which are accelerated toward the anode in a strong electric field produced by a large potential difference (high voltage) between the cathode and anode. The focusing cup concentrates the electrons onto a focal spot on the anode, usually made of tungsten. There the electrons are abruptly brought to rest, emitting continuous X-rays in all directions. Typically, less than 1% of the electrons' energy is converted into useful X-rays that emerge through a window in the tube. The other 99+% of the energy, lost in electronic collisions, is converted into heat, which must be removed from the anode. Anodes can be cooled by circulating oil or water. Rotating anodes are also used in X-ray tubes to keep the temperature lower.

Figure 2.9 shows typical continuous X-ray spectra generated from a tube operated at different voltages with the same current. The efficiency of bremsstrahlung production increases rapidly when the electron energy is raised. Therefore, the X-ray intensity increases considerably with tube voltage, even at constant current. The wavelength of an

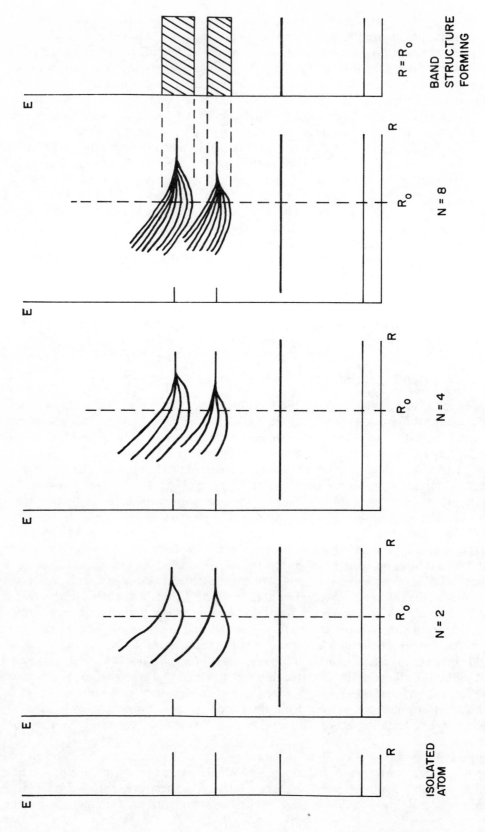

Figure 2.7. Energy-level diagram schematically showing the splitting of discrete atomic energy levels when $N = 2$, 4, and 8 atoms are brought together in a regular array. When N is very large, continuous energy bands result, as indicated at the right of the figure.

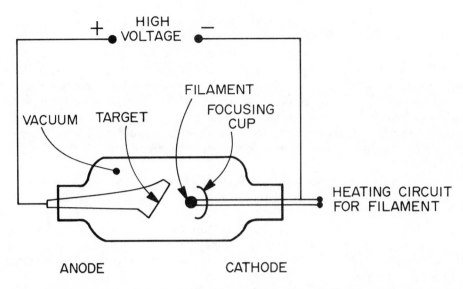

Figure 2.8. Schematic diagram of modern X-ray tube with fixed target anode.

X-ray photon with maximum energy can be computed from Eq. (2.26). For the top curve in Fig. 2.9, we find $\lambda_{min} = 12400/50000 = 0.248$ Å, where this curve intersects the abscissa. The X-ray energies are commonly referred to in terms of their peak voltages in kilovolts, denoted by kVp.

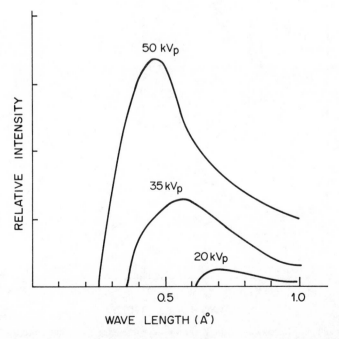

Figure 2.9. Typical continuous X-ray spectra from tube operating at three different peak voltages with the same current.

If the tube voltage is sufficient, electrons striking the target can eject electrons from the target atoms. (The K-shell binding energy is $E_K = 69.525$ keV for tungsten.) Discrete X-rays are then also produced. These are emitted when electrons from higher shells fill the inner-shell vacancies. The photon energies are characteristic of the element of which the target is made, just as the optical spectra are in the visible range. Characteristic X-rays appear superimposed on the continuous spectrum, as illustrated for tungsten in Fig. 2.10. They are designated K_α, K_β, etc. when the K-shell vacancy is filled by an electron from the L shell, M shell, etc. (In addition, when L-shell vacancies are filled, characteristic L_α, L_β, etc. X-rays are emitted. These have low energy and are usually absorbed in the tube housing.)

Because the electron energies in the other shells are not degenerate, the K X-rays have a fine structure, not shown in Fig. 2.10. The L shell, for example, consists of three subshells, in which for tungsten the electron binding energies in keV are $E_{LI} = 12.098$, $E_{LII} = 11.541$, and $E_{LIII} = 10.204$. The transition LIII → K gives a $K_{\alpha 1}$ photon with energy $E_K - E_{LIII} = 69.525 - 10.204 = 59.321$ keV; the transition LII → K gives a $K_{\alpha 2}$ photon with energy 57.984 keV. The optical transition LI → K is quantum mechanically forbidden and does not occur.

The first systematic study of characteristic X-rays was carried out in 1913 by the young British physicist, H. G. J. Moseley, working in Rutherford's laboratory. The diffraction of X-rays by crystals had been discovered by von Laue in 1912, and Moseley used this process to compare characteristic X-ray wavelengths. He found that the square root of the frequencies of corresponding lines (e.g., $K_{\alpha 1}$) in the characteristic X-ray spectra

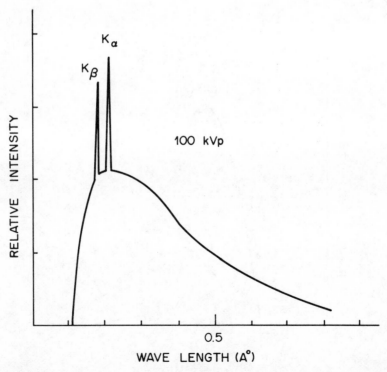

Figure 2.10. Spectrum showing characteristic K_α and K_β discrete X-rays in addition to the continuous X-rays. Characteristic K X-rays are present only when tube operating voltage is high enough to give electrons enough energy to ionize the K shell in the target atoms. Potential difference across tube in volts is then ≥ K-shell binding energy in eV.

increases by an almost constant amount from element to element in the periodic system. Alpha-particle scattering indicated that the number of charge units on the nucleus is about half the atomic weight. Moseley concluded that the number of positive nuclear charges and the number of electrons both increase by one from element to element. Starting with $Z = 1$ for hydrogen, the number of charge units Z determines the atomic number of an element, which gives its place in the periodic system.

The linear relationship between $\sqrt{\nu}$ and Z would be predicted if the electrons in many-electron atoms occupied orbits like those predicted by Bohr's theory for single-electron systems. As seen from Eq. (2.13), the frequencies of the photons for a given transition $i \rightarrow f$ in different elements is proportional to Z^2.

In view of Moseley's findings, the positions of cobalt and nickel had to be reversed in the periodic system. Although Co has the larger atomic weight, 58.93 compared with 58.70, its atomic number is 27, while that of Ni is 28. Moseley also predicted the existence of a new element with $Z = 43$. Technetium, which has no stable form, was discovered after nuclear fission.

2.11 AUGER ELECTRONS

An atom in which an L electron makes a transition to fill a vacancy in the K shell does not always emit a photon, particularly if it is an element of low Z. A different, nonoptical transition can occur in which an L electron is ejected from the atom, thereby leaving two vacancies in the L shell. The electron thus ejected from the atom is called an Auger electron.

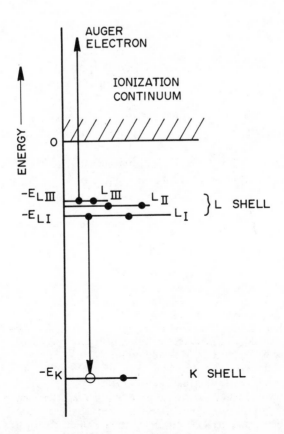

Figure 2.11. Schematic representation of an atomic transition that results in Auger-electron emission.

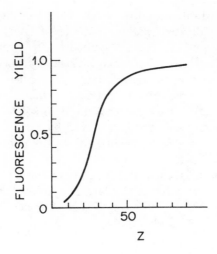

Figure 2.12. Fluorescence yield as a function of atomic number Z.

The emission of an Auger electron is illustrated in Fig. 2.11. The downward arrow indicates the transition of an electron from the L_I level into the K-shell vacancy, thus releasing an energy equal to the difference in binding energies, $E_K - E_{LI}$. As the alternative to photon emission, this energy can be transferred to an L_{III} electron, ejecting it from the atom with a kinetic energy

$$T = E_K - E_{LI} - E_{LIII}. \tag{2.37}$$

Two L-shell vacancies are thus produced. The Auger effect can occur with other combinations of the three L-shell levels. Equations analogous to (2.37) provide the possible Auger-electron energies.

The Auger process is not one in which a photon is emitted by one atomic electron and absorbed by another. In fact, the $L_I \to K$ transition shown in Fig. 2.11 is optically forbidden.

The fluorescence yield of an element is defined as the fraction of K-shell vacancies that are filled by photon emission rather than Auger-electron ejection. Figure 2.12 shows how the fluorescence yield goes from essentially zero for the low-Z elements to almost unity for high Z.

2.12 PROBLEMS

1. How many atoms are there in 3 L of N_2 at STP?
2. What is the volume occupied by 1 kg of methane (CH_4) at 23°C and 756 torr?
3. How many hydrogen atoms are there in the last problem?
4. What is the mass of a single atom of aluminum?
5. Estimate the number of atoms/cm² in an aluminum foil that is 1 mm thick.
6. Estimate the radius of a uranium nucleus. What is its cross-sectional area?
7. What is the density of the nucleus in a gold atom?
8. What was the minimum distance to which the 7.68 MeV alpha particles could approach the center of the gold nuclei in Rutherford's experiments?
9. How much energy would an alpha particle need in order to "just touch" the nuclear surface in a gold foil?
10. How much force acts on the electron in the ground state of the hydrogen atom?

11. What is the angular momentum of the electron in the $n = 5$ state of the H atom?
12. How does the angular momentum of the electron in the $n = 3$ state of H compare with that in the $n = 3$ state of He$^+$?
13. Calculate the ionization potential of Li^{2+}.
14. Calculate the radius of the $n = 2$ electron orbit in the Bohr hydrogen atom.
15. Calculate the orbital radius for the $n = 2$ state of Li^{2+}.
16. What are the energies of the photons with the two longest wavelengths in the Paschen series (Fig. 2.3)?
17. Calculate the wavelengths in the visible spectrum of the He$^+$ ion in which the electron makes transitions from higher states to states with quantum number $n = 1$, 2, 3, or 4.
18. Calculate the current of the electron in the ground state of the hydrogen atom.
19. What is the lowest quantum number of an H-atom electron orbit with a radius of at least 1 cm?
20. According to Bohr theory, how many bound states of He$^+$ have energies equal to bound-state energies in H?
21. How much energy is needed to remove an electron from the $n = 5$ state of He$^+$?
22. Calculate the reduced mass for the He$^+$ system.
23. What percentage error is made in the Rydberg constant for hydrogen if the electron mass is used instead of the reduced mass?
24. The negative muon is an elementary particle with a charge equal to that of the electron and a mass 207 times as large. A proton can capture a negative muon to form a hydrogen-like "mesic" atom. (The muon was formerly called the mu meson). For such a system, calculate (a) the radius of the first Bohr orbit and (b) the ionization potential. Do not assume a stationary nucleus.
25. What is the reduced mass for a system of two particles of equal mass, such as an electron and positron, orbiting about their center of mass?
26. What is meant by the fine structure in the spectrum of hydrogen and what is its physical origin?
27. Calculate the momentum of an ultraviolet photon of wave length 1000 Å.
28. What is the momentum of a photon of lowest energy in the Balmer series of hydrogen?
29. Calculate the de Broglie wavelength of the 7.68 MeV alpha particles used in Rutherford's experiment. Use nonrelativistic mechanics.
30. What is the energy of an electron having a wavelength of 0.123 Å?
31. Calculate the de Broglie wavelength of a 245 keV electron.
32. (a) What is the momentum of an electron with a de Broglie wavelength of 0.02 Å? (b) What is the momentum of a photon with a wavelength of 0.02 Å?
33. Calculate the kinetic energy of the electron and the energy of the photon in the last problem.
34. Estimate the uncertainty in the momentum of an electron whose location is uncertain by a distance of 2 Å. What is the uncertainty in the momentum of a proton under the same conditions?
35. What can one conclude about the relative velocities and energies of the electron and proton in the last problem? Are wave phenomena apt to be more apparent for light particles than for heavy ones?
36. The result given after Eq. (2.35) shows that an electron confined to nuclear dimensions, $\Delta x \sim 10^{-13}$ cm, could be expected to have a kinetic energy $T \sim 200$ MeV. What would be the value of Δx for $T \sim 100$ eV?
37. (a) Write the electron configuration of carbon. (b) How many s electrons does the C atom have? (c) How many p electrons?
38. The configuration of boron is $1s^2 2s^2 p^1$. (a) How many electrons are in the L shell? (b) How many electrons have orbital angular-momentum quantum number $l = 0$?
39. How many electrons does the nickel atom have with azimuthal quantum number $l = 2$?
40. What is the electron configuration of the magnesium ion, Mg^{2+}?
41. What is incorrect in the electron configuration $1s^2 2s^2 p^6 3s^2 p^8 d^{10}$?
42. (a) What are the largest and the smallest values that the magnetic quantum number m has in the Zn atom? (b) How many electrons have $m = 0$ in Zn?
43. Show that the total number of states available in a shell with principal quantum number n is $2n^2$.
44. What is the wavelength of a photon of maximum energy from an X-ray tube operating at a peak voltage of 80 keV?
45. If the operating voltage of an X-ray tube is doubled, by what factor does the wavelength of a photon of maximum energy change?

46. (a) How many electrons per second strike the target in an X-ray tube operating at a current of 50 mA? (b) If the potential difference between the anode and cathode is 100 keV, how much power is expended?

47. If the binding energies for electrons in the K, L, and M shells of an element are, respectively, 8979 eV, 951 eV, and 74 eV, what are the energies of the K_α and K_β characteristic X-rays? (These values are representative of Cu without the fine structure.)

48. The oxygen atom has a K-shell binding energy of 532 eV and L-shell binding energies of 23.7 eV and 7.1 eV. What are the possible energies of its Auger electrons?

CHAPTER 3
THE NUCLEUS AND NUCLEAR RADIATION

3.1 NUCLEAR STRUCTURE

The nucleus of an atom of atomic number Z and mass number A consists of Z protons and $N = A - Z$ neutrons. The atomic masses of all individual atoms are nearly integers, and A gives the total number of *nucleons* (i.e., protons and neutrons) in the nucleus. A species of atom, characterized by its nuclear constitution — its values of Z and A (or N) — is called a *nuclide*. It is conveniently designated by writing the appropriate chemical symbol with a subscript giving Z and superscript giving A. For example, ^1_1H, ^2_1H, and $^{238}_{92}U$ are nuclides. Nuclides of an element that have different A (or N) are called *isotopes*. Nuclides having the same number of neutrons are called *isotones*; e.g., $^{206}_{82}Pb$ and $^{204}_{80}Hg$ are isotones with $N = 124$. Hydrogen has three isotopes, ^1_1H, ^2_1H, and ^3_1H, all of which occur naturally. Deuterium, ^2_1H, is stable; tritium, ^3_1H, is radioactive. Fluorine has only a single naturally occurring isotope, $^{19}_9F$; all of its other isotopes are man made, radioactive, and short lived. The measured atomic weights of the elements reflect the relative abundances of the isotopes found in nature, as the next example illustrates.

Example
Chlorine is found to have two naturally occurring isotopes: $^{35}_{17}Cl$, which is 76% abundant, and $^{37}_{17}Cl$, 24% abundant. The atomic weights of the two isotopes are 34.97 and 36.97. Show that this isotopic composition accounts for the observed atomic weight of the element.

Solution
Taking the weighted average of the atomic weights of the two isotopes, we find for the atomic weight of Cl, $0.76 \times 34.97 + 0.24 \times 36.97 = 35.45$, as observed.

Since the electron configuration of the different isotopes of an element is the same, isotopes cannot be separated chemically. The existence of isotopes does cause a very slight perturbation in atomic energy levels, leading to an observed "isotopic shift" in some spectral lines. In addition, the different nuclear spins of different isotopes of the same element are responsible for hyperfine structure in the spectra of elements. As we mentioned at the end of Section 2.8, the existence of isotopes has a big effect on the vibration–rotation spectra of molecules.

Nucleons are bound together in a nucleus by the action of the strong, or nuclear, force. The range of this force is only of the order of nuclear dimensions, $\sim 10^{-13}$ cm, and it is powerful enough to overcome the Coulomb repulsion of the protons in the nucleus. Figure 3.1(a) schematically shows the potential energy of a proton as a function of the distance r separating its center and the center of a nucleus. The potential energy is zero at large separations. As the proton comes closer, its potential energy increases, due to the work done against the repulsive Coulomb force that acts between the two positive charges. Once the proton comes within range of the attractive nuclear force, though, its potential energy abruptly goes negative and it can react with the nucleus. If conditions are right, the proton's total energy can also become negative, and the proton will then occupy a bound state in the nucleus. As we learned in the Rutherford experiment in Section 2.2, a positively charged particle requires considerable energy in order to approach a nucleus closely. In contrast, the nucleus is accessible to a neutron of any energy. Because the neutron is uncharged, there is no Coulomb barrier for it to overcome. Figure 3.1(b) shows the potential-energy curve for a neutron and a nucleus.

Example

Estimate the minimum energy that a proton would have to have in order to react with the nucleus of a Cl atom.

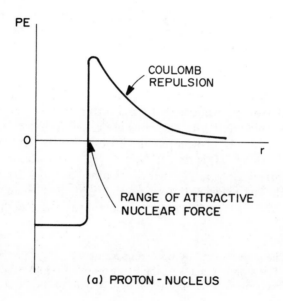

(a) PROTON - NUCLEUS

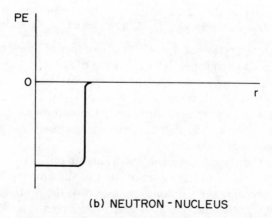

(b) NEUTRON - NUCLEUS

Figure 3.1. (a) Potential energy PE of a proton as a function of its separation r from the center of a nucleus. (b) Potential energy of a neutron and nucleus as a function of r. Uncharged neutron has no repulsive Coulomb barrier to overcome when approaching a nucleus.

Solution

In terms of Fig. 3.1(a), the proton would have to have enough energy to overcome the repulsive Coulomb barrier in a head-on collision. This would allow it to just reach the target nucleus. We can use Eq. (2.2) to estimate how far apart the centers of the proton and nucleus would then be, when they "just touch." With $A = 1$ and $A = 35$ in Eq. (2.2), we obtain for the radii of the proton (r_p) and the chlorine nucleus (r_{Cl})

$$r_p = 3 \times 1^{1/3} \times 10^{-13} = 3.0 \times 10^{-13} \text{ cm}, \tag{3.1}$$

$$r_{Cl} = 3 \times 35^{1/3} \times 10^{-13} = 9.8 \times 10^{-13} \text{ cm}. \tag{3.2}$$

The proton has unit positive charge, $e = 4.8 \times 10^{-10}$ esu, and the chlorine ($Z = 17$) nucleus has a charge $17e$. The potential energy of the two charges separated by the distance $r_p + r_{Cl} = 1.3 \times 10^{-12}$ cm is therefore

$$\text{PE} = \frac{17 \times (4.8 \times 10^{-10})^2}{1.3 \times 10^{-12}} = 3.0 \times 10^{-6} \text{ erg} = 1.9 \text{ MeV}. \tag{3.3}$$

(Problem 9 in Chapter 2 is worked like this example.)

Like an atom, a nucleus is itself a quantum-mechanical system of bound particles. However, the nuclear force, acting between nucleons, is considerably more complicated and more uncertain than the electromagnetic force that governs the structure and properties of atoms and molecules. In addition, wave equations describing nuclei cannot be solved with the same degree of numerical precision that atomic wave equations can. Nevertheless, many detailed properties of nuclei have been worked out and verified experimentally. Both the proton and the neutron are "spin-$\frac{1}{2}$" particles and hence obey the Pauli exclusion principle. Just as excited electron states exist in atoms, excited states can exist in nuclei. Whereas an atom has an infinite number of bound excited states, however, a nucleus has only a finite number, if any. This difference in atomic and nuclear structure is attributable to the infinite range of the Coulomb force as opposed to the short range and limited, though large, strength of the nuclear force. The energy-level diagram of the $^6_3$Li nucleus in Fig. 3.2 shows that it has a number of bound excited states.† The deuteron and alpha particle (nuclei of $^2_1$H and $^4_2$He) are examples of nuclei that have no bound excited states.

3.2 NUCLEAR BINDING ENERGIES

Changes can occur in atomic nuclei in a number of ways, as we shall see throughout this book. Nuclear reactions can be either exothermic (releasing energy) or endothermic (requiring energy in order to take place). The energies associated with nuclear changes are usually in the MeV range. They are thus $\sim 10^6$ times greater than the energies associated with the valence electrons that are involved in chemical reactions. This factor characterizes the enormous difference in the energy released when an atom undergoes a nuclear transformation as compared with a chemical reaction.

The energy associated with exothermic nuclear reactions comes from the conversion of mass into energy. If the mass loss is ΔM, then the energy released, Q, is given by Ein-

---

†The "level" at 4.52 MeV is very short lived and, therefore, does not have a sharp energy. All quantum-mechanical energy levels have a natural width, a manifestation of the uncertainty relation for energy and time, $\Delta E \Delta t \gtrsim \hbar$ [Eq. (2.30)]. The lifetimes of atomic states ($\sim 10^{-8}$ sec) are long and permit precise knowledge of their energies ($\Delta E \sim 10^{-7}$ eV). For many excited nuclear states, the lifetime Δt is so short that the uncertainty in their energy, ΔE, is large, as is the case here.

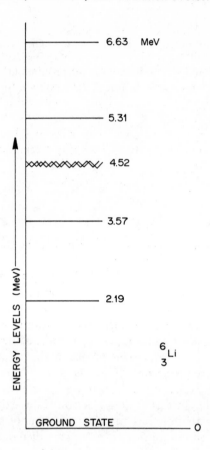

Figure 3.2. Energy levels of the $_3^6$Li nucleus, relative to the ground state of zero energy.

stein's relation, $Q = (\Delta M)c^2$, where c is the velocity of light. In this section we discuss the energetics of nuclear transformations.

We first establish the quantitative relationship between atomic mass units (AMU) and energy (MeV). By definition, the ^{12}C atom has a mass of exactly 12 AMU. Since its gram atomic weight is 12 g, it follows that 1 AMU, expressed in grams, is

$$1 \text{ AMU} = 1/(6.02 \times 10^{23}) = 1.66 \times 10^{-24} \text{ g}. \tag{3.4}$$

Using the Einstein relation and $c = 3 \times 10^{10}$ cm/sec, we obtain

$$1 \text{ AMU} = (1.66 \times 10^{-24})(3 \times 10^{10})^2$$

$$= 1.49 \times 10^{-3} \text{ erg} \tag{3.5}$$

$$= \frac{1.49 \times 10^{-3} \text{ erg}}{1.6 \times 10^{-6} \text{ erg MeV}^{-1}} = 931 \text{ MeV}. \tag{3.6}$$

More precisely, 1 AMU = 931.48 MeV.

We now consider one of the simplest nuclear reactions, the absorption of a thermal neutron by a hydrogen atom, accompanied by emission of a gamma ray. This reaction, which is very important for understanding the thermal-neutron dose to the body, can be represented by writing

$$\text{}^1_0\text{n} + \text{}^1_1\text{H} \rightarrow \text{}^2_1\text{H} + \text{}^0_0\gamma, \tag{3.7}$$

the photon having zero charge and mass. The reaction can also be designated $\text{}^1_1\text{H}(\text{n}, \gamma)\text{}^2_1\text{H}$. To find the energy released, we compare the total masses on both sides of the arrow. Appendix D contains data on nuclides which we shall frequently use. The atomic weight M of a nuclide of mass number A can be found from the mass difference, Δ, given in column 3. The quantity $\Delta = M - A$ gives the difference between the nuclide's atomic weight and its atomic mass number, expressed in MeV. (By definition, $\Delta = 0$ for the ^{12}C atom.) Since we are interested only in energy differences in the reaction (3.7), we obtain the energy released, Q, directly from the values of Δ, without having to calculate the actual masses of the neutron and individual atoms. Adding the Δ values for $\text{}^1_0\text{n}$ and $\text{}^1_1\text{H}$ and subtracting that for $\text{}^2_1\text{H}$, we find

$$Q = 8.0714 + 7.2890 - 13.1359 = 2.2245 \text{ MeV}. \tag{3.8}$$

This energy appears as a gamma photon emitted when the capture takes place (the thermal neutron has negligible kinetic energy).

The process (3.7) is an example of energy release by the fusion of light nuclei. The binding energy of the deuteron is 2.2245 MeV, which is the energy required to separate the neutron and proton again. As the next example shows, the binding energy of any nuclide can be calculated from a knowledge of its atomic weight (obtainable from Δ) together with the known individual masses of the proton, neutron, and electron.

Example
Find the binding energy of the nuclide $^{24}_{11}\text{Na}$.

Solution
One can work in terms of either AMU or MeV. The atom consists of 11 protons, 13 neutrons, and 11 electrons. The total mass in AMU of these separate constituents is, with the help of the data in Appendix A,

$$11(1.0073) + 13(1.0087) + 11(0.00055) = 24.199 \text{ AMU}. \tag{3.9}$$

From Appendix D, $\Delta = -8.418$ MeV gives the difference $M - A$. Thus, the mass of the $^{24}_{11}\text{Na}$ nuclide is less than 24 by the amount 8.418 MeV/(931.48 MeV AMU^{-1}) = 0.0090372 AMU. Therefore, the nuclide mass is $M = 23.991$ AMU. Comparison with (3.9) gives for the binding energy

$$\text{BE} = 24.199 - 23.991 = 0.208 \text{ AMU} = 194 \text{ MeV}. \tag{3.10}$$

This figure represents the total binding energy of the atom—nucleons plus electrons. However, the electron binding energies are small compared with nuclear binding, which accounts for essentially all of the 194 MeV. Thus the binding energy per nucleon in $^{24}_{11}\text{Na}$ is 194/24 = 8.08 MeV. [Had we worked in MeV, rather than AMU, the data from Appendix A give, in place of (3.9), 2.2541×10^4 MeV. Expressed in MeV, $A = 24 \times 931.48 = 2.2356 \times 10^4$ MeV. With $\Delta = -8$ MeV we have $M = A - \Delta = 2.2348 \times 10^4$ MeV. Thus the binding energy of the atom is $(2.2541 - 2.2348) \times 10^4 = 193$ MeV.]

The average binding energy per nucleon is plotted as a function of atomic mass number in Fig. 3.3. The curve has a broad maximum at about 8.5 MeV from $A = 40$ to 120.† It then drops off as one goes either to lower or higher A. The implication from this curve is that the *fusion* of light elements releases energy, as does the *fission* of heavy elements.

---

†The fact that the average nucleon binding energy is nearly constant over such a wide range of A is a manifestation of the saturation property of nuclear forces, mentioned at the end of Section 2.2.

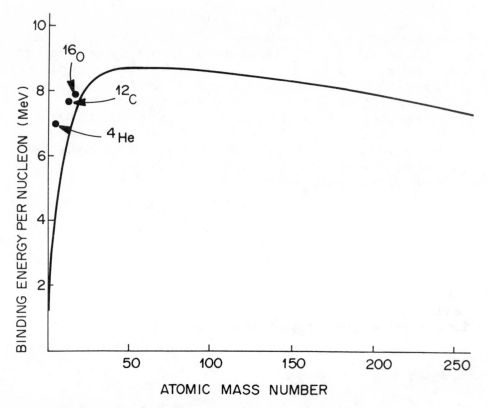

Figure 3.3. Average energy per nucleon as a function of atomic mass number.

Both transformations are made exothermic through the increased average nucleon binding energy that results. The $^1_1H(n, \gamma)^2_1H$ reaction considered above is an example of the release of energy through fusion. With a few exceptions, the average binding energies for all nuclides fall very nearly on the single curve shown. The nuclides ^4_2He, $^{12}_6C$, and $^{16}_8O$ show considerably tighter binding than their immediate neighbors. These nuclei are all "multiples" of the alpha particle, which appears to be a particularly stable nuclear subunit. (No nuclides with $A = 5$ exist for longer than $\sim 10^{-21}$ sec.†)

The loss of mass that accompanies the binding of particles is not a specifically nuclear phenomenon. The mass of the hydrogen atom is smaller than the sum of the proton and electron masses by 1.46×10^{-8} AMU. This is equivalent to an energy 1.46×10^{-8} AMU $\times$ 931 MeV/AMU $= 1.36 \times 10^{-5}$ MeV $= 13.6$ eV, the binding energy of the H atom.

We turn now to the subject of radioactivity, the property that some atomic species, called radionuclides, have of undergoing spontaneous nuclear disintegration. All of the heaviest elements are radioactive; $^{209}_{83}Bi$ is the only stable nuclide with $Z > 82$. All elements have radioactive isotopes, the majority being man made. The various kinds of

---

†Various forms of shell models have been studied for nuclei, analogous to an atomic shell model. The alpha particle consists of two spin-$\frac{1}{2}$ protons and two spin-$\frac{1}{2}$ neutrons in s states, forming the most tightly bound, "inner" nuclear shell. Generally, nuclei with even numbers of protons and neutrons ("even–even" nuclei) have the largest binding energies per nucleon.

radioactive decay and their associated nuclear energetics are described in the following sections.

3.3 ALPHA DECAY

Almost all naturally occurring alpha emitters are heavy elements with $Z \geq 83$. The principal features of alpha decay can be learned from the example of ^{226}Ra:

$$^{226}_{88}\text{Ra} \rightarrow {}^{222}_{86}\text{Rn} + {}^{4}_{2}\text{He}. \tag{3.11}$$

The energy Q released in the decay arises from a net loss in the masses $M_{\text{Ra,N}}$, $M_{\text{Rn,N}}$, and $M_{\text{He,N}}$, of the radium, radon, and helium nuclei:

$$Q = M_{\text{Ra,N}} - M_{\text{Rn,N}} - M_{\text{He,N}}. \tag{3.12}$$

This nuclear mass difference is very nearly equal the atomic mass difference, which, in turn, is equal to the difference in Δ values.† Letting Δ_{P}, Δ_{D}, and Δ_{He} denote the values of the parent, daughter, and helium atoms, we can write a general equation for obtaining the energy release in alpha decay:

$$Q_\alpha = \Delta_{\text{P}} - \Delta_{\text{D}} - \Delta_{\text{He}}. \tag{3.13}$$

Using the values in Appendix D for the decay of $^{226}_{88}$Ra to the ground state of $^{222}_{86}$Rn, we obtain

$$Q = 23.69 - 16.39 - 2.42 = 4.88 \text{ MeV}. \tag{3.14}$$

The Q value (3.14) is shared by the alpha particle and the recoil radon nucleus, and we can calculate the portion that each acquires. Since the radium nucleus was at rest, the momenta of the two decay products must be equal and opposite. Letting m and v represent the mass and initial velocity of the alpha particle and M and V those of the recoil nucleus, we write

$$mv = MV. \tag{3.15}$$

Since the initial kinetic energies of the products must be equal to the energy released in the decay, we have

$$\frac{1}{2}mv^2 + \frac{1}{2}MV^2 = Q. \tag{3.16}$$

Substituting $V = mv/M$ from Eq. (3.15) into (3.16) and solving for v^2, one finds

$$v^2 = \frac{2MQ}{m(m + M)}. \tag{3.17}$$

One thus obtains for the alpha-particle energy

$$E_\alpha = \frac{1}{2}mv^2 = \frac{MQ}{m + M}. \tag{3.18}$$

With the roles of the two masses interchanged, it follows that the recoil energy of the nucleus is

---

†Specifically, the relatively slight difference in the binding energies of the 88 electrons on either side of the arrow in (3.11) is neglected when atomic mass loss is equated to nuclear mass loss. In principle, nuclear masses are needed; however, atomic masses are much better known. These small differences are negligible for most purposes.

$$E_N = \frac{1}{2} MV^2 = \frac{mQ}{m + M}. \tag{3.19}$$

As a check, we see that $E_\alpha + E_N = Q$. Because of its much smaller mass, the alpha particle, having the same momentum as the nucleus, has much more energy. For ^{226}Ra, it follows from (3.14) and (3.18) that

$$E_\alpha = \frac{222 \times 4.88}{4 + 222} = 4.79 \text{ MeV}. \tag{3.20}$$

The radon nucleus recoils with an energy of only 0.09 MeV.

The conservation of momentum and energy, Eqs. (3.15) and (3.16), fix the energy of an alpha particle uniquely for given values of Q and M. Alpha particles therefore occur with discrete values of energy.

Appendix D gives the principal radiations emitted by various nuclides. We consider each of those listed for ^{226}Ra. Two alpha-particle energies are shown: 4.78 MeV,† occurring with a frequency of 95% of all decays and 4.60 MeV, occurring the other 5% of the time. The Q value for the less frequent alpha particle can be found from Eq. (3.18):

$$Q = \frac{(m + M)E_\alpha}{M} = \frac{226 \times 4.60}{222} = 4.68 \text{ MeV}. \tag{3.21}$$

The decay in this case goes to an excited state of the ^{222}Rn nucleus. Like excited atomic states, excited nuclear states can decay by photon emission. Photons from the nucleus are called gamma rays, and their energies are generally in the range from tens of keV to several MeV. Under the gamma rays listed in Appendix D for ^{226}Ra we find a 0.186 MeV photon emitted in 4% of the decays, in addition to another that occurs very infrequently (following alpha decay to still another excited level of higher energy in the daughter nucleus). We conclude that emission of the higher-energy alpha particle ($E = 4.79$ MeV) leaves the daughter ^{222}Rn nucleus in its ground state. Emission of the 4.60 MeV alpha particle leaves the nucleus in an excited state with energy $4.79 - 4.60 = 0.19$ MeV above the ground state. A photon of this energy can then be emitted from the nucleus, and, indeed, one of energy 0.186 MeV is listed for 4% of the decays. As an alternative to photon emission, under certain circumstances an excited nuclear state can decay by ejecting an atomic electron, usually from the K or L shell. This process, which produces the electrons listed (e^-), is called internal conversion, and will be discussed in Section 3.6.‡ For ^{226}Ra, since the excited state occurs in 5% of the total disintegrations and the 0.186 MeV photon is emitted only 4% of the time, it follows that internal conversion occurs in about 1% of the total decays. As we show in more detail in Section 3.6, the energy of the conversion electron is equal to the excited-state energy (in this case 0.186 MeV) minus the atomic-shell binding energy. The listing in Appendix D shows one of the e^- energies to be 0.170 MeV. In addition, since internal conversion leaves a K- or L-shell vacancy in the daughter atom, one also finds among the photons emitted the characteristic X-rays of Rn. Finally, as noted in the radiations listed in Appendix D for ^{226}Ra, various kinds of radiation are emitted from the radioactive daughters, in this case ^{222}Rn, ^{218}Po, ^{214}Pb, ^{214}Bi, and ^{214}Po.

---

†We obtained 4.79 MeV above; the difference is attributable to roundoff and/or to experimental differences and uncertainties.

‡In atoms, an Auger electron can be ejected from a shell in place of a photon, accompanying an electronic transition (Sect. 2.11).

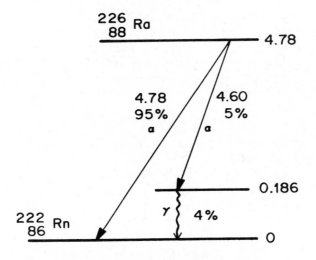

Figure 3.4. Nuclear decay scheme of $^{226}_{88}$Ra.

Decay-scheme diagrams, such as that shown in Fig. 3.4 for ^{226}Ra, conveniently summarize the nuclear transformations. The two arrows slanting downward to the left† show the two modes of alpha decay along with the alpha-particle energies and frequencies. Either changes the nucleus from that of ^{226}Ra to that of ^{222}Rn. When the lower-energy particle is emitted, the radon nucleus is left in an excited state with energy 0.186 MeV above the ground state. (The vertical distances in Fig. 3.4 are not to scale.) The subsequent gamma ray of this energy, which is emitted almost immediately, is shown by the vertical wavy line. The frequency 4% associated with this photon emission implies that an internal-conversion electron is emitted in the other 1% of the total number of disintegrations. Radiations not emitted directly from the nucleus (i.e., the Rn X-rays and the internal-conversion electron) are not shown on such a diagram, which represents the nuclear changes. Relatively infrequent modes of decay could also be shown, but are not included in Fig. 3.4. Most alpha emitters have a larger number of important alpha and gamma energies than the relatively simple spectrum of the ^{226}Ra nuclide.

The most energetic alpha particles are found to come from radionuclides having relatively short half-lives. An early empirical finding, known as the Geiger–Nuttall law, implies that there is a linear relationship between the logarithm of the range R of an alpha particle in air and the logarithm of the emitter's half-life T. The relation can be expressed in the form

$$-\ln T = a + b \ln R, \tag{3.22}$$

where a and b are empirical constants.

To conclude this section, we briefly consider the possible radiation-protection problems that alpha emitters can present. As we shall see in the next chapter, alpha particles have very short ranges and cannot even penetrate the outer, dead layer of skin. Therefore, they generally pose no direct external hazard to the body. Inhaled, ingested, or entering through a wound, however, an alpha source can present a hazard as an internal

---

†By convention, going left represents a decrease in Z and right, an increase in Z. Photon emission is represented by a vertical wavy line.

emitter. Depending upon the element, internal emitters tend to seek various organs and irradiate them. Radium, for example, seeks bone, where it can become lodged and irradiate an individual over his or her lifetime. In addition to the internal hazard, one can expect gamma rays along with an alpha source, as is the case with radium. Also, many alpha emitters have radioactive daughters that present radiation-protection problems.

3.4 BETA DECAY

In beta decay, a nucleus simultaneously emits an electron, or negative beta particle, $_{-1}^{0}\beta$, and an antineutrino, $_{0}^{0}\bar{\nu}$. Both of these particles are created at the moment of nuclear decay. The antineutrino, like its antiparticle† the neutrino, $_{0}^{0}\nu$, has no charge and little or no mass‡; they have been detected only in rather elaborate experiments.

As an example of beta decay, we consider ^{60}Co:

$$_{27}^{60}\text{Co} \rightarrow {}_{28}^{60}\text{Ni} + {}_{-1}^{0}\beta + {}_{0}^{0}\bar{\nu}. \tag{3.23}$$

In this case, the value of Q is equal to the difference between the mass of the ^{60}Co nucleus, $M_{\text{Co,N}}$, and that of the ^{60}Ni nucleus, $M_{\text{Ni,N}}$, plus one electron (m):

$$Q = M_{\text{Co,N}} - (M_{\text{Ni,N}} + m). \tag{3.24}$$

The nickel atom has one more electron than the cobalt atom. Therefore, if we neglect differences in atomic-electron binding energies, Eq. (3.24) implies that Q is simply equal to the difference in the masses of the ^{60}Co and ^{60}Ni *atoms*.§ Therefore, it follows that one can compute the energy released in beta decay from the difference in the values Δ_P and Δ_D, of the parent and daughter atoms:

$$Q_{\beta^-} = \Delta_\text{P} - \Delta_\text{D}. \tag{3.25}$$

Using the data from Appendix D, we find for the energy released in a β^- transformation of ^{60}Co to the ground state of ^{60}Ni

$$Q = -61.651 - (-64.471) = 2.820 \text{ MeV}. \tag{3.26}$$

In accordance with (3.23), this energy is shared by the beta particle, antineutrino, and recoil ^{60}Ni nucleus. The latter, because of its relatively large mass, receives negligible energy, and so

$$E_{\beta^-} + E_{\bar{\nu}} = Q, \tag{3.27}$$

where E_{β^-} and $E_{\bar{\nu}}$ are the initial kinetic energies of the electron and antineutrino. Depending on the relative directions of the momenta of the three decay products (β^-, $\bar{\nu}$, and recoil nucleus), E_{β^-} and $E_{\bar{\nu}}$ can each have any value between zero and Q, subject to the condition (3.27) on their sum. Thus the spectrum of beta-particle energies E_{β^-} is

---

†The Dirac equation predicts the existence of an antiparticle for every spin-$\frac{1}{2}$ particle and describes its relationship to the particle. Other examples include the positron, $_{+1}^{0}\beta$, antiparticle to the electron; the antiproton; and the antineutron. Creation of a spin-$\frac{1}{2}$ particle is always accompanied by creation of a related particle, which can be the antiparticle, such as happens in the creation of an electron–positron pair. Particle–antiparticle pairs can annihilate, as electrons and positrons do. A bar over a symbol is used to denote an antiparticle: e.g., ν, $\bar{\nu}$.

‡Experimentally, the neutrino and antineutrino masses cannot be larger than about 30 eV.

§We can think of adding and subtracting 27 electron masses in Eq. (3.24), giving $Q = (M_{\text{Co,N}} + 27m) - (M_{\text{Ni,N}} + 28m)$. Neglecting the difference in electron binding energies, then, we have $Q = M_{\text{Co},A} - M_{\text{Ni},A}$, where the subscript A denotes the atomic masses. It follows that Q is equal to the difference in Δ values for the two atoms.

continuous, with $0 \le E_{\beta^-} \le Q$, in contrast to the discrete spectra of alpha particles, as required by Eq. (3.18). Alpha particles are emitted in a decay into two bodies, which must share energy and momentum in a unique way, giving rise to discrete alpha spectra. Beta particles are emitted in a decay into three bodies, which can share energy and momentum in a continuum of ways, resulting in continuous beta spectra. The shape of a typical spectrum is shown in Fig. 3.5. The maximum beta-particle energy is always equal to the Q value for the nuclear transition. As a rule of thumb, the average beta energy is about one-third of Q: $\bar{E}_{\beta^-} \sim Q/3$.

To construct the decay scheme for ^{60}Co we consult Appendix D. We see that 99+% of the decays occur with $Q = 0.314$ MeV and that both of the gamma photons (listed with 100% frequency) occur with every disintegration. Therefore, almost every decay must go through an excited state of the daughter ^{60}Ni nucleus with an energy at least $1.173 + 1.332 = 2.505$ MeV above the ground state. Adding the maximum beta energy to this gives $2.505 + 0.314 = 2.819$ MeV, the value [Eq. (3.26)] calculated for a transition all the way to the ground state of the ^{60}Ni nucleus. Therefore, we conclude that the ^{60}Co nucleus first emits a beta particle, with $Q = 0.314$ MeV, which is followed successively by the two gamma rays. It remains to determine the energy of the nuclear excited state from which the second photon is emitted: 1.173 MeV or 1.332 MeV? Appendix D lists a rare beta particle with $Q = 1.48$ MeV. This decay must go to a level in the daughter nucleus having an energy $2.819 - 1.48 = 1.339$ MeV above the ground state. Thus we can conclude that the 1.332 MeV photon is emitted last in the transition to the ground state. The decay scheme is shown in Fig. 3.6. The arrows drawn slanting toward the right indicate the increase in atomic number that results from β^- decay. The rare mode is shown with a dashed line. No significant internal conversion occurs with this radionuclide.

A number of beta emitters have no accompanying gamma rays. Examples of such pure beta emitters are ^{3}H, ^{14}C, ^{32}P, ^{90}Sr, and ^{90}Y. Mixed beta–gamma emitters include ^{60}Co, ^{137}Cs, and many others. A number of radionuclides emit beta particles in decaying to several levels of the daughter nucleus, thus giving rise to complex beta spectra. A few radioisotopes can decay by emission of either an alpha or a beta particle. For example, $^{212}_{83}$Bi decays by alpha emission 36% of the time and by beta emission 64% of the time.

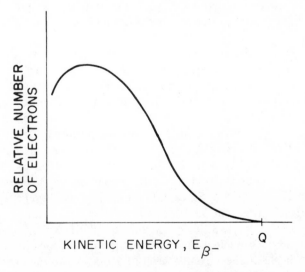

Figure 3.5. Shape of typical beta-particle energy spectrum.

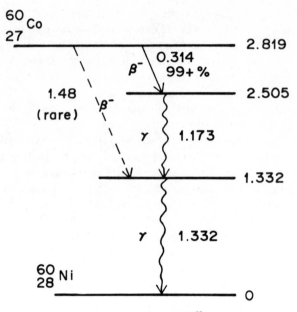

Figure 3.6. Decay scheme of $^{60}_{27}$Co.

Beta rays can have sufficient energy to penetrate the skin and thus be an external radiation hazard. Internal beta emitters are also a hazard. As is the case with ^{60}Co, many beta radionuclides also emit gamma rays. High-energy beta particles (i.e., in the MeV range) can emit bremsstrahlung, particularly in heavy-metal shielding. The bremsstrahlung from a beta source may be the only radiation that escapes the containment.

3.5 GAMMA-RAY EMISSION

As we have seen, one or more gamma photons can be emitted from the excited states of daughter nuclei following radioactive decay. Transitions that result in gamma emission leave Z and A unchanged and are called *isomeric*; nuclides in the initial and final states are called *isomers*.

As the examples in the last two sections illustrate, the gamma-ray spectrum from a radionuclide is discrete. Furthermore, just as optical spectra are characteristic of the chemical elements, a gamma-ray spectrum is characteristic of the particular radionuclides that are present. By techniques of gamma-ray spectroscopy (Chapter 9), the intensities of photons at various energies can be measured to determine the distribution of radionuclides in a sample. When ^{60}Co is present, for example, photons of energy 1.173 MeV and 1.332 MeV are observed with equal frequency. (Although these are called "^{60}Co gamma rays," we note from Fig. 3.6 that they are actually emitted by the daughter ^{60}Ni nucleus.) Radium can also be easily detected by its gamma-ray spectrum, which is more complex than indicated by Fig. 3.4. Since individual photons are registered in a spectrometer, gamma rays from infrequent modes of radioactive decay can often be readily measured. Figure 3.7 shows a detailed decay scheme for ^{226}Ra, which involves three excited states of the daughter ^{222}Rn nucleus and the emission of photons of four different energies. Transitions from the highest excited level (0.610 MeV) to the next (0.447 MeV) and from there to ground are forbidden by selection rules.

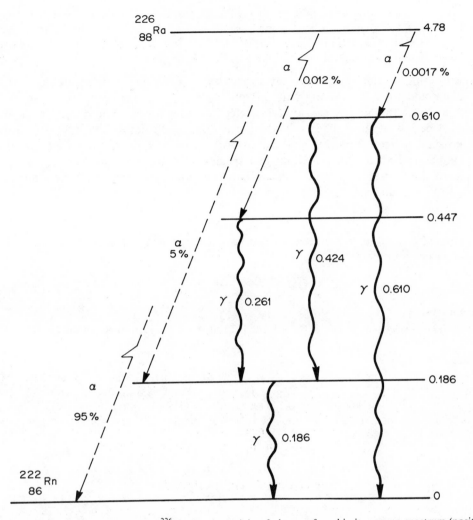

Figure 3.7. Detailed decay scheme for $^{226}_{88}$Ra showing origin of photons found in its gamma spectrum (position of initial $^{226}_{88}$Ra energy level not to scale).

Example

Like $^{60}_{27}$Co, another important gamma-ray source is the radioisotope $^{137}_{55}$Cs. Consult Appendix D and work out its decay scheme.

Solution

Also like $^{60}_{27}$Co, $^{137}_{55}$Cs is a β^- emitter that leaves its daughter, stable $^{137}_{56}$Ba, in an excited state that results in gamma emission. The decay is represented by

$$^{137}_{55}\text{Cs} \rightarrow {}^{137}_{56}\text{Ba} + {}^{0}_{-1}\beta + {}^{0}_{0}\bar{\nu}. \tag{3.28}$$

From the Δ values in Appendix D, we obtain for decay to the daughter ground state $Q = -86.9 + 88.0 = 1.1$ MeV. Comparison with the radiations listed in the Appendix indicates that decay by this mode takes place 7% of the time, releasing 1.176 MeV. Otherwise, the decay in 93% of the cases leaves the daughter nucleus in an excited state with energy $1.176 - 0.514 = 0.662$ MeV. A photon of this energy is shown with 85% frequency. Therefore, internal conversion occurs in $93 - 85 = 8\%$

of the disintegrations, giving rise to the conversion electrons, e⁻, with the energies shown. Characteristic Ba X-rays are emitted following the inner-shell vacancies created in the atom by internal conversion. The decay scheme of $^{137}_{55}$Cs is drawn in Fig. 3.8; the spectrum of electrons emitted by the source is shown schematically in Fig. 3.9.

The lifetimes of nuclear excited states vary, but $\sim 10^{-10}$ sec can be regarded as typical. Thus, gamma rays are usually emitted quickly after radioactive decay to an excited daughter state. In some cases, however, selection rules prevent photon emission for an extended period of time. The molybdenum isotope $^{99}_{42}$Mo decays by β^- emission to an excited state of technetium, which has an average lifetime of 8.7 hr before emitting a gamma photon. Such a long-lived nuclear species is termed *metastable* and is designated by use of the symbol m: $^{99m}_{43}$Tc. The nucleus then makes an isomeric transition (IT) in going to the ground state:

$$^{99m}_{43}\text{Tc} \rightarrow {}^{99}_{43}\text{Tc} + {}^{0}_{0}\gamma. \tag{3.29}$$

The energy released in an isomeric transition is simply equal to the difference in Δ values of the parent and daughter atoms:

$$Q_{\text{IT}} = \Delta_\text{P} - \Delta_\text{D}. \tag{3.30}$$

Example
Work out the decay scheme of $^{99m}_{43}$Tc with the help of the data given in Appendix D.

Solution
Using Eq. (3.30) and the given values of Δ, we obtain for the energy released in going to the ground state in the transition (3.29), $Q = 87.33 - 87.18 = 0.15$ MeV. This transition is responsible for the gamma photon listed in Appendix D, 0.140 (90%). By implication, internal conversion must occur the other 10% of the time, and one finds two electron energies (e⁻), one a little less than the pho-

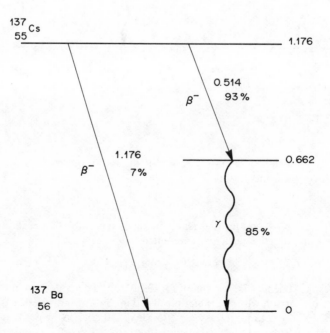

Figure 3.8. Decay scheme of $^{137}_{55}$Cs.

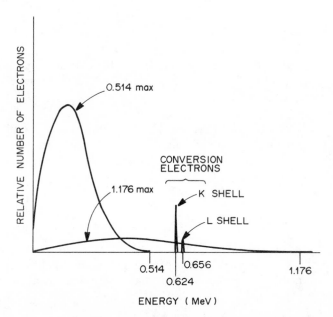

Figure 3.9. Sources of electrons from $^{137}_{55}$Cs and their energy spectra. There are two modes of β^- decay, with maximum energies of 0.514 MeV (93%) and 1.176 MeV (7%). Internal conversion electrons also occur at discrete energies of 0.624 MeV (from K shell) and 0.656 MeV (L shell) with a total frequency of 8%. See decay scheme in Fig. 3.8. The total spectrum of emitted electrons is the sum of the curves shown here.

ton energy (by an amount that equals the L-shell electron binding energy in the Tc atom). Because internal conversion leaves inner-shell vacancies, a $^{99m}_{}$Tc source also emits characteristic Tc X-rays, as listed. Since $^{99}_{43}$Tc decays by β^- emission into stable $^{99}_{44}$Ru, no daughter radiations occur.

The penetration of gamma rays in matter is fundamentally different in nature from that of alpha and beta particles. A given type of charged particle with a given initial energy has a definite range of penetration associated with it; a certain thickness of matter will completely absorb a beam of such particles. In contrast, the intensity of a beam of gamma rays is steadily attenuated in matter, but some photons can be expected to traverse even very thick shields. Gamma emitters present a radiation hazard for which the dose rate cannot always be reduced to zero. Protection from gamma (and X-ray) sources is discussed in Chapter 13.

3.6 INTERNAL CONVERSION

Internal conversion is the process in which the energy of an excited nuclear state is transferred to an atomic electron, most likely a K- or L-shell electron, ejecting it from the atom. It is an alternative to emission of a gamma photon from the nucleus.† We had examples of internal conversion in discussing the decay of ^{137}Cs and ^{99m}Tc.

The internal conversion coefficient α for a nuclear transition is defined as the ratio of the number of conversion electrons N_e and the number of competing gamma photons N_γ for that transtion:

---

†Internal conversion does not occur as a two-step process in which a photon is emitted by the nucleus and then absorbed by the atomic electron. The mechanism is entirely different. A similar observation was made in regard to Auger electrons in Section 2.11.

$$\alpha = \frac{N_e}{N_\gamma}. \tag{3.31}$$

The kinetic energy E_e of the ejected atomic electron is very nearly equal to the excitation energy E^* of the nucleus minus the binding energy E_B of the electron in its atomic shell:

$$E_e = E^* - E_B. \tag{3.32}$$

The conversion coefficients α increase as Z^3, the cube of the atomic number, and decrease with E^*, especially favoring small E^*. Internal conversion is thus prevalent in heavy nuclei, especially in the decay of low-lying excited states (small E^*). Gamma decay predominates in light nuclei.

3.7 ORBITAL ELECTRON CAPTURE

Some nuclei undergo a radioactive transformation by capturing an atomic electron, usually from the K shell, and emitting a neutrino. An isotope of palladium undergoes this process of electron capture (EC), going to a metastable state of the nucleus of the daughter rhodium:

$$^{103}_{46}\text{Pd} + {}_{-1}^{0}\text{e} \rightarrow {}^{103\text{m}}_{45}\text{Rh} + {}_{0}^{0}\nu. \tag{3.33}$$

The neutrino acquires the entire energy Q released by the reaction.

To find Q, we note that the captured electron releases its total mass, $m - E_B$, to the nucleus when it is absorbed there, E_B being the mass equivalent of the binding energy of the electron in the atomic shell. Therefore, in terms of the masses $M_{\text{Pd,N}}$ and $M_{\text{mRh,N}}$ of the parent and daughter nuclei, the energy released by the reaction (3.33) is given by

$$Q = M_{\text{Pd,N}} + m - E_B - M_{\text{mRh,N}}. \tag{3.34}$$

Since the palladium atom has one more electron than the rhodium atom, it follows that (neglecting the difference in the atomic binding energies) Q is equal to the difference in the two *atomic* masses, less the energy E_B.† Since the difference in the atomic masses is equal to the difference in the parent and daughter Δ values, we can write the general expression for the energy release by electron capture:

$$Q_{\text{EC}} = \Delta_P - \Delta_D - E_B. \tag{3.35}$$

Orbital electron capture thus cannot take place unless $Q > E_B$. For the K shell of palladium, $E_B = 0.024$ MeV. Using the Δ values from Appendix D in Eq. (3.35), we find that, for the decay to $^{103\text{m}}$Rh,

$$Q = -87.46 - (-87.974) - 0.024 = 0.490 \text{ MeV}. \tag{3.36}$$

In subsequently decaying to the ground state, $^{103\text{m}}$Rh releases an energy $-87.974 + 88.014 = 0.040$ MeV, as found from the values in Appendix D.

A decay scheme for ^{103}Pd is given in Fig. 3.10. Since electron capture decreases the atomic number of the nucleus, it is symbolized by an arrow pointing downward toward the left. The solid arrow represents the transition to $^{103\text{m}}$Rh that we analyzed above. The presence of the three gamma rays listed for ^{103}Pd in Appendix D implies that EC sometimes leaves the nucleus in other excited levels, as shown. These transitions, as well as one directly to the ground state, are indicated by the dashed arrows. It is not possible from

---

†A similar argument was given in the footnote after Eq. (3.24), except that E_B was not involved there.

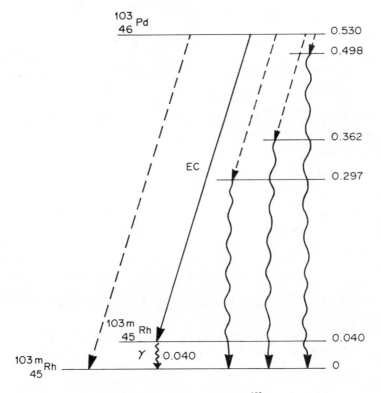

Figure 3.10. Decay scheme of $^{103}_{46}$Pd.

the information given in Appendix D to specify the frequency of these transitions relative to that represented by the solid arrow. Since electron capture necessarily leaves an inner-atomic-shell vacancy, characteristic X-rays of the daughter are always emitted. (Electron capture is detected through the observation of characteristic X-rays and Auger electrons as well as the recoil of the daughter nucleus.)

The radiations listed for ^{103m}Rh in Appendix D can also be explained. The photon with energy 0.040 MeV is shown in Fig. 3.10. Its 0.4% frequency implies that 99.6% of the time the excited nuclei decay to the ground state by internal conversion, resulting in the ejection of atomic electrons (e^-) with the energies shown. (The present instance affords an example of internal conversion being favored over gamma emission in the decay of low-lying excited states in heavy nuclei, mentioned at the end of the last section.) Internal conversion also leaves a vacancy in an atomic shell, and hence the characteristic X-rays of Rh are also found in a ^{103m}Rh source.

The neutrino emitted in electron capture has a negligible interaction with matter and offers no radiation hazard, as far as is known. Characteristic X-rays of the daughter will always be present. In addition, if the capture does not leave the daughter in its ground state, gamma rays will occur.

3.8 POSITRON DECAY

Some nuclei, such as $^{22}_{11}$Na, disintegrate by emitting a positively charged electron (positron, β^+) and a neutrino:

$$^{22}_{11}\text{Na} \rightarrow {}^{22}_{10}\text{Ne} + {}^{0}_{1}\beta + {}^{0}_{0}\nu. \qquad (3.37)$$

Positron decay has the same net effect as electron capture, reducing Z by one unit and leaving A unchanged. The energy released is given in terms of the masses $M_{\text{Na,N}}$ and $M_{\text{Ne,N}}$ of the sodium and neon nuclei by

$$Q = M_{\text{Na,N}} - M_{\text{Ne,N}} - m. \qquad (3.38)$$

Thus the mass of the parent nucleus must be greater than that of the daughter nucleus by at least the mass m of the positron it creates. As before, we need to express Q in terms of atomic masses, $M_{\text{Na,}A}$ and $M_{\text{Ne,}A}$. Since Na has 11 electrons and Ne 10, we write

$$Q = M_{\text{Na,N}} + 11m - (M_{\text{Ne,N}} + 10m) - 2m \qquad (3.39)$$
$$= M_{\text{Na,}A} - M_{\text{Ne,}A} - 2m, \qquad (3.40)$$

where the difference in the atomic binding of the electrons has been neglected in writing the last equality. In terms of the values Δ_{P} and Δ_{D} of the parent and daughter, the energy released in positron decay is given by

$$Q_{\beta^+} = \Delta_{\text{P}} - \Delta_{\text{D}} - 2mc^2. \qquad (3.41)$$

Therefore, for positron emission to be possible, the mass of the parent atom must be greater than that of the daughter by at least $2mc^2 = 1.022$ MeV. Using the information from Appendix D, we find for the energy released via positron emission in the decay (3.37) to the ground state of $^{22}_{10}\text{Ne}$

$$Q_{\beta^+} = -5.182 - (-8.025) - 1.022 = 1.821 \text{ MeV}. \qquad (3.42)$$

Electron capture, which results in the same net change as positron decay, can compete with (3.37):

$$^{0}_{-1}\text{e} + {}^{22}_{11}\text{Na} \rightarrow {}^{22}_{10}\text{Ne} + {}^{0}_{0}\nu. \qquad (3.43)$$

Neglecting the electron binding energy in the $^{22}_{11}\text{Na}$ atom, we obtain from Eq. (3.35) for the energy released by electron capture

$$Q_{\text{EC}} = -5.182 + 8.025 = 2.843 \text{ MeV}. \qquad (3.44)$$

[Comparison of Eqs. (3.35) and (3.41) shows that the Q value for EC is greater than that for β^+ decay by 1.022 MeV when E_{B} is neglected.]

We next develop the decay scheme for $^{22}_{11}\text{Na}$. Appendix D indicates that β^+ emission occurs 90.6% of the time and EC 9.4%. A gamma ray with energy 1.275 MeV occurs with essentially 100% frequency, indicating that either β^+ emission or EC leaves the daughter nucleus in an excited state with this energy. [An exception is the infrequent positron decay with $Q = 1.820$, which, by (3.42), goes directly to the ground state of the daughter.] The positron decay scheme is shown in Fig. 3.11(a) and that for electron capture in (b). The two are combined in (c) to show the complete decay scheme for $^{22}_{11}\text{Na}$. The energy levels are drawn relative to the ground state of $^{22}_{10}\text{Ne}$ as having zero energy. The starting EC level is $2mc^2$ higher than the starting level for β^+ decay.

Additional radiations are given in Appendix D for $^{22}_{11}\text{Na}$. Gamma rays of energy 0.511 MeV are shown with 180% frequency. These are annihilation photons that are present with all positron emitters. A positron slows down in matter and then annihilates with an atomic electron, giving rise to two photons, each having energy $mc^2 = 0.511$ MeV and traveling in opposite directions. Since a positron is emitted in about 90% of the decay processes, the frequency of an annihilation photon is 1.8 per disintegration of a $^{22}_{11}\text{Na}$

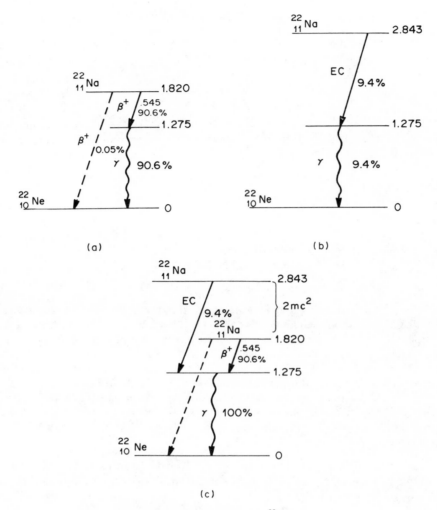

Figure 3.11. Decay scheme of $^{22}_{11}$Na.

atom. The remaining radiation shown, Ne X-rays, comes as the result of the atomic-shell vacancy following electron capture.

As this example shows, electron capture and positron decay are competitive processes. However, whereas positron emission cannot take place when the parent–daughter atomic mass difference is less than $2mc^2$, electron capture can, the only restriction being $Q > E_B$, as implied by Eq. (3.35). The nuclide $^{126}_{53}$I can decay by three routes: EC (55%), β^- (44%), or β^+ (1.3%).†

The radiation-protection problems associated with positron emitters include all those of β^- emitters (direct radiation and possible bremsstrahlung) and then some. As already mentioned, the 0.511 MeV annihilation photons are always present. In addition, because of the competing process of electron capture, characteristic X-rays can be expected.

---

†In general, the various possible decay modes for a nuclide are those for which $Q > 0$ for a transition to the daughter ground state.

Table 3.1. Formulas for Energy Release, Q, in Terms of Mass Differences, Δ_P and Δ_D, of Parent and Daughter Atoms

Type of decay	Formula	Reference
α	$Q_\alpha = \Delta_P - \Delta_D - \Delta_{He}$	Eq. (3.13)
β^-	$Q_{\beta^-} = \Delta_P - \Delta_D$	Eq. (3.25)
γ	$Q_{IT} = \Delta_P - \Delta_D$	Eq. (3.30)
EC	$Q_{EC} = \Delta_P - \Delta_D - E_B$	Eq. (3.35)
β^+	$Q_{\beta^+} = \Delta_P - \Delta_D - 2mc^2$	Eq. (3.41)

Example

Refer to Appendix D and deduce the decay scheme of $^{26}_{13}\text{Al}$.

Solution

This nuclide decays by β^+ emission (85%) and EC (15%). The energy release for EC with a transition to the daughter ground state is, from the Δ values,

$$Q_{EC} = -12.211 + 16.214 = 4.003 \text{ MeV.} \tag{3.45}$$

Here we have neglected the small binding energy of the atomic electron. The corresponding value for β^+ decay to the ground state is $Q_{\beta^+} = 4.003 - 1.022 = 2.981$ MeV. A 1.81 MeV gamma photon is emitted with 100% frequency, and so we can assume that both EC and β^+ decay modes proceed via an excited daughter state of this energy. Adding this to the maximum β^+ energy, we have $1.81 + 1.17 = 2.98$ MeV $= Q_{\beta^+}$. Therefore, the positron decay occurs as shown in Fig. 3.12(a). Its 85% frequency accounts for the annihilation photons listed with 170% frequency in Appendix D. The other 15% of the decays via EC also go through the level at 1.81 MeV. An additional photon of energy 1.12 MeV and frequency 4% is listed in Appendix D. This can arise if a fraction of the EC transformations go to a level with energy $1.81 + 1.12 = 2.93$ MeV above ground. The complete decay scheme is shown in Fig. 3.12(b).

This completes the description of the various types of radioactive decay. The formulas for finding the energy release Q from the mass differences Δ of the parent and daughter atoms are summarized in Table 3.1.

3.9 NATURAL RADIOACTIVITY

All of the heavy elements ($Z > 83$) found in nature are radioactive and decay by alpha or beta emission. The heaviest ones decay into successive radioactive daughters that form a series of radionuclides, ending in the production of a stable nuclide. Since the mass number can change only by four units (when alpha emission occurs), there are four possible series. The atomic mass numbers A of the members of each series are identifiable by the remainders left when A is divided by 4. The mass number of members of the thorium series, which begins with $^{232}_{90}\text{Th}$, are exactly divisible by 4. This series ends with the decay of $^{208}_{81}\text{Tl}$ into stable $^{208}_{82}\text{Pb}$. The series with remainder one after division of A by 4 is called the neptunium series. This series, which begins with $^{241}_{94}\text{Pu}$ and ends with $^{209}_{83}\text{Bi}$, is not found in nature; it is man-made. The neptunium isotope, $^{237}_{93}\text{Np}$, with a half-life of 2.2×10^6 years is the longest-lived member. The uranium series has remainder 2. It begins with $^{238}_{92}\text{U}$ and goes to stable $^{206}_{82}\text{Pb}$. The actinium series, with remainder 3, begins with $^{235}_{92}\text{U}$ and ends with $^{207}_{82}\text{Pb}$.

Several lighter elements have naturally occurring radioisotopes. One of the most important from the standpoint of human exposure is ^{40}K, which is about 0.01% abundant and has a half-life of 1.26×10^9 years. The nuclide can decay by β^- emission

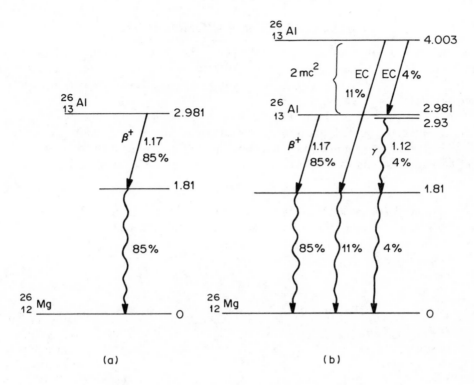

Figure 3.12. Decay scheme of $^{26}_{13}$Al (see example in text).

(89%), EC (11%), or β^+ emission (~0.001%). The maximum β^- energy is 1.314 MeV. This isotope is an important source of human internal radiation exposure, because potassium is a natural constituent of plants and animals.

Another important naturally occurring radioisotope is ^{14}C, used in carbon dating. It is produced by bombardment of nitrogen in the atmosphere by cosmic rays, causing the reaction $^{14}_{7}$N(n,p)$^{14}_{6}$C. Its half-life is 5730 yr. The radioisotope, existing as CO_2 in the atmosphere, is utilized by plants and becomes fixed in their structure through photosynthesis. The time at which ^{14}C was assimilated in a previously living specimen, used to make furniture or paper, for example, can be inferred from the relative amount of the isotope remaining in it today. Thus the age of such objects can be determined. The atmospheric testing of nuclear weapons has added a significant amount to the world's inventory of ^{14}C.

Example

How many alpha and beta particles are emitted by a nucleus of an atom of the uranium series, which starts as $^{238}_{92}$U and ends as stable $^{206}_{82}$Pb?

Solution

Nuclides of the four heavy-element radioactive series decay either by alpha or beta emission. A single disintegration, therefore, either (1) reduces the atomic number by 2 and the mass number by 4 or (2) increases the atomic number by 1 and leaves the mass number unchanged. Since the atomic mass numbers of $^{238}_{92}$U and $^{206}_{82}$Pb differ by 32, it follows that 8 alpha particles are emitted in the series. Since this alone would reduce the atomic number by 16, as compared with the actual reduction of 10, a total of 6 beta particles must also be emitted.

3.10 ACTIVITY AND EXPONENTIAL DECAY

The rate of decay of a radioactive sample is described by its activity, i.e., by the number of atoms that disintegrate per unit time. An activity of 3.7×10^{10} sec^{-1} is defined as 1 curie (Ci), this being the decay rate of 1 g of ^{226}Ra. In SI units, activity is expressed in becquerels (1 Bq = 1 sec^{-1}), with 1 Ci = 3.7×10^{10} Bq.

The activity of a pure radionuclide decreases exponentially with time, as we now show. If N represents the number of atoms of a radionuclide in a sample at any given time, then the change dN in the number during a short time dt is proportional to N and to dt. Letting λ be the constant of proportionality, we write

$$dN = -\lambda N\,dt. \tag{3.46}$$

The negative sign is needed because N decreases as the time t increases. The quantity λ is called the decay, or transformation, constant; it has the dimensions of inverse time (e.g., sec^{-1}). The decay rate, or activity, A, is given by

$$A = -\frac{dN}{dt} = \lambda N. \tag{3.47}$$

We separate the variables in Eq. (3.46) by writing

$$\frac{dN}{N} = -\lambda\,dt. \tag{3.48}$$

Integration of both sides gives

$$\ln N = -\lambda t + c, \tag{3.49}$$

where c is an arbitrary constant of integration, fixed by the initial conditions. If we specify that N_0 atoms of the radionuclide are present at time $t = 0$, then Eq. (3.49) implies that $c = \ln N_0$. In place of (3.49) we write

$$\ln N = -\lambda t + \ln N_0, \tag{3.50}$$

$$\ln \frac{N}{N_0} = -\lambda t, \tag{3.51}$$

or

$$\frac{N}{N_0} = e^{-\lambda t}. \tag{3.52}$$

Equation (3.52) describes the exponential radioactive decay law. Since the activity of a sample and the number of atoms present are proportional, activity follows the same rate of decrease,

$$\frac{A}{A_0} = e^{-\lambda t}, \tag{3.53}$$

where A_0 is the activity at time $t = 0$. The dose rate at a given location in the neighborhood of a fixed radionuclide source also falls off at the same exponential rate.

The function (3.53) is plotted in Fig. 3.13. During successive times T, called the half-life of the radionuclide, the activity drops by factors of one-half, as shown. To find T in terms of λ, we write from Eq. (3.53) at time $t = T$,

$$\frac{1}{2} = e^{-\lambda T}. \tag{3.54}$$

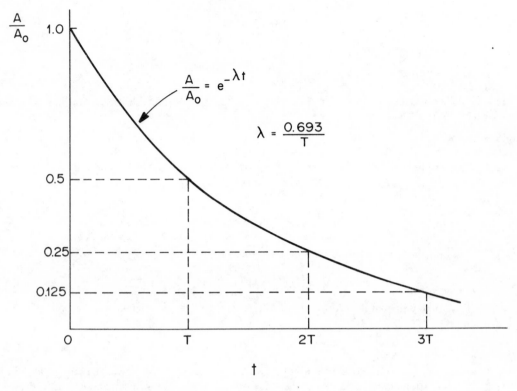

Figure 3.13. Exponential radioactivity decay law, showing relative activity, A/A_0, as function of time t; λ is the decay constant and T the half-life.

Taking the natural logarithm of both sides gives

$$-\lambda T = \ln\left(\frac{1}{2}\right) = -\ln 2, \tag{3.55}$$

and therefore

$$T = \frac{\ln 2}{\lambda} = \frac{0.693}{\lambda}. \tag{3.56}$$

Written in terms of the half-life, the exponential decay laws (3.52) and (3.53) become

$$\frac{N}{N_0} = \frac{A}{A_0} = e^{-0.693t/T}. \tag{3.57}$$

The decay law (3.57) can be derived simply on the basis of the half-life. If, for example, the activity decreases to a fraction A/A_0 of its original value after passage of time t/T half-lives, then we can write

$$\frac{A}{A_0} = \left(\frac{1}{2}\right)^{t/T}. \tag{3.58}$$

For example, for integral n, when $t = nT$, $A/A_0 = (\frac{1}{2})^n$. Taking the logarithm of both sides of Eq. (3.58) gives

$$\ln \frac{A}{A_0} = -\frac{t}{T} \ln 2 = -\frac{0.693t}{T}, \tag{3.59}$$

from which Eq. (3.57) follows.

Example

Calculate the activity of a 30 mCi source of $^{24}_{11}$Na after 2.5 days. What is the decay constant of this radionuclide?

Solution

The problem can be worked in several ways. We first find λ from Eq. (3.56) and then the activity from Eq. (3.53). The half-life $T = 15.0$ hr of the nuclide is given in Appendix D. From (3.56),

$$\lambda = \frac{0.693}{T} = \frac{0.693}{15.0 \text{ hr}} = 0.0462 \text{ hr}^{-1}. \tag{3.60}$$

With $A_0 = 30$ mCi and $t = 2.5$ days $\times$ 24 hr/day $= 60.0$ hr,

$$A = 30e - 0.0462 \text{ hr}^{-1} \times 60 \text{ hr} = 1.88 \text{ mCi}. \tag{3.61}$$

Note that the time units employed for λ and t must be the same in order that the exponential be dimensionless.

Example

A solution contains 0.10 μCi of ^{198}Au and 0.04 μCi of ^{131}I at time $t = 0$. What is the total beta activity in the solution at $t = 21$ days? At what time will the total activity decay to one-half its original value?

Solution

Both isotopes decay to stable daughters, and so the total beta activity is due to these isotopes alone. (A small fraction of ^{131}I decays into ^{131m}Xe, which does not contribute to the beta activity.) From Appendix D, the half-lives of ^{198}Au and ^{131}I are, respectively, 2.70 days and 8.05 days. At the end of 21 days, the activities A_{Au} and A_{I} of the nuclides are, from Eq. (3.57),

$$A_{\text{Au}} = 0.10e^{-0.693 \times 21/2.70} = 4.56 \times 10^{-4} \ \mu\text{Ci} \tag{3.62}$$

and

$$A_{\text{I}} = 0.04e^{-0.693 \times 21/8.05} = 6.56 \times 10^{-3} \ \mu\text{Ci}. \tag{3.63}$$

The total activity at $t = 21$ days is the sum of these two activities, $7.02 \times 10^{-3} \ \mu$Ci. To find the time t in days at which the activity has decayed to one-half its original value of $0.10 + 0.04 = 0.14$ Ci, we write

$$0.07 = 0.1e^{-0.693t/2.70} + 0.04e^{-0.693t/8.05}. \tag{3.64}$$

This is a transcendental equation, which cannot be solved in closed form for t. The solution can be found either graphically or by trial and error, focusing in between two values of t that make the right-hand side of (3.64) >0.07 and <0.07. We present a combination of both methods. The decay constants of the two nuclides are, for Au, $0.693/2.70 = 0.257$ day^{-1} and, for I, $0.693/8.05 = 0.0861$ day^{-1}. The activities in μCi, as functions of time t, are

$$A_{\text{Au}}(t) = 0.10e^{-0.257t} \tag{3.65}$$

and

$$A_{\text{I}}(t) = 0.04e^{-0.0861t}. \tag{3.66}$$

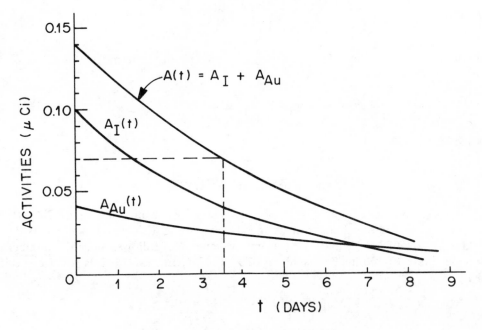

Figure 3.14. Graphical solution to example in text.

Figure 3.14 shows a plot of these two activities and the total activity, $A(t) = A_{Au} + A_1$, calculated as functions of t from these two equations. Plotted to scale, the total activity $A(t)$ is found to reach the value 0.07 μCi near $t = 3.50$ days. We can improve on this approximate graphical solution. Direct calculation from Eqs. (3.65) and (3.66) shows that $A(3.50) = 0.0703$ and $A(3.60) = 0.0689$. Linear interpolation suggests the solution $t = 3.52$ days, and, indeed, one can verify that $A(3.52) = 0.0700 \mu$Ci.

The average, or mean, life τ of a radionuclide is defined as the average of all of the individual lifetimes that the atoms in a sample of the radionuclide experience. It is equal to the mean value of t under the exponential curve in Fig. 3.15. Therefore, τ defines a rectangle, as shown, with area equal to the area under the exponential curve:

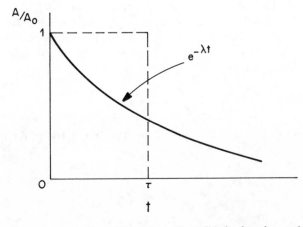

Figure 3.15. The average life τ of a radionuclide is given by $\tau = 1/\lambda$.

$$1 \times \tau = \int_0^\infty e^{-\lambda t} \, dt = -\frac{1}{\lambda} e^{-\lambda t}\big|_0^\infty = \frac{1}{\lambda}. \qquad (3.67)$$

Thus the mean life is the reciprocal of the decay constant. In terms of the half-life, we have

$$\tau = \frac{1}{\lambda} = \frac{T}{0.693}, \qquad (3.68)$$

showing that $\tau > T$.

3.11 SPECIFIC ACTIVITY

The specific activity of a sample is defined as its activity per unit mass, e.g., Ci/g. If the sample is a pure radionuclide, then its specific activity, SA, is determined by its decay constant λ, or half-life T, and by its atomic weight M as follows. Since the number of atoms per gram of the nuclide is $N = 6.02 \times 10^{23}/M$, Eq. (3.47) gives for the specific activity

$$SA = \frac{6.02 \times 10^{23} \, \lambda}{M} = \frac{4.17 \times 10^{23}}{MT}. \qquad (3.69)$$

If T is in seconds, then this formula gives the specific activity in Bq/g. In practice, using the atomic mass number A in place of M usually gives sufficient accuracy.

Example
Calculate the specific activity of ^{226}Ra in Bq/g.

Solution
From Appendix D, $T = 1602$ yr and $M = A = 226$. Converting T to seconds, we have

$$SA = \frac{4.17 \times 10^{23}}{226 \times 1602 \times 365 \times 24 \times 3600} \qquad (3.70)$$

$$= 3.65 \times 10^{10} \, \text{sec}^{-1} \, \text{g}^{-1} = 3.7 \times 10^{10} \, \text{Bq/g}. \qquad (3.71)$$

This, by definition, is an activity of 1 Ci.

The fact that ^{226}Ra has unit specific activity in terms of Ci/g can be used in place of Eq. (3.69) to find SA for other radionuclides. Compared with ^{226}Ra, a nuclide of shorter half-life and smaller atomic mass number A will have, in direct proportion, a higher specific activity than ^{226}Ra. The specific activity of a nuclide of half-life T and atomic mass number A is therefore given by

$$SA = \frac{1600}{T} \times \frac{226}{A} \, \text{Ci/g}, \qquad (3.72)$$

where T is expressed in years. (The equation gives $SA = 1$ for ^{226}Ra.)

Example
What is the specific activity of ^{14}C?

Solution
With $T = 5730$ yr and $A = 14$, Eq. (3.72) gives

$$SA = \frac{1600}{5730} \times \frac{226}{14} = 4.51 \ \text{Ci/g}. \qquad (3.73)$$

Alternatively, we can use Eq. (3.69) with $T = 5730 \times 365 \times 24 \times 3600 = 1.81 \times 10^{11}$ sec, obtaining

$$SA = \frac{4.17 \times 10^{23}}{14 \times 1.81 \times 10^{11}} = 1.65 \times 10^{11} \ \text{Bq/g} \qquad (3.74)$$

$$= \frac{1.65 \times 10^{11} \ \text{Bq/g}}{3.7 \times 10^{10} \ \text{Bq/Ci}} = 4.46 \ \text{Ci/g}, \qquad (3.75)$$

in agreement with (3.73).

Specific activity need not apply to a pure radionuclide. For example, ^{14}C produced by the $^{14}N(n,p)^{14}C$ reaction can be extracted chemically as a "carrier-free" radionuclide, i.e., without the presence of nonradioactive carbon isotopes. Its specific activity would be that calculated in the previous example. A different example is afforded by ^{60}Co, which is produced by neutron absorption in a sample of ^{59}Co (100% abundant), the reaction being $^{59}Co(n,\gamma)^{60}Co$. The specific activity of the sample depends on its radiation history, which determines the fraction of cobalt atoms that are made radioactive. Specific activity is also used to express the concentration of activity in solution; e.g., μCi/ml.

3.12 SERIAL RADIOACTIVE DECAY

In this final section we describe the activity of a sample in which one radionuclide produces one or more radioactive offspring in a chain. Several important cases will be discussed.

Secular Equilibrium ($T_1 \gg T_2$)

First, we calculate the total activity present at any time when a long-lived parent (1) decays into a relatively short-lived daughter (2), which, in turn, decays into a stable nuclide. The half-lives of the two radionuclides are such that $T_1 \gg T_2$; and we consider intervals of time that are short compared with T_1, so that the activity A_1 of the parent can be treated as constant. The total activity at any time is A_1 plus the activity A_2 of the daughter, on which we now focus. The rate of change, dN_2/dt, in the number of daughter atoms N_2 per unit time is equal to the rate at which they are produced, A_1, minus their rate of decay, $\lambda_2 N_2$:

$$\frac{dN_2}{dt} = A_1 - \lambda_2 N_2. \qquad (3.76)$$

To solve for N_2, we first separate variables by writing

$$\frac{dN_2}{A_1 - \lambda_2 N_2} = dt, \qquad (3.77)$$

where A_1 can be regarded as constant. Introducing the variable $u = A_1 - \lambda_2 N_2$, we have $du = -\lambda_2 dN_2$ and, in place of Eq. (3.77),

$$\frac{du}{u} = -\lambda_2 dt. \qquad (3.78)$$

Integration gives

$$\ln (A_1 - \lambda_2 N_2) = -\lambda_2 t + c, \qquad (3.79)$$

where c is an arbitrary constant. If N_{20} represents the number of atoms of nuclide (2) present at $t = 0$, then we have $c = \ln(A_1 - \lambda_2 N_{20})$. Equation (3.79) becomes

$$\ln \frac{A_1 - \lambda_2 N_2}{A_1 - \lambda_2 N_{20}} = -\lambda_2 t, \tag{3.80}$$

or

$$A_1 - \lambda_2 N_2 = (A_1 - \lambda_2 N_{20})e^{-\lambda_2 t}. \tag{3.81}$$

Since $\lambda_2 N_2 = A_2$, the activity of nuclide (2), and $\lambda_2 N_{20} = A_{20}$ is its initial activity, Eq. (3.81) implies that

$$A_2 = A_1(1 - e^{-\lambda_2 t}) + A_{20}e^{-\lambda_2 t}. \tag{3.82}$$

In many practical instances one starts with a pure sample of nuclide (1) at $t = 0$, so that $A_{20} = 0$, which we now assume. The activity A_2 then builds up as shown in Fig. 3.16. After about seven daughter half-lives ($t \gtrsim 7T_2$), $e^{-\lambda_2 t} \ll 1$ and Eq. (3.82) reduces to the condition $A_1 = A_2$, at which time the daughter activity is equal to that of the parent. This condition is called secular equilibrium. The total activity is $2A_1$. In terms of the numbers of atoms, N_1 and N_2, of the parent and daughter, secular equilibrium can be also expressed by writing

$$\lambda_1 N_1 = \lambda_2 N_2. \tag{3.83}$$

A chain of n short-lived radionuclides can all be in secular equilibrium with a long-lived parent. Then the activity of each member of the chain is equal to that of the parent and the total activity is $n + 1$ times the activity of the original parent.

General Case

When there is no restriction on the relative magnitudes of T_1 and T_2, we write in place of Eq. (3.76)

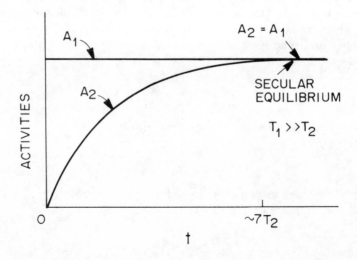

Figure 3.16. Activity A_2 of relatively short-lived radionuclide daughter ($T_2 \ll T_1$) as function of time t with initial condition $A_{20} = 0$. Activity of daughter builds up to that of the parent in about seven half-lives (T_2). Thereafter, daughter decays at the same rate it is produced ($A_2 = A_1$), and secular equilibrium is said to exist.

$$\frac{dN_2}{dt} = \lambda_1 N_1 - \lambda_2 N_2. \tag{3.84}$$

With the initial condition $N_{20} = 0$, the solution to this equation is

$$N_2 = \frac{\lambda_1 N_{10}}{\lambda_2 - \lambda_1} (e^{-\lambda_1 t} - e^{-\lambda_2 t}), \tag{3.85}$$

as can be verified by direct substitution into (3.84). This general formula yields Eq. (3.83) when $\lambda_2 \gg \lambda_1$ and $A_{20} = 0$, and hence also describes secular equilibrium.

Transient Equilibrium ($T_1 \gtrsim T_2$)

Another practical situation arises when $N_{20} = 0$ and the half-life of the parent is greater than that of the daughter, but not greatly so. According to Eq. (3.85), N_2 and hence the activity $A_2 = \lambda_2 N_2$ of the daughter initially build up steadily. With the continued passage of time, $e^{-\lambda_2 t}$ eventually becomes negligible with respect to $e^{-\lambda_1 t}$, since $\lambda_2 > \lambda_1$. Then Eq. (3.85) implies, after multiplication of both sides by λ_2, that

$$\lambda_2 N_2 = \frac{\lambda_2 \lambda_1 N_{10} e^{-\lambda_1 t}}{\lambda_2 - \lambda_1}. \tag{3.86}$$

Since $A_1 = \lambda_1 N_1 = \lambda_1 N_{10} e^{-\lambda_1 t}$ is the activity of the parent as a function of time, this relation says that

$$A_2 = \frac{\lambda_2 A_1}{\lambda_2 - \lambda_1}. \tag{3.87}$$

Thus, after initially increasing, the daughter activity A_2 goes through a maximum and then decreases at the same rate as the parent activity. Under this condition, illustrated in Fig. 3.17, transient equilibrium is said to exist. The total activity also reaches a maximum, as shown in the figure, at a time earlier than that of the maximum daughter activity. The time at which transient equilibrium is established depends on the individual magnitudes of T_1 and T_2. Secular equilibrium can be viewed as a special case of transient equilibrium in which $\lambda_2 \gg \lambda_1$ and the time of observation is so short that the decay of the activity A_1 is negligible. Under these conditions, the curve for A_1 in Fig. 3.17 would be flat, A_2 would approach A_1, and the figure would resemble Fig. 3.16.

No Equilibrium ($T_1 < T_2$)

When the daughter, initially absent ($N_{20} = 0$), has a longer half-life than the parent, its activity builds up to a maximum and then declines. Because of its shorter half-life, the parent eventually decays away and only the daughter is left. No equilibrium occurs. The activities in this case exhibit the patterns shown in Fig. 3.18.

Example
Starting with a 10.0 mCi sample of pure ^{90}Sr at time $t = 0$, how long will it take for the total activity (^{90}Sr + ^{90}Y) to build up to 17.5 mCi?

Solution
Appendix D shows that $^{90}_{38}$Sr β^- decays with a half-life of 27.7 yr into $^{90}_{39}$Y, which β^-decays into stable $^{90}_{40}$Zr with a half-life of 64.0 hr. These two isotopes illustrate a long-lived parent ($T_1 = 27.7$ yr) decaying into a short-lived daughter ($T_2 = 64.0$ hr). Secular equilibrium is reached in about seven daughter half-lives, i.e., in $7 \times 64 = 448$ hr. At the end of this time, the ^{90}Sr activity A_1 has not

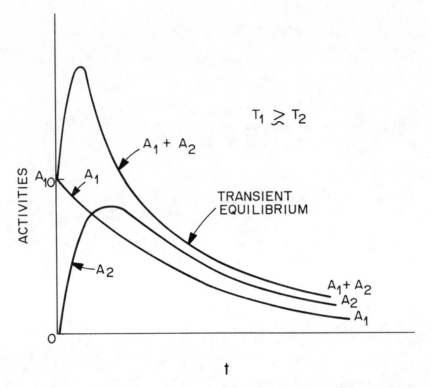

Figure 3.17. Activities as functions of time when T_1 is somewhat larger than T_2 ($T_1 \gtrsim T_2$) and $N_{20} = 0$. Transient equilibrium is eventually reached, in which all activities decay with the half-life T_1 of the parent.

diminished appreciably, the ^{90}Y activity A_2 has increased to the level $A_2 = A_1 = 10.0$ mCi, and the total activity is 20.0 mCi. In the present problem we are asked, in effect, to find the time at which the ^{90}Y activity reaches 7.5 mCi. The answer will be less than 448 hr. Equation (3.82) with $A_{20} = 0$ applies here.† The decay constant for ^{90}Y is $\lambda_2 = 0.693/T_2 = 0.693/64.0$ hr $= 0.0108$ hr^{-1}. With $A_1 = 10.0$ mCi, $A_2 = 7.5$ mCi, and $A_{20} = 0$, Eq. (3.82) gives

$$7.5 = 10.0(1 - e^{-0.0108t}), \tag{3.88}$$

where t is in hours. Rearranging, we have

$$e^{-0.0108t} = \frac{1}{4}, \tag{3.89}$$

giving $t = 128$ hr. (In this example note that the ^{90}Y activity increases in an inverse fashion to the way a pure sample of ^{90}Y would decay. It takes two half-lives, $2T_2 = 128$ hr, for the activity to build up to three-fourths its final value at secular equilibrium.)

Example
How many grams of ^{90}Y are in secular equilibrium with 1 mg of ^{90}Sr?

---

†Equation (3.85), describing the general case without restriction on the relative magnitudes of T_1 and T_2, can always be applied. To the degree of accuracy with which we are working, one will obtain the same numerical answer from the simplified Eq. (3.82), which already contains the appropriate approximations.

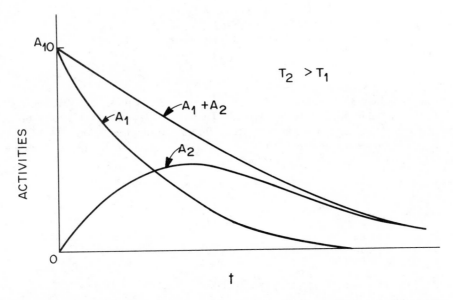

Figure 3.18. Activities as functions of time when $T_2 > T_1$ and $N_{20} = 0$. No equilibrium conditions occur. Eventually, only the daughter activity remains.

Solution

The amount of ^{90}Y will be that having the same activity as 1 mg of ^{90}Sr. The specific activity, SA, of ^{90}Sr ($T_1 = 27.7$ yr) is [from Eq. (3.72)]

$$SA_1 = \frac{1600}{27.7} \times \frac{226}{90} = 145 \text{ Ci/g}. \tag{3.90}$$

Therefore, the activity of the 1 mg sample of ^{90}Sr is

$$A_1 = 10^{-3} \text{ g} \times 145 \text{ Ci/g} = 0.145 \text{ Ci}, \tag{3.91}$$

which is also equal to the activity A_2 of the ^{90}Y. The latter has a specific activity

$$SA_2 = \frac{1600 \text{ yr}}{64.0 \text{ hr} \times \dfrac{1}{24} \dfrac{\text{day}}{\text{hr}} \times \dfrac{1}{365} \dfrac{\text{yr}}{\text{day}}} \times \frac{226}{90} \tag{3.92}$$

$$= 5.50 \times 10^5 \text{ Ci/g}. \tag{3.93}$$

Therefore, the mass of ^{90}Y in secular equilibrium with 1 mg of ^{90}Sr is

$$\frac{0.145 \text{ Ci}}{5.50 \times 10^5 \text{ Ci/g}} = 2.64 \times 10^{-7} \text{ g} = 0.264 \,\mu\text{g}. \tag{3.94}$$

Example

A sample contains 1 mCi of ^{191}Os at time $t = 0$. The isotope decays by β^- emission into metastable ^{191m}Ir, which then decays by γ emission into ^{191}Ir. The decay and half-lives can be represented by writing

$$^{191}_{76}\text{Os} \xrightarrow[15.0 \text{ days}]{\beta^-} {}^{191m}_{77}\text{Ir} \xrightarrow[4.9 \text{ sec}]{\gamma} {}^{191}_{77}\text{Ir}. \tag{3.95}$$

(a) How many grams of ^{191}Os are present at $t = 0$?
(b) How many millicuries of ^{191m}Ir are present at $t = 25$ days?
(c) How many atoms of ^{191m}Ir decay between $t = 100$ sec and $t = 102$ sec?
(d) How many atoms of ^{191m}Ir decay between $t = 30$ days and $t = 40$ days?

Solution

As in the last two examples, the parent half-life is large compared with that of the daughter. Secular equilibrium is reached in about $7 \times 4.9 = 34.3$ sec. Thereafter, the activities A_1 and A_2 of the ^{191}Os and ^{191m}Ir remain equal, as they are in secular equilibrium. During the periods of time considered in (b) and in (d), however, the osmium will have decayed appreciably; and so one deals with an example of transient equilibrium. The problem can be solved as follows.

(a) The specific activity of ^{191}Os is, from Eq. (3.72),

$$\text{SA}_1 = \frac{1600 \times 365}{15.0} \times \frac{226}{191} = 4.61 \times 10^4 \text{Ci/g}. \tag{3.96}$$

The mass of the sample, therefore, is

$$\frac{10^{-3} \text{ Ci}}{4.61 \times 10^4 \text{ Ci/g}} = 2.17 \times 10^{-8} \text{ g}. \tag{3.97}$$

(b) At $t = 25$ days,

$$A_2 = A_1 = 1 \times e^{-0.693 \times 25/15} = 0.315 \text{ mCi}. \tag{3.98}$$

(c) Between $t = 100$ sec and 102 sec secular equilibrium exists with the osmium source essentially still at its original activity. Thus the ^{191m}Ir decay rate at $t = 100$ sec is $A_2 = 1$ mCi $= 3.7 \times 10^7$ sec^{-1}. During the next 2 sec the number of ^{191m}Ir atoms that decay is $2 \times 3.7 \times 10^7 = 7.4 \times 10^7$.

(d) This part is like (c), except that the activities A_1 and A_2 do not stay constant during the time between 30 and 40 days. Since transient equilibrium exists, the number of atoms of ^{191m}Ir and ^{191}Os that decay are equal. The number of ^{191m}Ir atoms that decay, therefore, is equal to the integral of the ^{191}Os activity during the specified time (t in days):

$$3.7 \times 10^7 \int_{30}^{40} e^{-0.693t/15.0} \, dt = \frac{3.7 \times 10^7}{-0.0462} e^{-0.0462t} \Big|_{30}^{40} \tag{3.99}$$

$$= -8.01 \times 10^8 (0.158 - 0.250) = 7.37 \times 10^7. \tag{3.100}$$

3.13 PROBLEMS

1. Gallium occurs with two natural isotopes, ^{69}Ga (60.2% abundant) and ^{71}Ga (39.8%), having atomic weights 68.96 and 70.92. What is the atomic weight of the element?
2. The atomic weight of lithium is 6.941. It has two natural isotopes, ^{6}Li and ^{7}Li, with atomic weights of 6.015 and 7.016. What are the relative abundances of the two isotopes?
3. What minimum energy would an alpha particle need in order to react with a ^{238}U nucleus?
4. Calculate the energy released when a thermal neutron is absorbed by deuterium.
5. Calculate the total binding energy of the alpha particle.
6. How much energy is released when a ^{6}Li atom absorbs a thermal neutron in the reaction $^6_3\text{Li}(n, \alpha)^3_1\text{H}$?
7. What is the mass of a ^{6}Li atom in grams?
8. Calculate the average binding energy per nucleon for the nuclide $^{40}_{19}$K.
9. The atomic weight of ^{32}P is 31.973910. What is the value of Δ in MeV?
10. Show that 1 AMU $= 1.49 \times 10^{-3}$ erg.
11. Calculate the gamma-ray threshold for the reaction $^{12}\text{C}(\gamma, n)^{11}\text{C}$.
12. (a) Calculate the energy released by the alpha decay of $^{222}_{86}$Rn.

(b) Calculate the energy of the alpha particle.

(c) What is the energy of the recoil polonium atom?

13. The $^{238}_{92}$U nucleus emits a 4.20 MeV alpha particle. What is the total energy released in this decay?

14. The $^{226}_{88}$Ra nucleus emits a 4.60 MeV alpha particle 5% of the time when it decays to $^{222}_{86}$Rn.

    (a) Calculate the Q value for this decay.

    (b) What is the recoil energy of the ^{222}Rn atom?

15. The Q value for alpha decay of $^{239}_{94}$Pu is 5.25 MeV. Given the masses of the ^{239}Pu and ^{4}He atoms, 239.052175 AMU and 4.002603 AMU, calculate the mass of the $^{235}_{92}$U atom in AMU.

16. Calculate the Q value for the beta decay of the free neutron into a proton, $^1_0n \rightarrow {^1_1}p + {_{-1}^0}\beta + {^0_0}\bar{\nu}$.

17. (a) Calculate the energy released in the beta decay of $^{32}_{15}$P.

    (b) If a beta particle has 650 keV, how much energy does the antineutrino have?

18. Calculate the Q value for tritium beta decay.

19. Draw the decay scheme for $^{42}_{19}$K.

20. (a) Draw the decay scheme for $^{203}_{80}$Hg.

    (b) Estimate the K-shell electron binding energy from the data given in Appendix D.

21. Draw the decay scheme for $^{59}_{26}$Fe, labeling energies and frequencies (percentages) for each transition.

22. Draw the decay scheme for $^{198}_{79}$Au.

23. Calculate the recoil energy of the technetium atom as a result of photon emission in the isomeric transition ^{99m}Tc $\rightarrow$ ^{99}Tc $+ \gamma$.

24. Refer to the decay scheme of $^{137}_{55}$Cs in Fig. 3.8. The binding energies of the K- and L-shell electrons of the daughter $^{137}_{56}$Ba atom are 38 keV and 6 keV.

    (a) What are the energies of the internal-conversion electrons ejected from these shells?

    (b) What is the wavelength of the barium K_α X-ray emitted when an L-shell electron makes a transition to the K shell?

    (c) What is the value of the internal-conversion coefficient?

    (d) If the ionization potential of barium is 5.21 eV, what is the maximum energy of an Auger electron produced following internal conversion?

25. (a) Calculate the Q value for K orbital-electron capture by the $^{37}_{18}$Ar nucleus, neglecting the electron binding energy.

    (b) Repeat (a), including the binding energy, 3.20 keV, of the K-shell electron in argon.

    (c) What becomes of the energy released as a result of this reaction?

26. What is the maximum positron energy in the decay of $^{35}_{18}$Ar?

27. Explain the origins of the radiations listed in Appendix D for $^{85}_{39}$Y. Draw the decay scheme.

28. The nuclide $^{65}_{30}$Zn decays by electron capture (98.3%) and by positron emission (1.7%).

    (a) Calculate the Q value for both modes of decay.

    (b) Draw the decay scheme for ^{65}Zn.

    (c) What are the physical processes responsible for each of the major radiations listed in Appendix D?

    (d) Estimate the binding energy of a K-shell electron in copper.

29. Does $^{26m}_{13}$Al decay to the ground state of its daughter $^{26}_{12}$Mg?

30. Show that $^{55}_{26}$Fe, which decays by electron capture, cannot decay by positron emission.

31. The isotope $^{126}_{53}$I can decay by EC, β^-, and β^+ transitions.

    (a) Calculate the Q values for the three modes of decay to the ground states of the daughter nuclei.

    (b) Draw the decay scheme.

    (c) What kinds of radiation can one expect from a ^{126}I source?

32. To which of the natural series do the following heavy radionuclides belong: $^{213}_{83}$Bi, $^{215}_{84}$Po, $^{230}_{90}$Th, $^{233}_{92}$U, and $^{224}_{88}$Ra?

33. What is the value of the decay constant of ^{40}K?

34. What is the decay constant of tritium?

35. The activity of a radioisotope is found to decrease by 30% in 1 wk. What are the values of its (a) decay constant, (b) half-life, and (c) mean life?

36. What percentage of the original activity of a radionuclide remains after (a) 5 half-lives and (b) 10 half-lives?
37. The isotope ^{132}I decays by β^- emission into stable ^{132}Xe with a half-life of 2.3 hr.
    (a) How long will it take for $\frac{7}{8}$ of the original ^{132}I atoms to decay?
    (b) How long will it take for a sample of ^{132}I to lose 95% of its activity?
38. A very old specimen of wood contains 10^{12} atoms of ^{14}C.
    (a) How many ^{14}C atoms did it contain in the year 9474 B.C.?
    (b) How many ^{14}C atoms did it contain in 1986 B.C.?
39. A radioactive sample consists of a mixture of ^{35}S and ^{32}P. Initially, 5% of the activity is due to the ^{35}S and 95% to the ^{32}P. At what subsequent time will the activities of the two nuclides in the sample be equal?
40. The gamma exposure rate at the surface of a shielded ^{198}Au source is 10 R/hr (roentgen/hour, Sec. 10.2). What will be the exposure rate in this position after 2 wk?
41. Compute the specific activity of (a) ^{238}U, (b) ^{90}Sr, and (c) ^{3}H.
42. How many grams of ^{32}P are there in a 5 mCi source?
43. An encapsulated ^{210}Po radioisotope is to be used as a heat source, in which an implanted thermocouple junction converts heat into electricity with an efficiency of 15% to power a small transmitter for a space probe.
    (a) How many curies of ^{210}Po are needed at launch time if the transmitter is to be supplied with 100 W of electricity 1 yr after launch?
    (b) Calculate the number of grams of ^{210}Po needed.
    (c) If the transmitter shuts off when the electrical power to it falls below 1 W, how long can it be expected to operate after launch?
    (d) What health physics precautions would you recommend during fabrication, encapsulation, and handling of the device?
44. A 0.2 g sample of $^{85}_{36}$Kr gas, which decays into stable $^{85}_{37}$Rb, is accidentally broken and escapes inside a sealed warehouse measuring $40 \times 30 \times 20$ m. What is the specific activity of the air inside in Ci/cm^3?
45. A 6.2 mg sample of ^{90}Sr is in secular equilibrium with its daughter ^{90}Y.
    (a) How many curies of ^{90}Sr are present?
    (b) How many curies of ^{90}Y are present?
    (c) What is the mass of ^{90}Y present?
    (d) What will the activity of the ^{90}Y be after 100 yr?
46. What is the mean life of a ^{226}Ra atom?
47. At time $t = 0$ a sample consists of 2 Ci of ^{90}Sr and 8 Ci of ^{90}Y.
    (a) What will the activity of ^{90}Y be in the sample after 100 hr?
    (b) At what time will the ^{90}Y activity be equal to 3 Ci?
48. Show that Eq. (3.85) leads to secular equilibrium, $A_1 = A_2$, under the appropriate conditions.
49. Show by direct substitution that the solution given by Eq. (3.85) satisfies Eq. (3.84).

CHAPTER 4
INTERACTION OF HEAVY CHARGED PARTICLES WITH MATTER

This chapter and the next four describe the mechanisms by which different types of ionizing radiation interact with matter. Knowledge of the basic physics of radiation interaction and energy transfer is fundamental to radiation detection, measurement, and control, as well as to understanding the biological effects of radiation on living tissue. We consider "heavy" charged particles first, i.e., charged particles other than the electron and positron.

4.1 ENERGY-LOSS MECHANISMS

A heavy charged particle traversing matter loses energy primarily through the ionization and excitation of atoms. (Except at low velocities, a heavy charged particle loses a negligible amount of energy in nuclear collisions.) The moving charged particle exerts electromagnetic forces on atomic electrons and imparts energy to them. The energy transferred may be sufficient to knock an electron out of an atom and thus ionize it, or it may leave the atom in an excited, nonionized state. As we show in the next section, a heavy charged particle can transfer only a small fraction of its energy in a single electronic collision. Its deflection in the collision is negligible. Thus, a heavy charged particle travels an almost straight path through matter, losing energy almost continuously in small amounts through collisions with atomic electrons, leaving ionized and excited atoms in its wake.

In contrast, beta particles (β^- or β^+), which are discussed in the next chapter, can lose a large fraction of their energy and undergo large deflections in single collisions with atomic electrons. Thus they do not travel in straight lines. A beta particle can also be sharply deflected by an atomic nucleus, causing it to emit photons in the process called bremsstrahlung (braking radiation). Figure 4.1 shows the contrast between the straight tracks of two alpha particles and the tortuous track of a beta particle in photographic emulsion.

4.2 MAXIMUM ENERGY TRANSFER IN A SINGLE COLLISION

In this section we calculate the maximum energy that a heavy charged particle can lose in colliding with an atomic electron. We assume that the particle is moving rapidly compared with the electron and that the energy transferred is large compared with the binding energy of the electron in the atom. Under these conditions the electron can be considered to be initially free and at rest. We treat the problem classically and then give the relativistic results.

Figure 4.2(a) shows schematically a heavy particle (mass M and velocity V) approaching an electron (mass m, at rest). After the collision, which for maximum energy transfer

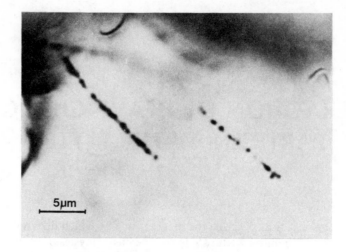

5μm

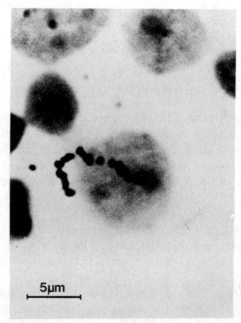

5μm

Figure 4.1. (Top) Alpha-particle autoradiography of rat bone after inhalation of ^{241}Am. Biological preparation by R. Masse and N. Parmentier. (Bottom) Beta-particle autoradiograpy of isolated rat-brain nucleus. The ^{14}C-thymidine incorporated in the nucleolus is located at the track origin of the electron emitted by the tracer element. Biological preparation by M. Wintzerith and P. Mandel. (Courtesy R. Rechenmann and E. Wittendorp–Rechenmann, Laboratoire de Biophysique des Rayonnements et de Methodologie INSERM U.220, Strasbourg, France)

is head-on, the particles in (b) move with speeds V_1 and v_1 along the initial line of travel of the incident particle. Since the total kinetic energy and momentum are conserved in the collision, we have the two relationships

$$\frac{1}{2} MV^2 = \frac{1}{2} MV_1^2 + \frac{1}{2} mv_1^2 \tag{4.1}$$

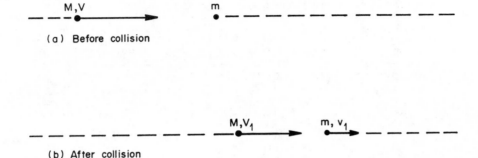

(a) Before collision

(b) After collision

Figure 4.2. Representation of head-on collision of a particle of mass M and speed V with an electron of mass m, initially free and at rest.

and

$$MV = MV_1 + mv_1. \tag{4.2}$$

If we solve Eq. (4.2) for v_1 and substitute the result into (4.1), we obtain

$$V_1 = \frac{(M - m)V}{M + m}. \tag{4.3}$$

Using this expression for V_1, we find for the maximum energy transfer

$$Q_{max} = \frac{1}{2} MV^2 - \frac{1}{2} MV_1^2 = \frac{4mME}{(M + m)^2}, \tag{4.4}$$

where $E = MV^2/2$ is the initial kinetic energy of the heavy particle. Note that when the masses are equal ($M = m$), Eq. (4.4) gives $Q_{max} = E$; and so the incident particle can transfer all of its energy in a billiard-ball-type collision.

Example
Calculate the maximum energy that a 10 MeV proton can lose in a single electronic collision.

Solution
For a proton of this energy the nonrelativistic formula (4.4) is accurate. Neglecting m compared with M, we have $Q_{max} = 4mE/M = 4 \times 1 \times 10/1836 = 2.18 \times 10^{-2}$ MeV = 21.8 keV, which is only 0.22% of the proton's energy.

The exact relativistic expression for the maximum energy transfer, with m and M denoting the rest masses of the electron and the heavy particle, is

$$Q_{max} = \frac{2\gamma^2 mV^2}{1 + 2\gamma m/M + m^2/M^2}, \tag{4.5}$$

where $\gamma = 1/\sqrt{1 - \beta^2}$, $\beta = V/c$, and c is the speed of light (Appendix C). Except at extreme relativistic energies, $\gamma m/M \ll 1$, in which case (4.5) reduces to

$$Q_{max} = 2\gamma^2 mV^2 = 2\gamma^2 mc^2\beta^2, \tag{4.6}$$

which is the usual relativistic result.

Table 4.1. Maximum Possible Energy Transfer, Q_{max}, in Proton Collision with Electron

Proton kinetic energy T (MeV)	Q_{max} (MeV)	Maximum percentage energy transfer $100 Q_{max}/T$
0.1	0.00022	0.22
1	0.0022	0.22
10	0.0219	0.22
100	0.229	0.23
10^3	3.33	0.33
10^4	136.	1.4
10^5	1.06×10^4	10.6
10^6	5.38×10^5	53.8
10^7	9.21×10^6	92.1

Example

Use the relativistic formula (4.6) to calculate the maximum possible energy loss in a single collision of the 10 MeV proton in the last example.

Solution

We first find γ. Since the proton rest energy is $Mc^2 = 938$ MeV, we can use the formula in Appendix C for the relativistic kinetic energy, $T = 10$ MeV, to write $10 = 938(\gamma - 1)$. It follows that $\gamma = 1.01066$ and $\beta^2 = 0.02099$. Since the electron rest energy is $mc^2 = 0.511$ MeV (Appendix A), Eq. (4.6) yields $Q_{max} = 21.9$ keV.

Table 4.1 gives numerical results for a range of proton energies. Except at extremely high energies, where Eq. (4.5) must be used, a heavy charged particle can lose only a relatively small fraction of its energy in a single collision with an electron. Encounters in which an amount of energy comparable to Q_{max} is transferred are very rare, particularly at high energies.

4.3 THE BETHE FORMULA FOR STOPPING POWER

The linear rate of energy loss to atomic electrons along the path of a heavy charged particle in a medium (expressed, for example, in MeV/cm) is the basic physical quantity that determines the dose that the particle delivers in the medium. This quantity, designated $-dE/dx$, is called the stopping power of the medium for the particle. In 1913 Bohr derived an explicit formula giving the stopping power for heavy charged particles. Because quantum mechanics had not yet been discovered, Bohr relied on intuition and insight to obtain the proper semiclassical representation of atomic collisions. He calculated the energy loss of a heavy particle in a collision with an electron at a given distance of passing and then averaged over all possible distances and energy losses up to the maximum, Q_{max}. The nonrelativistic formula that Bohr obtained gave the correct physical features of stopping power as borne out by experiment and by the later quantum mechanical theory of Bethe.†

Using relativistic quantum mechanics, Bethe derived the following expression for the stopping power of a uniform medium for heavy charged particles:

---

†For derivations of the Bohr and Bethe stopping-power formulas, the reader is referred to J. E. Turner, *Health Phys.* **13**, 1255 (1967) and references therein.

$$-\frac{dE}{dx} = \frac{4\pi z^2 e^4 n}{mc^2\beta^2}\left[\ln\frac{2mc^2\beta^2}{I(1-\beta^2)} - \beta^2\right]. \tag{4.7}$$

In this relation

z = atomic number of the heavy particle,

e = magnitude of electron charge,

n = number of electrons per unit volume in the medium,

m = electron rest mass,

c = speed of light in vacuum,

$\beta = V/c$ = speed of the particle relative to c, and

I = mean excitation energy of the medium.

The stopping power depends only on the charge ze and velocity β of the heavy particle. The relevant properties of the medium are its mean excitation energy I (next section) and the electronic density n, to which the stopping power is proportional. The mass m in the formula is that of the target atomic electrons. Plots of the mass stopping power ($-dE/\rho\,dx$, the stopping power divided by the density) of water for a number of particles are shown in Fig. 4.3. The ln term in (4.7) causes the rise in these curves at high energies as $\beta \to 1$. At low energies, the factor in front of the bracket in (4.7) increases, but the ln term decreases, causing a peak (called the Bragg peak) to occur. The rate of energy loss is a maximum there.

Stopping power, $-dE/dx$, is often expressed in the units MeV/cm. The corresponding

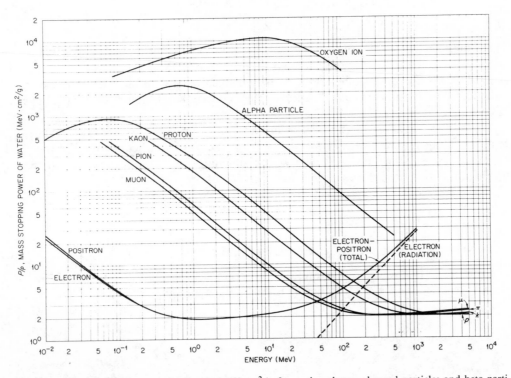

Figure 4.3. Mass stopping power of water in MeV cm²/g for various heavy charged particles and beta particles. The muon, pion, and kaon are elementary particles with rest masses equal, respectively, to about 207, 270, and 967 electron rest masses. (Courtesy Oak Ridge National Laboratory, operated by Martin Marietta Energy Systems, Inc., for the Department of Energy)

units for mass stopping power, $-\mathrm{d}E/\rho\,\mathrm{d}x$, are MeV cm^2/g. The mass stopping power is a useful quantity because it expresses the rate of energy loss of the charged particle per g/cm^2 of the medium traversed. In a gas, for example, $-\mathrm{d}E/\mathrm{d}x$ depends on pressure, but $-\mathrm{d}E/\rho\,\mathrm{d}x$ does not, because dividing by the density exactly compensates for the pressure. In addition, the mass stopping power does not differ greatly for materials with similar atomic composition. For example, for 10 MeV protons the mass stopping power of H_2O is 45.9 MeV cm^2/g and that of anthracene ($C_{14}H_{10}$) is 44.2 MeV cm^2/g. The curves in Fig. 4.3 for water can be used for tissue, plastics, hydrocarbons, and other materials that consist primarily of light elements. For Pb ($Z = 82$), on the other hand, $-\mathrm{d}E/\rho\,\mathrm{d}x = 17.5$ MeV cm^2/g for 10 MeV protons. Generally, heavy atoms are less efficient on a g/cm^2 basis for slowing down heavy charged particles, because many of their electrons are too tightly bound in the inner shells to participate effectively in the absorption of energy.

4.4 MEAN EXCITATION ENERGIES

Mean excitation energies I for a number of elements have been calculated from the quantum-mechanical definition obtained in the derivation of Eq. (4.7). They can also be measured in experiments in which all of the quantities in Eq. (4.7) except I are known. The following approximate empirical formulas can be used to estimate the I value in eV for an element with atomic number Z:

$$I \cong \begin{cases} 19.0 \text{ eV}, \ Z = 1 \text{ (hydrogen)} & (4.8) \\ 11.2 + 11.7Z \text{ eV}, \ 2 \leq Z \leq 13 & (4.9) \\ 52.8 + 8.71Z \text{ eV}, \ Z > 13. & (4.10) \end{cases}$$

Since only the logarithm of I enters the stopping-power formula, values obtained by using these formulas are accurate enough for most applications. The value of I for an element depends only to a slight extent on the chemical compound in which the element is found and on the state of condensation of the material, solid, liquid, or gas (Bragg additivity rule).

When the material is a compound or mixture, the stopping power can be calculated by simply adding the separate contributions from the individual constituent elements. If there are N_i atoms/cm^3 of an element with atomic number Z_i and mean excitation energy I_i, then in formula (4.7) one makes the replacement

$$n \ln I = \sum_i N_i Z_i \ln I_i, \tag{4.11}$$

where n is the total number of electrons/cm^3 in the material $\left(n = \sum_i N_i Z_i \right)$. In this way the composite $\ln I$ value for the material is obtained from the individual elemental $\ln I_i$ values weighted by the electron densities $N_i Z_i$ of the various elements.

Example
Calculate the mean excitation energy of H_2O.

Solution
We obtain the I values for H and O from Eqs. (4.8) and (4.9), and then apply (4.11). For H, $I_H = 19.0$ eV, and for O, $I_0 = 11.2 + 11.7 \times 8 = 105$ eV. The electronic densities $N_i Z_i$ and n can be computed in a straightforward way. However, only the ratios $N_i Z_i/n$ are needed to find I, and

these are much simpler to use. Since the H_2O molecule has 10 electrons, 2 of which belong to H ($Z = 1$) and 8 to O ($Z = 8$), we may write from Eq. (4.11)

$$\ln I = \frac{2 \times 1}{10} \ln 19.0 + \frac{1 \times 8}{10} \ln 105 = 4.312, \tag{4.12}$$

giving $I = 74.6$ eV.

4.5 TABLE FOR COMPUTATION OF STOPPING POWERS

In this section we develop a numerical table to facilitate the computation of stopping power for a heavy charged particle in any material. In the next section we use the table to calculate the proton stopping power of H_2O as a function of energy.

In cgs units, the multiplicative factor in Eq. (4.7) can be written with the help of the constants given in Appendix A as

$$\frac{4\pi z^2 e^4 n}{mc^2\beta^2} = \frac{4\pi z^2 (4.80 \times 10^{-10})^4 n}{9.11 \times 10^{-28}(3.00 \times 10^{10})^2\beta^2} = 8.14 \times 10^{-31}\frac{z^2 n}{\beta^2}\frac{\text{erg}}{\text{cm}}. \tag{4.13}$$

The units are those of $e^4 n/mc^2$:

$$\frac{(\text{esu})^4\,\text{cm}^{-3}}{\text{erg}} = \frac{(\text{erg cm})^2\text{cm}^{-3}}{\text{erg}} = \frac{\text{erg}}{\text{cm}}, \tag{4.14}$$

where we have used erg cm to replace esu^2 (Appendix B). Converting to MeV, we have

$$\frac{8.14 \times 10^{-31}z^2 n}{\beta^2}\frac{\text{erg}}{\text{cm}} \times \frac{1}{1.60 \times 10^{-6}}\frac{\text{MeV}}{\text{erg}} = \frac{5.09 \times 10^{-25}z^2 n}{\beta^2}\frac{\text{MeV}}{\text{cm}}. \tag{4.15}$$

In the dimensionless ln term in (4.7) we can express the energies I and $2mc^2$ in eV. The stopping power is then

$$-\frac{dE}{dx} = \frac{5.09 \times 10^{-25}z^2 n}{\beta^2}\left[\ln\frac{1.02 \times 10^6\beta^2}{I_{\text{ev}}(1 - \beta^2)} - \beta^2\right]\frac{\text{MeV}}{\text{cm}}. \tag{4.16}$$

This general formula for any heavy charged particle in any medium can be written

$$-\frac{dE}{dx} = \frac{5.09 \times 10^{-25}z^2 n}{\beta^2}\,[F(\beta) - \ln I_{\text{ev}}]\,\frac{\text{MeV}}{\text{cm}}, \tag{4.17}$$

where

$$F(\beta) = \ln\frac{1.02 \times 10^6\beta^2}{1 - \beta^2} - \beta^2. \tag{4.18}$$

Example
Compute $F(\beta)$ for a proton with kinetic energy $T = 10$ MeV.

Solution
In the last example in Section 4.2 we found that $\beta^2 = 0.02099$. Substitution of this value into Eq. (4.18) gives $F(\beta) = 9.973$.

The quantities β^2 and $F(\beta)$ are given for protons of various energies in Table 4.2. Since, for a given value of β, the kinetic energy of a particle is proportional to its rest mass, the table can also be used for other heavy particles as well. For example, the ratio of the kinetic energies T_d and T_p of a deuteron and a proton traveling at the same speed is

Table 4.2. Data for Computation of Stopping Power for Heavy Charged Particles

Proton kinetic energy (MeV)	β^2	$F(\beta)$, Eq. (4.18)
0.01	0.000021	2.179
0.02	0.000043	3.775
0.04	0.000085	4.468
0.06	0.000128	4.873
0.08	0.000171	5.161
0.10	0.000213	5.384
0.20	0.000426	6.077
0.40	0.000852	6.771
0.60	0.001278	7.175
0.80	0.001703	7.462
1.00	0.002129	7.685
2.00	0.004252	8.376
4.00	0.008476	9.066
6.00	0.01267	9.469
8.00	0.01685	9.753
10.00	0.02099	9.973
20.00	0.04133	10.65
40.00	0.08014	11.32
60.00	0.1166	11.70
80.00	0.1510	11.96
100.0	0.1834	12.16
200.0	0.3205	12.77
400.0	0.5086	13.36
600.0	0.6281	13.73
800.0	0.7088	14.02
1000.	0.7658	14.26

$$\frac{T_d}{T_p} = \frac{M_d}{M_p} = 2. \tag{4.19}$$

The value of $F(\beta) = 9.973$ that we computed above for a 10 MeV proton applies, therefore, to a 20 MeV deuteron. Linear interpolation can be used where needed in the table.

4.6 STOPPING POWER OF WATER FOR PROTONS

For protons $z = 1$, and for water $n = (10/18) \times 6.02 \times 10^{23} = 3.34 \times 10^{23}$ cm^{-3}. Also, as found at the end of Section 4.4, $\ln I_{eV} = 4.312$. From Eq. (4.17) it follows that the stopping power of water for a proton of speed β is given by

$$-\frac{dE}{dx} = \frac{0.170}{\beta^2} [F(\beta) - 4.31] \text{MeV/cm}. \tag{4.20}$$

At 1 MeV, for example, we find in Table 4.2 that $\beta^2 = 0.00213$ and $F(\beta) = 7.69$; therefore Eq. (4.20) gives

$$-\frac{dE}{dx} = \frac{0.170}{0.00213} (7.69 - 4.31) = 270 \text{ MeV/cm}. \tag{4.21}$$

This is numerically equal to the value of the mass stopping power plotted in Fig. 4.3 for unit-density water. The curves in the figure were obtained by such calculations.

4.7 RANGE

The range of a charged particle is the distance it travels before coming to rest. The reciprocal of the stopping power gives the distance traveled per unit energy loss. Therefore, the range $R(T)$ of a particle of kinetic energy T is the integral of this quantity down to zero energy:

$$R(T) = \int_0^T \left(-\frac{dE}{dx} \right)^{-1} dE. \tag{4.22}$$

Table 4.3 gives the mass stopping power and range of protons in water. The latter is expressed in g/cm^2; i.e., the range in cm multiplied by the density of water ($\rho = 1$ g/cm^3). Like mass stopping power, the range in g/cm^2 applies to all materials of similar atomic composition.

Although the integral in (4.22) cannot be evaluated in closed form, the explicit functional form of (4.7) enables one to scale the proton ranges in Table 4.3 to obtain the ranges of other heavy charged particles. Inspection of Eqs. (4.7) and (4.22) shows that the range of a heavy particle is given by an equation of the form

$$R(T) = \frac{1}{z^2} \int_0^T \frac{dE}{G(\beta)}, \tag{4.23}$$

where z is the particle's charge and $G(\beta)$ depends only on the particle's velocity β. Since $E = Mc^2/\sqrt{1 - \beta^2}$, where M is the particle's rest mass, we may write $dE = Mg(\beta)\, d\beta$, where g is another function of velocity. It follows that Eq. (4.23) can be written in the form

$$R(\beta) = \frac{M}{z^2} \int_0^\beta \frac{g(\beta)}{G(\beta)} \, d\beta = \frac{M}{z^2} f(\beta), \tag{4.24}$$

where the function $f(\beta)$ depends only on the velocity of the heavy charged particle. The structure of Eq. (4.24) enables one to scale ranges for different particles in the following manner. Since $f(\beta)$ is the same for two heavy charged particles at the same initial speed β, the ratio of their ranges is simply

$$\frac{R_1(\beta)}{R_2(\beta)} = \frac{z_2^2 M_1}{z_1^2 M_2}, \tag{4.25}$$

where M_1 and M_2 are the rest masses and z_1 and z_2 are the charges. If particle number 2 is a proton ($M_2 = 1$ and $z_2 = 1$), then we can write for the range R of the other particle (mass $M_1 = M$ proton masses and charge $z_1 = z$)

$$R(\beta) = \frac{M}{z^2} R_p(\beta), \tag{4.26}$$

where $R_p(\beta)$ is the proton range.

Example
Use Table 4.3 to find the range of an 80 MeV $^3He^{2+}$ ion in soft tissue.

Solution
Applying (4.26), we have $z^2 = 4$, $M = 3$, and $R(\beta) = 3R_p(\beta)/4$. Thus the desired range is three-quarters that of a proton traveling with the speed of an 80 MeV $^3He^{2+}$ ion. To find the speed, we use the relationship from Appendix C: $T = Mc^2(\gamma - 1)$. With $T = 80$ MeV and $Mc^2 = 3$ AMU $= 3 \times 931 = 2790$ MeV, we find $\gamma = 1.029$ and $\beta^2 = 0.0550$. In Table 4.3 this value lies

Table 4.3. Mass Stopping Power $-dE/\rho\,dx$ and Range R_p for Protons in Water

Kinetic energy (MeV)	β^2	$-dE/\rho\,dx$ (MeV cm²/g)	R_p (g/cm²)
0.01	.000021	500.	3×10^{-5}
0.04	.000085	860.	6×10^{-5}
0.05	.000107	910.	7×10^{-5}
0.08	.000171	920.	9×10^{-5}
0.10	.000213	910.	1×10^{-4}
0.50	.001065	428.	8×10^{-4}
1.00	.002129	270.	0.002
2.00	.004252	162.	0.007
4.00	.008476	95.4	0.023
6.00	.01267	69.3	0.047
8.00	.01685	55.0	0.079
10.0	.02099	45.9	0.118
12.0	.02511	39.5	0.168
14.0	.02920	34.9	0.217
16.0	.03327	31.3	0.280
18.0	.03731	28.5	0.342
20.0	.04133	26.1	0.418
25.0	.05126	21.8	0.623
30.0	.06104	18.7	0.864
35.0	.07066	16.5	1.14
40.0	.08014	14.9	1.46
45.0	.08948	13.5	1.80
50.0	.09867	12.4	2.18
60.0	.1166	10.8	3.03
70.0	.1341	9.55	4.00
80.0	.1510	8.62	5.08
90.0	.1675	7.88	6.27
100.	.1834	7.28	7.57
150.	.2568	5.44	15.5
200.	.3207	4.49	25.5
300.	.4260	3.52	50.6
400.	.5086	3.02	80.9
500.	.5746	2.74	115.
600.	.6281	2.55	152.
700.	.6721	2.42	192.
800.	.7088	2.33	234.
900.	.7396	2.26	277.
1000.	.7658	2.21	321.
2000.	.8981	2.05	795.
4000.	.9639	2.09	1780.

between the proton ranges $R_p = 0.623$ and 0.864 g/cm². Linear interpolation in β^2 gives $R_p = 0.715$ g/cm². It follows that the range of the 80 MeV ^{3}He^{2+} particle is $3(0.715)/4$ g/cm², or 0.536 cm in unit-density soft tissue.

Figure 4.4 shows the ranges in g/cm² of protons, alpha particles, and electrons in water or muscle (virtually the same), bone, and lead. For a given proton energy, the range in g/cm² is greater in Pb than in H_2O, consistent with the smaller mass stopping power of Pb, as mentioned at the end of Section 4.3. The same comparison is true for electrons in Pb and water at the lower energies in Fig. 4.4 ($\leq$20 MeV). At higher energies, bremsstrahlung greatly increases the rate of energy loss for electrons in Pb, reducing the range in g/cm² below that in H_2O.

Figure 4.5 gives the ranges in cm of protons, alpha particles, and electrons in air at standard temperature and pressure. For alpha particles in air at 15°C and 1 atm pressure,

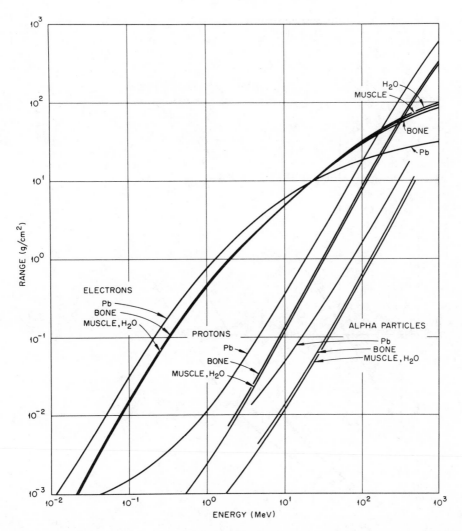

Figure 4.4. Ranges of protons, alpha particles, and electrons in water, muscle, bone, and lead, expressed in g/cm^2. (Courtesy Oak Ridge National Laboratory, operated by Martin Marietta Energy Systems, Inc., for the Department of Energy)

the following approximate empirical relations† fit the observed range R in cm as a function of energy E in MeV:

$$R = 0.56E, \qquad E < 4; \tag{4.27}$$

$$R = 1.24E - 2.62, \qquad 4 < E < 8. \tag{4.28}$$

As discussed at the end of Section 3.3, alpha rays from sources external to the body present little danger because their range is less than the minimum thickness of the outermost, dead layer of cells of the skin (epidermis, minimum thickness ~7 mg/cm^2). The

---

†U.S. Public Health Service, *Radiological Health Handbook*, Publ. No. 2016, Bureau of Radiological Health, Rockville, MD (1970).

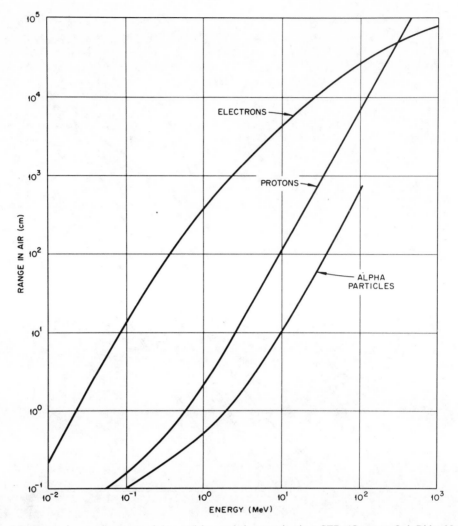

Figure 4.5. Ranges in cm of protons, alpha particles, and electrons in air at STP. (Courtesy Oak Ridge National Laboratory, operated by Martin Marietta Energy Systems, Inc., for the Department of Energy)

next example illustrates the nature of the potential hazard from one important alpha emitter when inhaled and trapped in the lung.

Example

The radon daughter $^{214}_{84}$Po, which emits a 7.69 MeV alpha particle, is present in the atmosphere of uranium mines. What is the range of this particle in soft tissue? Describe briefly the nature of the radiological hazard from inhalation of this nuclide.

Solution

We use the proton range in Table 4.3 to find the alpha particle range in tissue. Applied to alpha rays, Eq. (4.26) gives ($z^2 = 4$ and $M = 4$) $R_\alpha(\beta) = R_p(\beta)$. Thus, the ranges of an alpha particle and a proton with the same velocity are the same. The ratio of the kinetic energies at the same speed is $T_\alpha/T_p = M_\alpha/M_p = 4$, and so $T_p = T_\alpha/4 = 7.69/4 = 1.92$ MeV. The alpha-particle range is, therefore, equal to the range of a 1.92 MeV proton. Interpolation in Table 4.3 gives $R_p = R_\alpha = 0.0066$

Table 4.4. Calculated Slowing-Down Rates, $-dE/dt$, and Estimated
Stopping Times τ for Protons in Water

Proton energy T (MeV)	Slowing-down rate $-dE/dt$ (MeV/sec)	Estimated stopping time τ (sec)
0.5	4.21×10^{11}	1.2×10^{-12}
1.0	3.74×10^{11}	2.7×10^{-12}
10.0	2.00×10^{11}	5.0×10^{-11}
100.0	9.37×10^{10}	1.1×10^{-9}
1000.0	5.81×10^{10}	1.7×10^{-8}

cm in tissue of unit density. The ^{214}Po alpha particles thus cannot penetrate the 0.007 cm minimum epidermal thickness from outside the body to reach living cells. On the other hand, inhaled particulate matter containing ^{214}Po can be deposited in the lung. There the range of the alpha particles is sufficient to reach the basal cells of the bronchial epithelium. The increase in lung-cancer incidence among uranium miners over that normally expected has been linked to the alpha-particle dose from inhaled radon daughters.

Because of the statistical nature of energy losses by atomic collisions, all particles of a given type and initial energy do not travel exactly the same distance before coming to rest in a medium. Range as used here refers to the mean range, or average distance traveled. The phenomenon of range straggling will be discussed in Section 6.6.

4.8 SLOWING-DOWN RATE

We can use the stopping-power formula to calculate the rate at which a heavy charged particle slows down. The time rate of energy loss, $-dE/dt$, can be expressed in terms of the stopping power by using the chain rule of differentiation: $-dE/dt = (-dE/dx)(dx/dt) = V(-dE/dx)$, where $V = dx/dt$ is the velocity of the particle. For a proton with kinetic energy $T = 0.5$ MeV in water, for example, the rate of energy loss is $-dE/dt = 4.21 \times 10^{11}$ MeV/sec. We obtain a rough estimate of the time it takes a proton of kinetic energy T to stop by simply taking the ratio $T/[V(-dE/dx)]$. For a 0.5 MeV proton in water this estimate gives 1.2×10^{-12} sec. Slowing-down rates and stopping times for protons of other energies in water are given in Table 4.4. Electrons of a given energy stop much faster than protons.

4.9 LIMITATIONS OF BETHE'S STOPPING-POWER FORMULA

The stopping-power formula (4.7) is valid at high energies as long as the inequali'
$\gamma m/M \ll 1$, mentioned before Eq. (4.6), holds (e.g., up to ~10^6 MeV for proto'
Other physical factors, not included in Bethe's theory, come into play at higher ene'
These include forces on the atomic electrons due to the particle's spin and m'
moment as well as its internal electric and magnetic structures (particle form'
Bethe's formula is also based on the assumption that the particle moves much f'
atomic electrons. At low energies it fails because the term $\ln 2mc^2\beta^2/J$
becomes negative, giving a negative value for the stopping power.

In the low-energy region, also, a positively charged particle captures and'
as it moves, thus reducing its net charge and stopping power. Electron'
important when the speed V of the heavy particle is comparable to or'

that an electron needs in order to orbit about the particle as a nucleus. Based on Eq. (2.9) of the discussion of Bohr's theory, the orbital speed of an electron in the ground state about a nucleus of charge ze is $ze^2/\hbar$. Thus, as a condition for electron capture and loss one has $ze^2/\hbar V \gtrsim 1$. For electron capture by protons $(z = 1)$, we see from Eq. (2.9) that $V = 2.2 \times 10^8$ cm/sec, corresponding to a kinetic energy of ~25 keV.

The dependence of the Bethe formula on z^2, the square of the charge of the heavy particle, implies that pairs of particles with the same mass and energy but opposite charge, such as pions, $\pi^\pm$, and muons, $\mu^\pm$, have the same stopping power and range. Departures from this prediction have been measured and theoretically explained by the inclusion of z^3 and higher powers of the charge in the stopping-power formula.

4.10 PROBLEMS

1. Derive Eq. (4.4) from Eqs. (4.1) and (4.2).
2. (a) Calculate the maximum energy that a 3 MeV alpha particle can transfer to an electron in a single collision. (b) Repeat for a 100 MeV pion.
3. According to Eq. (4.6), what would be the relationship between the kinetic energies T_p and T_d of a proton and a deuteron that could transfer the same maximum energy to an atomic electron?
4. Which can transfer more energy to an electron in a single collision — a proton or an alpha particle? Explain.
5. Calculate the maximum fraction of its energy that a 10 MeV muon can lose in a single collision with an electron.
6. Compute the mean excitation energy of (a) Be, (b) Al, (c) Cu, (d) Pb.
7. Calculate the mean excitation energy of C_6H_6.
8. Compute the mean excitation energy of SiO_2.
9. What is the I value of air? Assume a composition of 4 parts N_2 to 1 part O_2 by volume.
10. Show that the stopping-power formula (4.7) gives $-dE/dx$ in the dimensions of energy/length.
11. (a) Calculate $F(\beta)$ directly from Eq. (4.18) for a 52 MeV proton. (b) Use Table 4.2 to obtain $F(\beta)$ by interpolation.
12. Find $F(\beta)$ from Table 4.2 for a 500 MeV alpha particle.
13. (a) Use Table 4.2 to determine $F(\beta)$ for a 5 MeV deuteron. (b) What is the stopping power of water for a 5 MeV deuteron?
14. (a) What is $F(\beta)$ for a 100 MeV muon? (b) Calculate the stopping power of copper for a 100 MeV muon.
15. Using Eq. (4.20), calculate the stopping power of water for (a) a 7 MeV proton, (b) a 7 MeV pion, (c) a 7 MeV alpha particle. Compare answers with Fig. 4.3.
16. Using Table 4.3 for the proton mass stopping power of water, estimate the stopping power of Lucite (density = 1.19 g/cm^3) for a 35 MeV proton.
17. Refer to Fig. 4.3. (a) By what factor can the stopping power of water for alpha particles exceed that for protons? (b) By what factor does the maximum alpha-particle stopping power exceed the maximum proton stopping power? (c) Why is the answer to (b) not 4, the ratio of the square of their charges? (d) What is the value of the maximum stopping power for pions?
18. From Table 4.3 determine the minimum energy that a proton must have to penetrate 30 cm of tissue, the approximate thickness of the human body.
19. What is the range of a 15 MeV $^3He^{2+}$ particle in water?
20. Write a formula that gives the range of a π^+ at a given velocity in terms of the range of a proton at that velocity.
21. How much energy does an alpha particle need to penetrate the minimal protective epidermal layer of skin (thickness ~7 mg/cm^2)?
22. Use Table 4.3 to determine the range in cm of an 11 MeV proton in air at STP.
23. A proton and an alpha particle with the same velocity are incident on a soft-tissue target. Which will penetrate to a greater depth?

24. What is the range of a 5 MeV deuteron in soft tissue?
25. (a) What is the range of a 4 MeV alpha particle in tissue? (b) Using the answer for (a), estimate the range in cm in air at STP. Compare with Fig. 4.5.
26. ^{239}Pu emits a 5.16 MeV alpha particle. What is its range in cm in (a) muscle, (b) bone of density 1.9 g/cm^3, (c) air at 22°C and 750 mm Hg?
27. Convert the formulas (4.27) and (4.28), which apply to alpha particles in air at 15°C and 1 atm, to air at STP.
28. Estimate the time it takes for a 6 MeV proton to stop in water.
29. (a) How does the stopping time for a 6 MeV proton in water (Problem 28) compare with the stopping time in lead? (b) How does the stopping time in water compare with that in air?
30. Estimate the time required for a 2.5 MeV alpha particle to stop in tissue.
31. (a) Estimate the slowing down time for a 2 MeV pion in water. (b) Repeat for a 2 MeV muon. (c) Give a physical reason for the difference in the times.
32. Estimate the energy at which an alpha particle begins to capture and lose electrons when slowing down.
33. Estimate the energy at which electron capture and loss become important when a positive pion slows down.

CHAPTER 5
INTERACTION OF BETA PARTICLES WITH MATTER

5.1 ENERGY-LOSS MECHANISMS

We treat electron and positron energy-loss processes together, referring to both simply as beta particles. Their stopping powers and ranges are virtually the same, except at low energies, as can be seen from Fig. 4.3. Energetic gamma photons produced by the annihilation of positrons with atomic electrons (Sect. 7.5) present a radiation problem with β^+ sources that does not occur with β^- emitters.

Like heavy charged particles, beta particles can excite and ionize atoms. In addition, they can also radiate energy by bremsstrahlung. As seen from Fig. 4.3, the radiative contribution to the stopping power (shown by the dashed line) becomes important only at high energies. At 100 MeV, for example, radiation accounts for about half the total rate of energy loss in water. We consider separately the collisional stopping power $(-dE/dx)_{col}$ and the radiative stopping power $(-dE/dx)_{rad}$ for beta particles. Beta particles can also be scattered elastically by atomic electrons, a process that has a significant effect on beta-particle penetration and diffusion in matter at low energies.

5.2 COLLISIONAL STOPPING POWER

The collisional stopping power for beta particles is different from that of heavy charged particles because of two physical factors. First, as mentioned in Section 4.1, a beta particle can lose a large fraction of its energy in a single collision with an atomic electron, which has equal mass. Second, a β^- particle is identical to the atomic electron with which it collides and a β^+ is the electron's antiparticle. In quantum mechanics, the identity of the particles implies that one cannot distinguish experimentally between the incident and struck electron after a collision. Energy loss is defined in such a way that the electron of lower energy after collision is treated as the struck particle. Unlike heavy charged particles, the identity of β^- and the relation of β^+ to atomic electrons imposes certain symmetry requirements on the equations that describe their collisions with atoms.

The collisional stopping-power formulas for electrons and positrons can be written

$$\left(-\frac{dE}{dx}\right)_{col}^{\pm} = \frac{4\pi e^4 n}{mc^2\beta^2}\left[\ln\frac{mc^2\tau\sqrt{\tau+2}}{\sqrt{2}\,I} + F^{\pm}(\beta)\right], \tag{5.1}$$

where

$$F^-(\beta) = \frac{1-\beta^2}{2} + \frac{1}{2(\tau+1)^2}\left[\frac{\tau^2}{8} - (2\tau+1)\ln 2\right] \tag{5.2}$$

is used for electrons and

$$F^+(\beta) = \ln 2 - \frac{\beta^2}{24}\left[23 + \frac{14}{\tau+2} + \frac{10}{(\tau+2)^2} + \frac{4}{(\tau+2)^3}\right] \tag{5.3}$$

for positrons. Here $\tau = T/mc^2$ is the kinetic energy T of the β^- or β^+ particle expressed in multiples of the electron rest energy mc^2. The other symbols in these equations, including I, are the same as in Eq. (4.7). Similar to Eq. (4.17), we have from (5.1)

$$\left(-\frac{dE}{dx}\right)^{\pm}_{\text{col}} = \frac{5.09 \times 10^{-25}n}{\beta^2}\left[\ln\frac{3.61 \times 10^5 \tau\sqrt{\tau+2}}{I_{\text{eV}}} + F^{\pm}(\beta)\right] \text{MeV/cm}. \tag{5.4}$$

As with heavy charged particles, this can be put into a general form:

$$\left(-\frac{dE}{dx}\right)^{\pm}_{\text{col}} = \frac{5.09 \times 10^{-25}n}{\beta^2}\left[G^{\pm}(\beta) - \ln I_{\text{eV}}\right] \text{MeV/cm}, \tag{5.5}$$

where

$$G^{\pm}(\beta) = \ln(3.61 \times 10^5 \tau\sqrt{\tau+2}) + F^{\pm}(\beta). \tag{5.6}$$

Example
Calculate the collisional stopping power of water for 1 MeV electrons.

Solution
This quantity is $(-dE/dx)^-_{\text{col}}$, given by Eq. (5.5). We need to compute β^2, τ, $F^-(\beta)$ and then $G^-(\beta)$. As in Section 4.6, we have $n = 3.34 \times 10^{23}$ cm^{-3} and $\ln I_{\text{eV}} = 4.31$. Using the relativistic formula for kinetic energy with $T = 1$ MeV and $mc^2 = 0.511$ MeV, we write

$$1 = 0.511\left(\frac{1}{\sqrt{1-\beta^2}} - 1\right), \tag{5.7}$$

giving $\beta^2 = 0.886$. Also, $\tau = T/mc^2 = 1/0.511 = 1.96$. From Eq. (5.2),

$$F^-(\beta) = \frac{1-0.886}{2} + \frac{1}{2(1.96+1)^2}\left[\frac{(1.96)^2}{8} - (2 \times 1.96 + 1)\ln 2\right], \tag{5.8}$$

giving $F^-(\beta) = -0.110$. From Eq. (5.6),

$$G^-(\beta) = \ln(3.61 \times 10^5 \times 1.96\sqrt{1.96+2}) - 0.110 = 14.0. \tag{5.9}$$

Finally, applying Eq. (5.5), we find

$$\left(-\frac{dE}{dx}\right)^-_{\text{col}} = \frac{5.09 \times 10^{-25} \times 3.34 \times 10^{23}}{0.886}[14.0 - 4.31] = 1.86 \text{ MeV/cm}. \tag{5.10}$$

It is of interest to compare this result with that for a 1 MeV positron. The quantities β^2 and τ are the same. Calculation gives $F^+(\beta) = -0.312$, which is a little larger in magnitude than $F^-(\beta)$. In place of (5.9) and (5.10) one finds $G^+(\beta) = 13.8$ and $(-dE/dx)^+_{\text{col}} = 1.82$ MeV/cm. The β^+ collisional stopping power is practically equal to that for β^- at 1 MeV in water.

The collisional, radiative, and total mass stopping powers of water as well as the radiation yield and range for electrons are given in Table 5.1. The total stopping power for β^- or β^+ particles is the sum of the collisional and radiative contributions:

$$\left(-\frac{dE}{dx}\right)^{\pm}_{\text{tot}} = \left(-\frac{dE}{dx}\right)^{\pm}_{\text{col}} + \left(-\frac{dE}{dx}\right)^{\pm}_{\text{rad}}, \tag{5.11}$$

Table 5.1. Electron Collisional, Radiative, and Total Mass Stopping Powers;
Radiation Yield; and Range in Water

Kinetic energy	β^2	$-\dfrac{1}{\rho}\left(\dfrac{dE}{dx}\right)^-_{col}$ (MeV cm^2/g)	$-\dfrac{1}{\rho}\left(\dfrac{dE}{dx}\right)^-_{rad}$ (MeV cm^2/g)	$-\dfrac{1}{\rho}\left(\dfrac{dE}{dx}\right)^-_{tot}$ (MeV cm^2/g)	Radiation yield	Range (g/cm^2)
10 eV	0.00004	4.0	—	4.0	—	4×10^{-8}
30	0.00012	44.	—	44.	—	2×10^{-7}
50	0.00020	170.	—	170.	—	3×10^{-7}
75	0.00029	272.	—	272.	—	4×10^{-7}
100	0.00039	314.	—	314.	—	5×10^{-7}
200	0.00078	298.	—	298.	—	8×10^{-7}
500 eV	0.00195	194.	—	194.	—	2×10^{-6}
1 keV	0.00390	126.	—	126.	—	5×10^{-6}
2	0.00778	77.5	—	77.5	—	2×10^{-5}
5	0.0193	42.6	—	42.6	—	8×10^{-5}
10	0.0380	23.2	—	23.2	0.0001	0.0002
25	0.0911	11.4	—	11.4	0.0002	0.0012
50	0.170	6.75	—	6.75	0.0004	0.0042
75	0.239	5.08	—	5.08	0.0006	0.0086
100	0.301	4.20	—	4.20	0.0007	0.0140
200	0.483	2.84	0.006	2.85	0.0012	0.0440
500	0.745	2.06	0.010	2.07	0.0026	0.174
700 keV	0.822	1.94	0.013	1.95	0.0036	0.275
1 MeV	0.886	1.87	0.017	1.89	0.0049	0.430
4	0.987	1.91	0.065	1.98	0.0168	2.00
7	0.991	1.93	0.084	2.02	0.0208	2.50
10	0.998	2.00	0.183	2.18	0.0416	4.88
100	0.999+	2.20	2.40	4.60	0.317	32.5
1000 MeV	0.999+	2.40	26.3	28.7	0.774	101.

with a similar relation holding for the mass stopping powers. Radiative stopping power, radiation yield, and range are treated in the next three sections. Table 5.1 can also be used for positrons with energies above about 10 keV.

The calculated mass stopping power of liquid water for electrons at low energies is shown in Fig. 5.1. (Measurement of this important quantity does not appear to be technically feasible.) The radiative stopping power is negligible here. This curve is very basic to radiation physics, because the end product of any form of ionizing radiation is a spatial distribution of low-energy secondary electrons, which slow down through the energy range shown in the figure. Since it takes an average of only about 25 eV to produce a secondary electron in liquid water, radiation produces low-energy electrons in abundance. A 10 keV electron, for example, produces a total of about 400 secondary electrons, a large fraction of which occur with initial energies of less than 150 eV. The details of electron transport and charged-particle track structure and their relation to chemical and biological effects will be considered in Chapter 11.

5.3 RADIATIVE STOPPING POWER

The acceleration of a heavy charged particle in an atomic collision is usually small, and except under extreme conditions negligible radiation occurs. A beta particle, on the other hand, having little mass can be accelerated strongly by the same electromagnetic force within an atom and thereby emit radiation, called bremsstrahlung. Bremsstrahlung occurs when a beta particle is deflected in the electric field of a nucleus and, to a lesser extent, in the field of an atomic electron. At high beta-particle energies, the radiation is

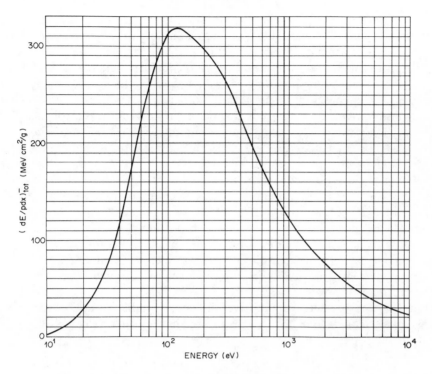

Figure 5.1. Mass stopping power of water for low-energy electrons.

emitted mostly in the forward direction, that is, in the direction of travel of the beta particle. As indicated in Fig. 5.2, this circumstance is observed in a betatron or synchrotron,
a device that accelerates electrons to high energies in circular orbits. Most of the synchrotron radiation, as it is called, is emitted in a narrow sweeping beam nearly in the direction
of travel of the electrons that produce it.

Energy loss by an electron in radiative collisions was studied quantum mechanically by
Bethe and Heitler. If the electron passes near a nucleus, the field in which it is accelerated
is essentially the bare Coulomb field of the nucleus. If it passes at a greater distance, the

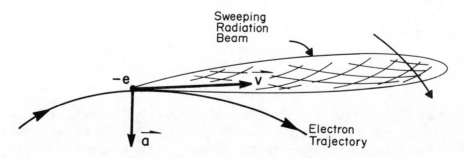

Figure 5.2. Synchroton radiation. At high energies, photons are emitted by electrons (charge $-e$) in circular orbits
in the direction (cross-hatched area) of their instantaneous velocity **v**. The direction of the electrons' acceleration **a** is also shown.

partial screening of the nuclear charge by the atomic electrons becomes important, and the field is no longer coulombic. Thus, depending on how close the electron comes to the nucleus, the effect of atomic-electron screening will be different. The screening and subsequent energy loss also depend on the energy of the incident beta particle. The maximum energy that a bremsstrahlung photon can have is equal to the kinetic energy of the beta particle. The photon energy spectrum is approximately flat out to this maximum.

Unlike collisional energy losses, no single analytic formula exists for calculating the radiative stopping power $(-dE/dx)_{rad}^{\pm}$. Instead, numerical procedures are used to obtain values, such as those in Table 5.1. Details of the analysis show that energy loss by radiation behaves quite differently from that by ionization and excitation. The efficiency of bremsstrahlung in elements of different atomic number Z varies nearly as Z^2. Thus, for beta particles of a given energy, bremsstrahlung losses are considerably greater in high-Z materials, such as lead, than in low-Z materials, such as water. As seen from Eq. (5.1), the collisional energy-loss rate in an element is proportional to n and hence to Z. In addition, the radiative energy-loss rate increases nearly linearly with beta-particle energy, whereas the collisional rate increases only logarithmically. At high energies, therefore, bremsstrahlung becomes the predominant mechanism of energy loss for beta particles, as can be seen from Table 5.1.

The following approximate formula gives the ratio of radiative and collisional stopping powers for an electron of total energy E, expressed in MeV, in an element of atomic number Z:

$$\frac{(-dE/dx)_{rad}^-}{(-dE/dx)_{col}^-} \cong \frac{ZE}{800}. \tag{5.12}$$

This formula shows that in lead ($Z = 82$), for example, the two rates of energy loss are approximately equal at a total energy given by

$$\frac{82E}{800} \cong 1. \tag{5.13}$$

Thus $E \cong 9.8$ MeV, and the electron's kinetic energy is $T = E - mc^2 \cong 9.3$ MeV. In oxygen ($Z = 8$), the two rates are equal when $E \cong 100$ MeV $\cong T$, an order-of-magnitude higher energy than in lead. The radiative stopping power $(-dE/dx)_{rad}^-$ for electrons is shown by the dashed curve in Fig. 4.3.

At very high energies the dominance of radiative over collisional energy losses gives rise to electron–photon cascade showers. Since the bremsstrahlung photon spectrum is approximately flat out to its maximum (equal to the electron's kinetic energy), high-energy beta particles emit high-energy photons. These, in turn, produce Compton electrons and electron–positron pairs, which then produce additional bremsstrahlung photons, and so on. These repeated interactions result in an electron–photon cascade shower, which can be initiated by either a high-energy beta particle or a photon.

5.4 RADIATION YIELD

We have discussed the relative rates of energy loss by collision and by radiation. Radiation yield is defined as the average fraction of its energy that a beta particle radiates as bremsstrahlung in slowing down completely. Radiation yields are given in Table 5.1 for electrons of various energies in water. At 100 MeV, for example, the rates of energy loss by collision and by radiation are approximately equal. As the electron slows down, however, the relative amount lost by radiation decreases steadily. In slowing down completely, a 100 MeV electron loses an average of 0.317 of its initial energy (i.e., 31.7 MeV) by radiation. Radiation yield increases with electron energy. A 1000 MeV electron stopping in water will lose an average of 0.774 of its energy by bremsstrahlung. For an electron of given energy, radiation yield also increases with atomic number.

An estimate of radiation yield can give an indication of the potential bremsstrahlung hazard of a beta-particle source. If electrons of initial kinetic energy T in MeV are stopped in an absorber of atomic number Z, then the radiation yield is given approximately by the formula[†]

$$Y \cong \frac{6 \times 10^{-4} ZT}{1 + 6 \times 10^{-4} ZT}. \tag{5.14}$$

To keep bremsstrahlung to a minimum, low-Z materials can be used as a shield to stop beta particles. Such a shield can, in turn, be surrounded by a material of high Z to efficiently absorb the bremsstrahlung photons.

Example
Estimate the fraction of the energy of a 2 MeV beta ray that is converted into bremsstrahlung when the particle is absorbed in aluminum and in lead.

Solution
For Al, $ZT = 13 \times 2 = 26$, and $Y \cong 0.016/1.016 = 0.016$. For Pb ($Z = 82$), $Y \cong 0.090$. Thus, about 1.6% of the electron kinetic energy is converted into photons in Al, while the corresponding figure for Pb is about 9.0%.

For radiation-protection purposes, conservative assumptions can be made in order to apply Eq. (5.14) to the absorption of beta particles from a radioactive source. To this end, the *maximum* beta-particle energy is used for T. This assumption overestimates the energy converted into radiation because bremsstrahlung efficiency is less at the lower electron energies. Furthermore, assuming that all bremsstrahlung photons have the energy T will also give a conservative estimate of the actual photon hazard.

Example
A small 10 mCi ^{90}Y source is enclosed in a lead shield just thick enough to absorb the beta particles, which have a maximum energy of 2.27 MeV and an average energy of 0.76 MeV. Estimate the rate at which energy is radiated as bremsstrahlung. For protection purposes, estimate the photon fluence rate at a distance of 1 m from the source.

Solution
Setting $T = 2.27$ and $Z = 82$ in Eq. (5.14) gives for the fraction of the beta-particle energy converted into photons $Y \cong 0.10$. The total beta-particle energy released per second from the 10 mCi source is $(3.7 \times 10^{8}/\text{sec}) (0.76 \text{ MeV}) = 2.81 \times 10^{8}$ MeV/sec. Multiplication by Y gives, for the rate of energy emission by bremsstrahlung, $\sim 2.81 \times 10^{7}$ MeV/sec. The energy fluence rate at a distance of 1 m is therefore $\sim (2.81 \times 10^{7} \text{ MeV/sec})/(4\pi \times 100^{2} \text{ cm}^{2}) = 2.24 \times 10^{2}$ MeV/(cm^{2} sec). For assessing the radiation hazard, we assume that the photons have an energy of 2.27 MeV. Therefore, the photon fluence rate at this distance is $\sim 2.24 \times 10^{2}/2.27 = 98.7$ photons/(cm^{2} sec). For comparison, we note that use of an aluminum ($Z = 13$) shield to stop the beta particles would give $Y \cong 0.0174$, reducing the bremsstrahlung by a factor of 5.75.

[†]H. W. Koch and J. W. Motz, *Rev. Mod. Phys.* **31**, 920 (1959).

5.5 RANGE

The range of beta particles can be defined like that of heavy particles by Eq. (4.22) in which the total stopping power $(-dE/dx)^{\pm}_{tot}$ is used. This definition assumes that the kinetic energy of the particle changes as a continuous function as it slows down and comes to rest. While this "continuous-slowing-down approximation" is physically appropriate for heavy charged particles, it is not always realistic for electrons, which can lose large fractions of their energy in single collisions. Nevertheless, the definition (4.22) is used to calculate electron range as a function of energy. One thus obtains a value for the range that is approximately equal to the average path length an electron travels, as distinct from the distance it penetrates in an absorber (cf. Fig. 4.1). Electrons of reasonably high energy ($\gtrsim$ 10 keV) tend to travel in such a way that the range calculated in the continuous-slowing-down approximation is a useful measure of average electron penetration distance. Table 5.1 gives electron ranges in water down to 10 eV. As with heavy charged particles, electron ranges expressed in g/cm^2 are approximately the same in different materials of similar atomic composition.

Electron ranges in H_2O, muscle, bone, Pb, and air are included in Figs. 4.4 and 4.5. For the same reasons as with heavy charged particles (discussed at the end of Sect. 4.3), the collisional mass stopping power for beta particles is smaller in high-Z materials, such as lead, than in water. In Fig. 4.4, this fact accounts for the greater range of electrons in Pb compared with H_2O at energies below about 20 MeV. At higher energies, the radiative energy-loss rate in Pb more than compensates for the difference in the collisional rate, and the electron range in Pb is less than in H_2O.

The following empirical equations for electrons in low-Z materials relate the range R in g/cm^2 to the kinetic energy T in MeV:
For $0.01 \leq T \leq 2.5$ MeV,

$$R = 0.412T^{1.27-0.0954 \ln T} \tag{5.15}$$

or

$$\ln T = 6.63 - 3.24(3.29 - \ln R)^{1/2}; \tag{5.16}$$

for $T > 2.5$ MeV,

$$R = 0.530T - 0.106, \tag{5.17}$$

or

$$T = 1.89R + 0.200. \tag{5.18}$$

These relations fit the curve plotted in Fig. 5.3.†

Example
How much energy does a 2.2 MeV electron lose in passing through 5 mm of Lucite (density $\rho =$ 1.19 g/cm^3)?

Solution
Lucite is a low-Z material, and so we may apply Eqs. (5.15)–(5.18) or use Fig. 5.3 directly. We shall employ the equations and check the results against the figure. First, we find how far the 2.2 MeV electron can travel in Lucite. From Eq. (5.15) with $T = 2.2$ MeV,

---

†Figure 5.3 and the relations (5.15)–(5.18) are taken from the U.S. Public Health Service, *Radiological Health Handbook*, Publ. No. 2016, Bureau of Radiological Health, Rockville, MD (1970).

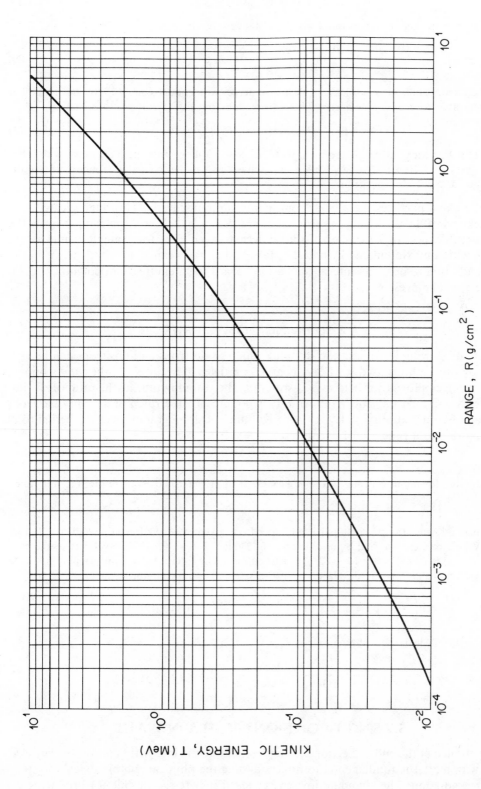

Figure 5.3. Beta-particle range–energy curve for materials of low atomic number. [From U.S. Public Health Service, *Radiological Health Handbook*, Pub. No. 2016, Bureau of Radiological Health, Rockville, MD (1970)]

$$R = 0.412(2.2)^{1.27-0.0954\ln 2.2} = 1.06 \text{ g/cm}^2,$$ (5.19)

in agreement with Fig. 5.3. This range gives a distance

$$d = \frac{R}{\rho} = \frac{1.06 \text{ g/cm}^2}{1.19 \text{ g/cm}^3} = 0.891 \text{ cm}.$$ (5.20)

Since the Lucite is only 0.5 cm thick, the electron emerges with enough energy T' to carry it another 0.391 cm, or 0.465 g/cm^2. The energy T' can be found from Eq. (5.16):

$$\ln T' = 6.63 - 3.24(3.29 - \ln 0.465)^{1/2} = 0.105,$$ (5.21)

and so $T' = 1.11$ MeV, again in agreement with Fig. 5.3. It follows that the energy lost by the electron is $T - T' = 2.20 - 1.11 = 1.09$ MeV. The analysis and numerical values found here are the same also for a 2.2 MeV positron.

Unlike alpha particles, beta rays from many radionuclides have a range greater than the thickness of the epidermis. As seen from Fig. 5.3, a 70 keV electron can penetrate the minimum thickness of 7 mg/cm^2 of the epidermal layer. ^{90}Y, for example, emits a beta particle with a maximum energy of 2.27 MeV, which has a range of over 1 g/cm^2 in tissue. In addition to being an internal radiation hazard, beta emitters can potentially damage the skin and eyes.

5.6 SLOWING-DOWN TIMES

The rate of slowing down and the stopping time for electrons and positrons can be estimated by the methods used for heavy charged particles in Section 4.8, the *total* stopping power being employed for beta rays. Estimating the stopping time as the ratio of the initial energy and the total slowing-down rate is not grossly in error. For a 1 MeV electron, for example, this ratio is $\tau = 1.9 \times 10^{-11}$ sec; numerical integration over the actual total stopping rate as a function of energy gives 1.3×10^{-11} sec.

Example
Calculate the slowing-down rate of an 800 keV electron in water and estimate the stopping time.

Solution
The slowing-down rate for a beta particle of speed v is given by $-dE/dt = v(-dE/dx)^-_{tot}$. For an 800 keV beta particle, we find by interpolating in Table 5.1, $\beta^2 = 0.843$. The velocity of the electron is $v = \beta c = \sqrt{0.843} \times 3 \times 10^{10} = 2.75 \times 10^{10}$ cm/sec. The interpolated total stopping power at 800 keV from Table 5.1 is $(-dE/dx)^-_{tot} = 1.94$ MeV/cm. The slowing-down rate is

$$-\frac{dE}{dt} = v\left(-\frac{dE}{dx}\right)^-_{tot} = 2.75 \times 10^{10} \frac{\text{cm}}{\text{sec}} \times 1.94 \frac{\text{MeV}}{\text{cm}} = 5.34 \times 10^{10} \text{ MeV/sec}.$$ (5.22)

With $T = 0.800$ MeV, the stopping time is

$$\tau \cong \frac{T}{-dE/dt} = \frac{0.800 \text{ MeV}}{5.34 \times 10^{10} \text{ MeV/sec}} = 1.5 \times 10^{-11} \text{ sec}.$$ (5.23)

5.7 SINGLE-COLLISION SPECTRA IN WATER

As mentioned at the end of Section 5.2, understanding the interaction of low-energy electrons with matter is fundamental to understanding the physical and biological effects of ionizing radiation. The abundant low-energy electrons are responsible for producing the initial alterations that lead to chemical changes in tissue and tissue-like materials, such as water, which has been extensively studied experimentally and theoretically.

The interaction of an electron of kinetic energy T can be characterized physically by the probability $N(T,E)\,dE$ that it loses an amount of energy between E and $E + dE$ in a single collision. The distribution $N(T,E)$ is called the single-collision spectrum for an electron of energy T. As a probability function, it is normalized,

$$\int_0^{Q_{max}} N(T,E)\,dE = 1, \tag{5.24}$$

and has the dimensions of inverse energy.

Calculated single-collision spectra for electrons of energy $T = 30$ eV, 50 eV, 150 eV, and 10 keV in water are shown in Fig. 5.4. The information given by the curves can be illustrated by an example. For 10 keV electrons, the average value of the single-collision spectrum for energy losses between 45 and 50 eV is 0.01 eV^{-1}. Since the width of the interval is 5 eV, it follows that the relative number of collisions with energy losses between 45 and 50 eV is $(0.01\text{ eV}^{-1}) \times 5\text{ eV} = 0.05$. Thus, a 10 keV electron in water has about a 5% chance of experiencing an energy loss E between 45 and 50 eV in its next collision.

The structure of the curves in Fig. 5.4 reflects the basic physics of electron interactions with the liquid-water system. All curves start at an estimated threshold energy of 7.4 eV, which is the minimum energy required for electronic excitation. Excitation takes place to bound, discrete energy levels, as shown by the resonance peaks in the energy-loss spectra at the lower values of E. This structure is particularly pronounced at $T = 30$ eV, where the electron energy is so low that excitation is about as probable as ionization. The spec-

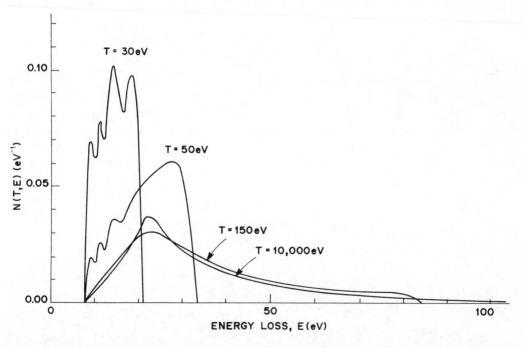

Figure 5.4. Single-collision spectra $N(T,E)$ for electrons of kinetic energies $T = 30$ eV, 50 eV, 150 eV, and 10 keV in water.

trum of energy losses to ionization is, of course, continuous, since the energy of the ejected electron is not restricted to discrete values. Figure 5.5 shows the probability that a given energy-loss event will cause ionization rather than excitation. The relative importance of ionization is seen to increase very rapidly with the energy of the electron. At $T > 150$ eV, almost 95% of the energy losses result in ionization, rather than excitation. The energy-loss spectra in Fig. 5.4 remain remarkably similar for electrons with energies above 150 eV. The spectrum at 10 keV—two orders of magnitude higher—is approximately the same as that at 150 eV except that it has a high-energy tail. Most of the energy losses by 10 keV electrons are therefore the same as those experienced by 150 eV electrons. Occasionally, however, the higher-energy electron will have a "hard" collision in which a very energetic secondary electron is produced.

The collisional stopping power is related to the single-collision spectrum $N(T,E)$. The average energy lost $\bar{E}(T)$ by an electron of energy T in a single collision is given by the weighted average over the energy-loss spectrum,

$$\bar{E}(T) = \int_0^{Q_{max}} EN(T,E)\,dE. \tag{5.25}$$

The stopping power at energy T is the product of $\bar{E}(T)$ and the probability $\mu(T)$ per unit distance that an inelastic collision occurs:

$$\left(-\frac{dE}{dx}\right)_{tot}^{-} = \mu(T)\bar{E}(T) = \mu(T)\int_0^{Q_{max}} EN(T,E)\,dE. \tag{5.26}$$

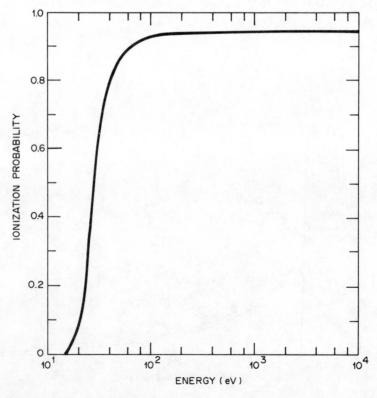

Figure 5.5. Probability that a given energy-loss event will result in ionization rather than excitation in water as a function of electron energy.

Curves for $N(T,E)$, such as those in Fig. 5.4, were combined with knowledge of the inelastic collision probability to calculate the stopping power shown in Fig. 5.1.

5.8 EXAMPLES OF ELECTRON TRACKS IN WATER

Figure 5.6 shows a three-dimensional representation of three electron "tracks" calculated by a Monte Carlo computer code to simulate electron transport in water. Each primary electron starts with an energy of 5 keV from the origin and moves initially toward the right along the horizontal axis. Each dot represents the location, at 10^{-11} sec, of a chemically active species produced by the action of the primary electron or one of its secondaries. The Monte Carlo code randomly selects collision events from specified distribu-

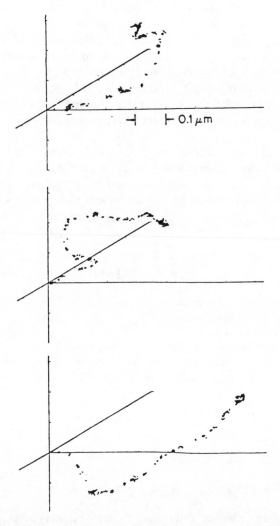

Figure 5.6. Three calculated tracks of 5 keV electrons in water. Each electron starts from the origin and initially travels along the horizontal axis toward the right. Each dot gives the position of a chemically active species at 10^{-11} sec. [From J. E. Turner, J. L. Magee, H. A. Wright, A. Chatterjee, R. N. Hamm, and R. H. Ritchie, Physical and chemical development of electron tracks in liquid water, *Rad. Res.* **96**, 437–449 (1983). Courtesy Oak Ridge National Laboratory, operated by Martin Marietta Energy Systems, Inc., for the Department of Energy]

tions of flight distance, energy loss, and angle of scatter in order to calculate the fate of individual electrons, simulating as nearly as is known what actually happens in nature. All electrons are transported in the calculation until their energies fall below the threshold of 7.4 eV for electronic excitation. Ticks along the axes are 0.1 μm = 1000 Å apart.

These examples illustrate a number of characteristics of the tracks of electrons that stop in matter. As we have mentioned already, the tracks tend to wander, due to large-angle deflections that electrons can experience in single collisions. The wandering is augmented at low energies, near the end of a track, by the greatly increased and almost isotropic elastic scattering that occurs there. In addition, energy-loss events are more sparsely distributed at the beginning of the track, where the primary electron is moving faster. This is generally true of charged-particle tracks, since the stopping power is smaller at high energies than near the end of the range. Note also the clustering of events, particularly in the first part of the tracks. Such groupings, called "spurs," are due to the production of a secondary electron with just enough energy to produce several additional ionizations and excitations. The range of the original secondary electron is not great enough to take it appreciably away from the region through which the primary electron passes. Clustering occurs as a result of the broad shape of the single-collision spectra in Fig. 5.4 with most of the area covering energy losses $\leq$70 eV.

Figure 5.7(a) gives a stereoscopic representation of another 5 keV electron track, traveling out of the page toward the reader, calculated in water. The same track is shown from the side in (b), except that the primary electron was forced to move straight ahead in the calculation.

The discussion here covers much of the basic physics underlying our understanding of the effects of ionizing radiation on matter. Studies with water can be partially checked by radiochemical measurements. They shed some light on the physical and chemical changes induced by radiation that ultimately lead to biological effects in living systems. The subsequent chemical evolution within the track of a charged particle in water is described in Chapter 11.

5.9 PROBLEMS

1. Calculate $F^-(\beta)$ for a 600 keV electron.
2. Calculate $F^+(\beta)$ for a 600 keV positron.
3. Derive Eq. (5.4) from Eq. (5.1).
4. Calculate the collisional stopping power of water for 600 keV electrons.
5. Calculate the collisional stopping power of water for 600 keV positrons.
6. Show that, when $\beta^2 \ll 1$, the collisional stopping power formula for an electron with kinetic energy T can be written

$$\left(-\frac{dE}{dx}\right)_{col}^- = \frac{2\pi e^4 n}{mv^2}\left(\ln\frac{mv^2 T}{4I^2} + 1\right).$$

7. Use the formula from the last problem to calculate the stopping power of CO_2 at STP for 9.5 keV electrons.
8. Estimate the energy at which the collisional and radiative stopping powers are equal in (a) Be, (b) Cu, (c) Pb.
9. What is the ratio of the collisional and radiative stopping powers of Al for electrons of energy (a) 10 keV, (b) 1 MeV, (c) 100 MeV?
10. Estimate the radiation yield for 10 MeV electrons in (a) Al, (b) Fe, (c) Au.
11. A small 0.05 Ci ^{198}Au source (maximum beta-particle energy 0.962 MeV, average 0.30 MeV) is enclosed in a lead shield just thick enough to absorb all the beta particles. (a) Estimate the

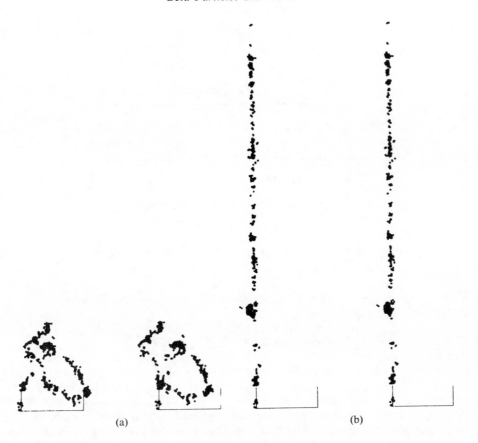

Figure 5.7. (a) Stereoscopic view of a 5 keV electron track in water. (b) Lateral view of same track in which the primary electron is forced to always move straight ahead. (Courtesy R. N. Hamm, Oak Ridge National Laboratory, operated by Martin Marietta Energy Systems, Inc., for the Department of Energy)

energy fluence rate at a point 50 cm away. (b) What is the estimated photon fluence rate at that point for the purpose of assessing radiation hazard?

12. A 12 mA electron beam is accelerated through a potential difference of 2×10^5 V in an X-ray tube with a tungsten target. X-rays are generated as bremsstrahlung from the electrons stopping in the target. Neglect absorption in the tube. (a) Estimate the fraction of the beam power that is emitted as radiation. (b) How much power is radiated as bremsstrahlung?

13. Calculate the range in cm of the maximum-energy beta ray (2.27 MeV) from ^{90}Y in bone of density 1.9 g/cm^3.

14. Use Eq. (5.15) to calculate the range of a 400 keV beta particle in water. How does the answer compare with Table 5.1?

15. Derive Eq. (5.16) from Eq. (5.15).

16. Use Table 5.1 to estimate the range in cm in air at STP for electrons of energy (a) 50 keV, (b) 830 keV, (c) 100 MeV.

17. A positron emerges normally from a 4 mm thick plastic slab (density 1.14 g/cm^3) with an energy of 1.62 MeV. What was its energy when it entered the slab?

18. Use the range curve in Fig. 4.4 and make up a formula or formulas like Eqs. (5.15) and (5.17) giving the range of electrons in lead as a function of energy.

19. A cell culture (Fig. 5.8) is covered with a 1 cm sheet of Lucite (density = 1.19 g/cm^3). What thickness of lead (in cm) is needed on top of the Lucite to prevent 10 MeV beta rays from

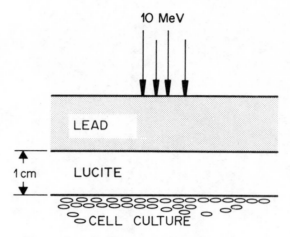

Figure 5.8. Cell culture covered with lucite and lead.

reaching the culture? Use the approximate empirical formulas relating range R in g/cm^2 to electron kinetic energy T in MeV:

$$\text{Lucite: } R = 0.334T^{1.48}, \qquad 0 \le T \le 4$$
$$\text{lead: } \quad R = 0.426T^{1.14}, \qquad 0.1 \le T \le 10.$$

20. As far as protecting the cell culture in Problem 19 from radiation, what advantage would be gained if the positions of the Lucite and the lead were swapped, so that the Lucite was on top?
21. Estimate the slowing-down time in water for positrons of energy (a) 100 keV, (b) 1 MeV.
22. (a) Use the information from Problems 6 and 7 to calculate the slowing-down rate of a 9.5 keV electron in CO_2 at STP. (b) Estimate the stopping time for the electron.
23. (a) Estimate the stopping time of a 9.5 keV electron in soft tissue. (b) Why is this time considerably shorter than the time in Problem 22(b)?
24. (a) Calculate the ratio of the slowing-down times of a 1 MeV proton and a 1 MeV electron in water. (b) Calculate the ratio for a 250 MeV proton and a 0.136 MeV electron ($\beta = 0.614$ for both). (c) Discuss physical reasons for the time difference in (a) and (b).
25. (a) Approximately how many secondary electrons are produced when a 5 MeV electron stops in water? (b) What is the average number of ions per cm (specific ionization) along its track?
26. For a 150 eV electron, the ordinate in Fig. 5.4 has an average value of about 0.03 in the energy-loss interval between 19 eV and 28 eV. What fraction of the collisions of 150 eV electrons in water result in energy losses between 19 and 28 eV?
27. From Fig. 5.5, estimate for a 100 eV electron the probability that a given energy-loss event will result in excitation, rather than ionization, in water.

CHAPTER 6
PHENOMENA ASSOCIATED WITH CHARGED-PARTICLE TRACKS

6.1 DELTA RAYS

A heavy charged particle or an electron traversing matter sometimes produces a secondary electron with enough energy to leave the immediate vicinity of the primary particle's path and produce a noticeable track of its own. Such a secondary electron is called a delta ray. Figure 6.1 shows a number of examples of delta rays along calculated tracks of protons and alpha particles at several energies with the same speeds. The 20 MeV alpha particle produces a very-high-energy delta ray, which itself produces another delta ray. There is no sharp distinction in how one designates one secondary electron along a track as a delta ray and another not, except that its track be noticeable or distinct from that of the primary charged particle.

6.2 RESTRICTED STOPPING POWER

As will be discussed in Chapter 10, radiation dose is defined as the energy absorbed per unit mass in an irradiated material. Absorbed energy thus plays a preeminent role in dosimetry and in radiation protection.

Stopping power gives the energy *lost* by a charged particle in a medium. This is not always equal to the energy *absorbed* in a target, especially if the target is small compared with the ranges of secondary electrons produced. On the biological scale, many living cells have diameters of the order of microns (10^{-4} cm). Subcellular structures can be many times smaller; the DNA double helix, for example, has a diameter of about 20 Å. Delta rays and other secondary electrons can effectively transport energy out of the original site in which it is lost by a primary particle.

The concept of restricted stopping power has been introduced to associate energy loss in a target more closely with the energy that is actually absorbed there. The restricted stopping power, written $-(dE/dx)_\Delta$, is defined as the linear rate of energy loss due only to collisions in which the energy transfer does not exceed a specified value Δ. To calculate this quantity, one integrates the weighted energy-loss spectrum only up to Δ, rather than Q_{max}. In place of Eq. (5.26) one defines the restricted stopping power for electrons as

$$-\left(\frac{dE}{dx}\right)_\Delta = \mu(T)\int_0^\Delta EN(T,E)\,dE, \tag{6.1}$$

a similar expression being applied to heavy charged particles. Depending on the application, different values of Δ can be selected, e.g., $-(dE/dx)_{100\,eV}$, $-(dE/dx)_{1\,keV}$, etc. In the context of restricted stopping power, the subscript Q_{max} or ∞ is used to designate the usual stopping power:

PROTONS   ALPHAS

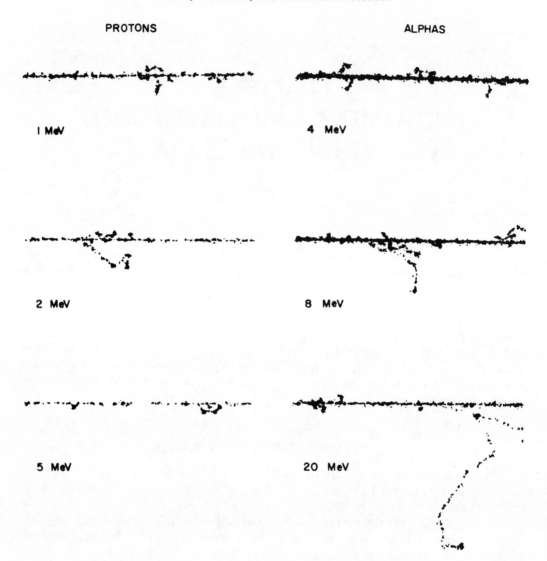

Figure 6.1. Calculated track segments (0.7 μm) of protons and alpha particles having the same velocities in water. (Courtesy Oak Ridge National Laboratory, operated by Martin Marietta Energy Systems, Inc., for the Department of Energy)

$$-\frac{dE}{dx} = -\left(\frac{dE}{dx}\right)_{Q_{max}} = -\left(\frac{dE}{dx}\right)_{\infty}. \tag{6.2}$$

We see from Table 5.1 that restricting single-collision energy losses by electrons in water to 100 eV or less, for example, limits the range of secondary electrons to ~5×10^{-7} cm, or about 40 Å. With $\Delta = 1$ keV, the maximum range of the secondary electrons contributing to the restricted stopping power is 5×10^{-6} cm, or about 500 Å.

Example

A sample of viruses, assumed to be in the shape of spheres of diameter 300 Å, is to be irradiated by a charged-particle beam in an experiment. Estimate a cutoff value that would be appropriate for determining a restricted stopping power that would be indicative of the actual energy *absorbed* in the individual virus particles.

Solution

As an approximation, one can specify that the range of the most energetic delta ray should not exceed 300 Å $= 3 \times 10^{-6}$ cm. We assume that the virus sample has unit density. Table 5.1 shows that this distance is approximately the range of a 700 eV secondary electron. Therefore, we choose $\Delta = 700$ eV and use the restricted stopping power $-(\mathrm{d}E/\mathrm{d}x)_{700\,\mathrm{eV}}$ as a measure of the average energy absorbed in an individual virus particle from a charged particle traversing it.

Restricted mass stopping powers of water for protons are given in Table 6.1. At energies of 0.05 MeV and below, collisions that transfer more than 100 eV do not contribute significantly to the total stopping power, and so $(-\mathrm{d}E/\rho\,\mathrm{d}x)_{100\,\mathrm{eV}} = (-\mathrm{d}E/\rho\,\mathrm{d}x)_{\infty}$. In fact, at 0.05 MeV, $Q_{max} = 109$ eV. At 0.10 MeV, on the other hand, $Q_{max} = 220$ eV; and so the restricted mass stopping power with $\Delta = 1$ keV is significantly larger than that with $\Delta = 100$ eV. At 1 MeV, a negligible number of collisions result in energy transfers of more than 10 keV; at 10 MeV, about 6% of the stopping power is due to collisions that transfer more than 10 keV. Corresponding data for the restricted collisional mass stopping power for electrons are presented in Table 6.2. Here, the restricted stopping powers are different at much lower energies than in Table 6.1.

6.3 LINEAR ENERGY TRANSFER (LET)

The concept of linear energy transfer, or LET, was introduced in the early 1950s to characterize the rate of energy transfer per unit distance along a charged-particle track. As

Table 6.1. Restricted Mass Stopping Power of Water, $-(\mathrm{d}E/\rho\,\mathrm{d}x)_{\Delta}$ in MeV cm^2/g, for Protons

Energy (MeV)	$-\left(\dfrac{\mathrm{d}E}{\rho\,\mathrm{d}x}\right)_{100\,\mathrm{eV}}$	$-\left(\dfrac{\mathrm{d}E}{\rho\,\mathrm{d}x}\right)_{1\,\mathrm{keV}}$	$-\left(\dfrac{\mathrm{d}E}{\rho\,\mathrm{d}x}\right)_{10\,\mathrm{keV}}$	$-\left(\dfrac{\mathrm{d}E}{\rho\,\mathrm{d}x}\right)_{\infty}$
0.05	910.	910.	910.	910.
0.10	711.	910.	910.	910.
0.50	249.	424.	428.	428.
1.00	146.	238.	270.	270.
10.0	24.8	33.5	42.2	45.9
100.	3.92	4.94	5.97	7.28

Table 6.2. Restricted Collisional Mass Stopping Power of Water, $-(\mathrm{d}E/\rho\,\mathrm{d}x)_{\Delta}$ in MeV cm^2/g, for Electrons

Energy (MeV)	$-\left(\dfrac{\mathrm{d}E}{\rho\,\mathrm{d}x}\right)_{100\,\mathrm{eV}}$	$-\left(\dfrac{\mathrm{d}E}{\rho\,\mathrm{d}x}\right)_{1\,\mathrm{keV}}$	$-\left(\dfrac{\mathrm{d}E}{\rho\,\mathrm{d}x}\right)_{10\,\mathrm{keV}}$	$-\left(\dfrac{\mathrm{d}E}{\rho\,\mathrm{d}x}\right)_{\infty}$
0.0002	298.	298.	298.	298.
0.0005	183.	194.	194.	194.
0.001	109.	126.	126.	126.
0.003	40.6	54.4	60.1	60.1
0.005	24.9	34.0	42.6	42.6
0.01	15.1	20.2	23.2	23.2
0.05	4.12	5.26	6.35	6.75
0.10	2.52	3.15	3.78	4.20
1.00	1.05	1.28	1.48	1.89

such, LET and stopping power were synonymous. In studying radiation effects in terms of LET, the distinction was made between the energy transferred from a charged particle in a target and the energy actually absorbed there. In 1962 the International Commission on Radiation Units and Measurements (ICRU) defined LET as the quotient $-dE_L/dx$, where dE_L is the "average energy locally imparted" to a medium by a charged particle in traversing a distance dx. The words "locally imparted," however, were not precisely specified, and LET was not always used with exactly the same meaning. In 1970, the ICRU defined LET_Δ as the restricted stopping power for energy losses not exceeding Δ:

$$LET_\Delta = -\left(\frac{dE}{dx}\right)_\Delta,$$ (6.3)

with the symbol LET_∞ denoting the usual (unrestricted) stopping power.

LET is often found in the literature with no subscript or other clarification. It can generally be assumed then that the unrestricted stopping power is implied.

Example
Use Table 6.1 to determine $LET_{1\,keV}$ and $LET_{5\,keV}$ for 1 MeV protons in water.

Solution
Note that Eq. (6.3) for LET involves stopping power rather than mass stopping power. Since $\rho = 1$ for water, the numbers in Table 6.1 also give $-(dE/dx)_\Delta$ in MeV/cm. We find $LET_{1\,keV} = 238$ MeV/cm given directly in the table. Linear interpolation gives $LET_{5\,keV} = 252$ MeV/cm.

LET is often expressed in units of keV/μm of water. Conversion of units shows that 1 keV/μm = 10 MeV/cm (Problem 10).

6.4 SPECIFIC IONIZATION

The average number of ion pairs that a particle produces per unit distance traveled is called the specific ionization. This quantity, which expresses the density of ionizations along a track, is often considered in studying the response of materials to radiation and in interpreting some biological effects. The specific ionization of a particle at a given energy is equal to the stopping power divided by the average energy required to produce an ion pair at that particle energy. The stopping power of air for a 5 MeV alpha particle is 1.23 MeV/cm and an average of about 36 eV is needed to produce an ion pair. Thus, the specific ionization of a 5 MeV alpha particle in air is $(1.23 \times 10^6$ eV/cm)/(36 eV/ip) = 34,200 ip/cm. For a 5 MeV alpha particle in water or soft tissue, $-dE/dx = 950$ MeV/cm (Fig. 4.3). Since about 25 eV is required to produce an ion pair, the specific ionization is 3.80×10^7 ip/cm.

6.5 ENERGY STRAGGLING

As a charged particle penetrates matter, statistical fluctuations occur in the number of collisions along its track and in the amount of energy lost in each collision. As a result, a number of identical particles starting out under identical conditions will show (1) a distribution of energies as they pass a given depth and (2) a distribution of path lengths traversed before they stop. The phenomenon of unequal energy losses under identical conditions is called energy straggling and the existence of different path lengths is referred to as range straggling. We examine these two forms of straggling in this and the next section.

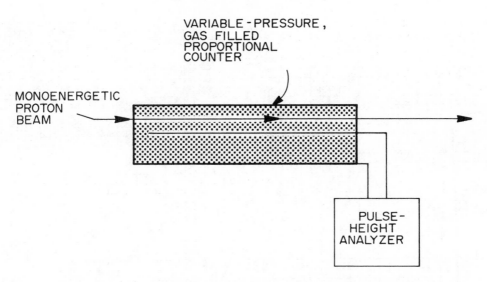

Figure 6.2. Schematic arrangement for studying energy straggling experimentally.

Energy straggling can be observed experimentally by the setup shown schematically in Fig. 6.2. A monoenergetic beam of protons (or other charged particles) is passed through a gas-filled cylindrical proportional counter parallel to its axis. The ends of the cylinder can be thin aluminum or other material which absorbs little energy. Each proton makes a number of electronic collisions and produces a single pulse in the counter, which is operated so that the pulse height is proportional to the total energy that the proton deposits in the counter gas. (Proportional counters are described in Chapter 9). Thus, by measuring the distribution of pulse heights, called the pulse-height spectrum, in an experiment one obtains the distribution of proton energy losses in the gas. By changing the gas pressure and repeating the experiment, one can study how the energy-loss distribution depends on the amount of matter traversed.

Some data are shown in Fig. 6.3, based on experiments reported by Gooding and Eisberg,[†] using 37 MeV protons and a 10 cm long counter filled with a mixture of 96% Ar and 4% CO_2 at pressures up to 1.2 atm. The average energy needed to produce an ion pair in the gas is about 25 eV. Data are provided for gas pressures of 0.2 atm and 1.2 atm. The ordinate shows the relative number of counts at different pulse heights given by the abscissa. We can examine both curves quantitatively. For reference, the value $Q_{max} = 80.6$ keV for a single collision of a 37 MeV proton with an electron is also shown.

At 0.2 atm, the most probable energy loss measured for a proton traversing the gas is $E_p = 27$ keV and the average loss is $\bar{E} = 34$ keV. Since about 25 eV is needed to produce an ion pair, the average number of secondary electrons is $34,000/25 = 1360$ per proton. Some of these electrons are produced directly by the proton and the others are produced by secondary electrons. If we assume that the proton directly ejects secondary electrons with a mean energy of ~60 eV, then the proton makes approximately $34,000/60 = 570$ collisions in traversing the gas. At this lower pressure, where the proton mean energy loss is considerably less than Q_{max} and only a few hundred collisions take place, the pulse-

[†] T. J. Gooding and R. M. Eisberg, Statistical fluctuations in energy losses of 37-MeV protons, *Phys. Rev.* **105**, 357 (1957).

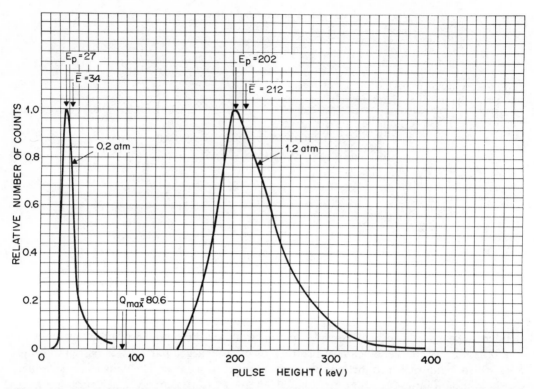

Figure 6.3. Pulse-height spectra for 37 MeV protons traversing proportional counter with gas at 0.2 atm and at 1.2 atm pressure. See text. [Based on T. J. Gooding and R. M. Eisberg, Statistical fluctuations in energy losses of 37-MeV protons, *Phys. Rev.* **105**, 357–360 (1957)].

height spectrum shows the skewed distribution characteristic of the single-collision spectrum. The relative separation of the peak and mean energies is $(\bar{E} - E_p)/\bar{E} = (34 - 27)/34 = 0.21$.

At 1.2 atm, an average of six times as many proton collisions takes place. The observed mean energy loss $\bar{E} = 212$ keV is some three times Q_{max}. The pulse-height spectrum, while still skewed, is somewhat more symmetric. The relative separation of the peak and mean energies is $(212 - 202)/212 = 0.05$, which is considerably less than that at 0.2 atm. At still higher gas pressures, the pulse-height spectrum shifts further to the right and becomes more symmetric, approaching a Gaussian shape with $E_p = \bar{E}$.

6.6 RANGE STRAGGLING

Range straggling can be measured with the experimental arrangement shown in Fig. 6.4(a). A monoenergetic beam of charged particles is incident on an absorber whose thickness can be varied by using additional layers of the material. A count-rate meter is used to measure the relative number of beam particles that emerge from the absorber as a function of its thickness. A plot of relative count rate vs. thickness for energetic heavy charged particles has the form shown in Fig. 6.4(b). At first the absorber serves only to reduce the energy of the particles traversing it, and therefore the count-rate curve is flat. When additional absorber material is added, the curve remains flat until the thickness approaches the range of the particles. Then the number of particles emerging from the

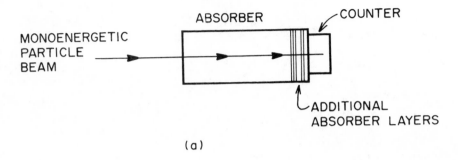

(a)

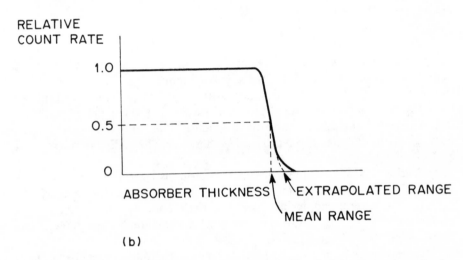

(b)

Figure 6.4. (a) Experimental arrangement for observing range straggling. (b) Plot of relative count rate vs. absorber thickness, showing the mean and extrapolated ranges.

absorber begins to decrease rapidly and almost linearly as more material is added until all of the particles are stopped in the absorber.

The mean range is defined as the absorber thickness at which the relative count rate is 0.50, as shown in Fig. 6.4(b). The extrapolated range is determined by extending the straight portion of the curve to the abscissa. The distribution of stopping depths about the mean range is nearly Gaussian in shape.

Range straggling is not large for heavy charged particles. For 100 MeV protons in tissue, for example, the root-mean-square fluctuation in pathlength is about 0.09 cm. The range is 7.57 cm (Table 4.3), and so the relative spread in stopping distances is $(0.09/7.57) \times 100 = 1.2\%$.

6.7 MULTIPLE COULOMB SCATTERING

We have seen how straggling affects the penetration of charged particles and introduces some fuzziness into the concept of range. Another phenomenon, elastic scattering from atomic nuclei via the Coulomb force, further complicates the analysis of particle penetration. The path of a charged particle in matter—even a fast electron or a heavy charged particle—deviates from a straight line because it undergoes frequent small-angle nuclear scattering events.

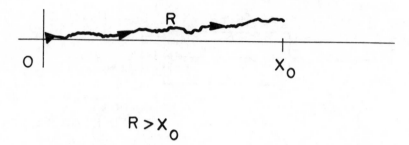

Figure 6.5. Schematic representation of the effect of multiple Coulomb scattering on the path of a heavy charged particle that starts moving from the origin O toward the right along the X axis. The displacement from the X axis is exaggerated for illustrative purposes.

Figure 6.5 illustrates how a heavy particle, starting out along the X axis at the origin O in an absorber might be deviated repeatedly by multiple Coulomb scattering until coming to rest at a depth X_0. The total pathlength traveled, R, which is the quantity calculated from Eq. (4.22) and given in the tables, is greater than the depth of penetration X_0. The latter is sometimes called the projected range. The difference between R and X_0 for heavy charged particles is typically $\leq 1\%$, and so R is usually considered to be the same as X_0.

Another effect of multiple Coulomb scattering is to spread a pencil beam of charged particles into a diverging beam as it penetrates a target, as illustrated in Fig. 6.6. The magnitude of the spreading increases with the atomic number of the material. When a pencil beam of 120 MeV protons penetrates 1 cm of water, for example, about 4% of the particles emerge outside an angle $\varphi = 1.5°$ in Fig. 6.6.

In radiotherapy with charged-particle beams, multiple Coulomb scattering often significantly diminishes the dose that can be concentrated in a tumor, particularly when it is located at some depth in the body.

6.8 PROBLEMS

1. A 4 MeV proton in water produces a delta ray with energy $0.1Q_{max}$. How large is the range of this delta ray compared with the range of the proton?

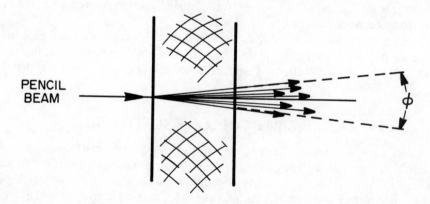

Figure 6.6. Multiple Coulomb scattering causes spread in a pencil beam of charged particles as they penetrate a target.

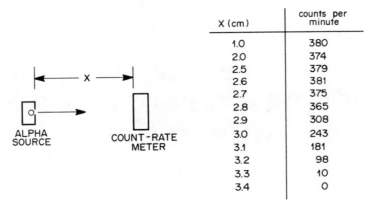

X (cm)	counts per minute
1.0	380
2.0	374
2.5	379
2.6	381
2.7	375
2.8	365
2.9	308
3.0	243
3.1	181
3.2	98
3.3	10
3.4	0

Figure 6.7. Count rate from a collimated, monoenergetic alpha-particle beam.

2. How does the maximum energy of a delta ray that can be produced by a 3 MeV alpha particle compare with that from a 3 MeV proton?

3. What is the energy of a proton that can produce a delta ray with enough energy to traverse a cell having a diameter of 2.5 μm?

4. Find $-(dE/dx)_{500\,ev}$ for a 500 keV proton in water.

5. (a) For 0.05 MeV protons, what is the smallest value of Δ for which $-(dE/dx)_{\Delta} = -(dE/dx)_{\infty}$?

   (b) Repeat for 0.10 MeV protons.

   (c) Are your answers consistent with Table 6.1?

6. Use Table 6.1 to estimate the restricted mass stopping power $-(dE/\rho\,dx)_{2\,kev}$ of water for 10 MeV protons.

7. Use Table 6.2 to estimate the ratio of the restricted stopping power of water $-(dE/dx)_{500\,ev}$ to the total stopping power for 50 keV electrons.

8. Find $LET_{1\,kev}$ for 10 MeV protons in

   (a) soft tissue,

   (b) bone of density 1.93 g/cm^3.

9. From Fig. 4.3 find LET_{∞} for a

   (a) 2 MeV alpha particle,

   (b) 100 MeV muon,

   (c) 100 keV positron.

10. Show that the common LET unit, keV/μm, is equal to 10 MeV/cm.

11. What is the specific ionization of a 12 MeV proton in tissue if an average of 25 eV is needed to produce an ion pair?

12. In Section 6.4 we calculated the specific ionization of a 5 MeV alpha particle in air (3.42×10^4 ip/cm) and in water (3.80×10^7 ip/cm). Why is the specific ionization so much greater in water?

13. How does the maximum specific ionization of an alpha particle in a medium compare with that of a proton?

14. (a) Define energy straggling.

    (b) Does energy straggling cause range straggling?

15. (a) From the numerical analysis of Fig. 6.3 given in Section 6.5, how many total ionizations per proton would be expected at a pressure of 0.7 atm?

    (b) How many of these ionizations would be produced by secondary electrons?

16. Figure 6.7 shows an experimental arrangement in which the count rate from a collimated, monoenergetic alpha-particle beam is measured at different separations x in air. From the given data, determine

    (a) the mean range,

    (b) the extrapolated range.

CHAPTER 7
THE INTERACTION OF PHOTONS
WITH MATTER

7.1 MECHANISMS

Unlike charged particles, photons are electrically neutral and do not steadily lose energy as they penetrate matter. Instead, they can travel some distance before interacting with an atom. How far a given photon will penetrate is governed statistically by a probability of interaction per unit distance traveled, which depends on the specific medium traversed and on the photon energy. When the photon interacts, it might be absorbed and disappear or it might be scattered, changing its direction of travel, with or without loss of energy.

The principal mechanisms of energy deposition by photons in matter are photoelectric absorption, Compton scattering, pair production, and photonuclear reactions.

7.2 PHOTOELECTRIC EFFECT

The ejection of electrons from a surface as a result of light absorption is called the photoelectric effect. The arrangement in Fig. 7.1 can be used to study this process experimentally. Monochromatic light passes into an evacuated glass tube through a quartz window (which allows ultraviolet light to be used) and strikes an electrode (1) causing photoelectrons to be ejected. Electrode 1 can be made of a metal to be studied or have its surface covered with such a metal. The current I that flows during illumination can be measured as a function of the variable potential difference V_{21} applied between the two electrodes, 1 and 2, of the tube. Curves (a) and (b) represent data obtained at two different intensities of the incident light. With the surface illuminated, there will be some current even with $V_{21} = 0$. When V_{21} is made positive and increased, the efficiency of collecting photoelectrons at electrode 2 increases; the current rises to a plateau when all of the electrons are being collected. The ratio of the plateau currents is equal to the relative light intensity used for curves (a) and (b).

When the polarity of the potential difference is reversed ($V_{21} < 0$), photoelectrons ejected from the illuminated electrode 1 now experience an attractive force back toward it. Making V_{21} more negative allows only the most energetic photoelectrons to reach electrode 2, thus causing the current I to decrease. Independently of the light intensity, the photoelectric current drops to zero when the reversed potential difference reaches a magnitude V_0, called the stopping potential. The potential energy eV_0, where e is the magnitude of the electronic charge, is equal to the maximum kinetic energy, T_{max}, of the photoelectrons:

$$T_{max} = eV_0. \tag{7.1}$$

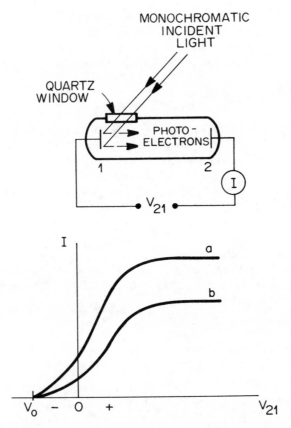

Figure 7.1. Experiment on photoelectric effect. With electrode 1 illuminated with monochromatic light of constant intensity, the current I is measured as a function of the potential difference V_{21} between electrodes 2 and 1. Curves (a) and (b) represent data at two different intensities of the incident light.

The stopping potential V_0 varies linearly with the frequency ν of the monochromatic light used. A threshold frequency ν_0 is found below which no photoelectrons are emitted, even with intense light. The value of ν_0 depends on the metal used for electrode 1.

The photoelectric effect is of special historical significance. The experimental findings are incompatible with the classical wave theory of light, which had been so successful in the latter part of the 19th century. Based on a wave concept, one would expect the maximum kinetic energy of photoelectrons, $T_{\max}$ in Eq. (7.1), to increase as the intensity of the incident light is increased. Yet the value of V_0 for a metal was found to be independent of the intensity (Fig. 7.1). Furthermore, one would also expect some photoelectrons to be emitted by light of any frequency, simply by making it intense enough. However, a threshold frequency ν_0 exists for every metal.

To explain the photoelectric effect, Einstein in 1905 proposed that the incident light arrives in discrete quanta (photons), having an energy given by $E = h\nu$, where h is Planck's constant. He further assumed that a photoelectron is produced when a single electron in the metal completely absorbs a single photon. The kinetic energy T with which the photoelectron is emitted from the metal is equal to the photon energy minus an energy φ that the electron expends in escaping the surface:

$$T = h\nu - \varphi. \tag{7.2}$$

The energy φ may result from collisional losses (stopping power) and from work done against the net attractive forces that normally keep the electron in the metal. A minimum energy, φ_0, called the work function of the metal, is required to remove the most loosely bound electron from the surface. The maximum kinetic energy that a photoelectron can have is given by

$$T_{max} = h\nu - \varphi_0. \tag{7.3}$$

Einstein received the Nobel Prize in 1921 "for his contributions to mathematical physics, and especially for his discovery of the law of the photoelectric effect."

The probability of producing a photoelectron when light strikes an atom is strongly dependent on the atomic number Z and the energy $h\nu$ of the photons. It is largest for high-Z materials and low-energy photons with frequencies above the threshold value ν_0. The probability varies as $Z^4/(h\nu)^3$.

Example

(a) What threshold energy must a photon have to produce a photoelectron from Al, which has a work function of 4.20 eV? (b) Calculate the maximum energy of a photoelectron ejected from Al by UV light with a wavelength of 1500 Å. (c) How does the maximum photoelectron energy vary with the intensity of the UV light?

Solution

(a) The work function $\varphi_0 = 4.20$ eV represents the minimum energy that a photon must have to produce a photoelectron. (b) The energy of the incident photons in eV is given in terms of the wavelength λ in angstroms by Eq. (2.26):

$$E = h\nu = \frac{12400}{\lambda} = \frac{12400}{1500} = 8.27 \text{ eV}. \tag{7.4}$$

From Eq. (7.3),

$$T_{max} = 8.27 - 4.20 = 4.07 \text{ eV}. \tag{7.5}$$

(c) T_{max} is independent of the light intensity.

7.3 ENERGY–MOMENTUM REQUIREMENTS FOR PHOTON ABSORPTION BY AN ELECTRON

As with charged particles, when the energy transferred by a photon to an atomic electron is large compared with its binding energy, then the electron can be treated as initially free and at rest. We now show that the conservation of energy and momentum prevents the *absorption* of a photon by an electron under these conditions. Thus, the binding of an electron and its interaction with the rest of the atom is essential for the photoelectric effect to occur. However, a photon can be *scattered* from a free electron, either with a reduction in its energy (Compton effect, next section) or with no change in energy (Thomson scattering).

If an electron, intially free and at rest (rest energy mc^2), absorbs a photon of energy $h\nu$ and momentum $h\nu/c$ (Appendix C), then the conservation of energy and momentum require, respectively, that

$$mc^2 + h\nu = \gamma mc^2 \tag{7.6}$$

and

$$h\nu/c = \gamma mc\beta. \tag{7.7}$$

Here $\gamma = (1 - \beta^2)^{-1/2}$ is the relativistic factor and $\beta = v/c$ is the ratio of the speed of the electron after absorbing the photon and the speed of light c. Multiplying both sides of Eq. (7.7) by c and subtracting from (7.6) gives

$$mc^2 = \gamma mc^2(1 - \beta).\tag{7.8}$$

This equation has only the trivial solution $\beta = 0$ and $\gamma = 1$, which, by Eq. (7.6), leads to the condition $h\nu = 0$. We conclude that the photoelectric effect occurs because the absorbing electron interacts with the nucleus and the other electrons in the atom to conserve the total energy and momentum of all interacting partners.

We now turn to the scattering of photons by electrons.

7.4 COMPTON EFFECT

Figure 7.2 illustrates the experimental arrangement used by Compton in 1922. Molybdenum K_α X-rays (photon energy 17.4 keV, wavelength $\lambda = 0.714$ Å) were directed at a graphite target and the wavelengths λ' of scattered photons were measured at various angles θ with respect to the incident photon direction. The intensities of the scattered radiation vs. λ' for three values of θ are sketched in Fig. 7.3. Each plot shows peaks at two values of λ': one at the wavelength λ of the incident photons and another at a longer wavelength, $\lambda' > \lambda$. The appearance of scattered radiation at a longer wavelength is called the Compton effect. The Compton shift in wavelength, $\Delta\lambda = \lambda' - \lambda$, was found to depend only on θ; it is independent of the incident-photon wavelength λ. In the crucial new experiment in 1922, Compton measured the shift $\Delta\lambda = 0.024$ Å at $\theta = 90°$.

The occurrence of scattered radiation at the same wavelength as that of the incident radiation can be explained by classical electromagnetic wave theory. The electric field of an incident wave accelerates atomic electrons back and forth at the same frequency $\nu = c/\lambda$ with which it oscillates. The electrons, therefore, emit radiation with the same wavelength. This Thomson scattering of radiation from atoms with no change in wavelength was known before Compton's work. The occurrence of the scattered radiation at longer wavelengths contradicted classical expectations.

To account for his findings, Compton proposed the following quantum model. In Fig. 7.4(a), a photon of energy $h\nu$ and momentum $h\nu/c$ (wavy line) is incident on a stationary,

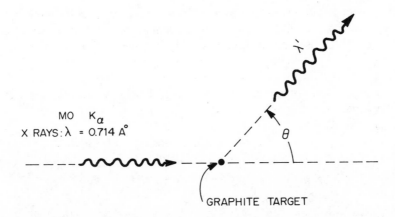

Figure 7.2. Compton measured intensity of scattered photons as a function of their wavelength λ' at various scattering angles θ. Incident radiation was molybdenum K_α X-rays, having a wavelength $\lambda = 0.714$ Å.

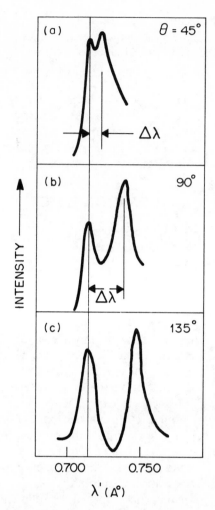

Figure 7.3. Intensity vs. wavelength λ' of photons scattered at angles (a) $\theta = 45°$, (b) 90°, and (c) 135°.

free electron. After the collision, the photon in (b) is scattered at an angle θ with energy $h\nu'$ and momentum $h\nu'/c$. The struck electron recoils at an angle φ with total energy E' and momentum P'. Conservation of total energy in the collision requires that

$$h\nu + mc^2 = h\nu' + E'. \tag{7.9}$$

Conservation of the components of momentum in the horizontal and vertical directions gives the two equations

$$\frac{h\nu}{c} = \frac{h\nu'}{c} \cos\theta + P' \cos\varphi \tag{7.10}$$

and

$$\frac{h\nu'}{c} \sin\theta = P' \sin\varphi. \tag{7.11}$$

Eliminating P' and φ from these three equations and solving for ν', one finds that

(a) BEFORE COLLISION

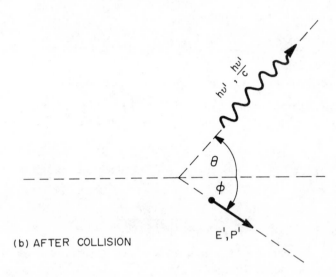

(b) AFTER COLLISION

Figure 7.4. Diagram illustrating Compton scattering of photon (energy $h\nu$, momentum $h\nu/c$) from electron, initially free and at rest with total energy mc^2. As a result of the collision, the photon is scattered at an angle θ with reduced energy $h\nu'$ and momentum $h\nu'/c$ and the electron recoils at angle φ with total energy E' and momentum P'.

$$h\nu' = \frac{h\nu}{1 + (h\nu/mc^2)(1 - \cos\theta)}. \tag{7.12}$$

With this result, the Compton shift is given by

$$\Delta\lambda = \lambda' - \lambda = c\left(\frac{1}{\nu'} - \frac{1}{\nu}\right) = \frac{h}{mc}(1 - \cos\theta). \tag{7.13}$$

Thus, as Compton found experimentally, the shift does not depend on the incident photon frequency ν. The magnitude of the shift at $\theta = 90°$ is

$$\Delta\lambda = \frac{h}{mc} = \frac{6.625 \times 10^{-27}}{9.11 \times 10^{-28} \times 3.00 \times 10^{10}} = 2.42 \times 10^{-10} \text{ cm}, \tag{7.14}$$

which agrees with the measured value. The quantity $h/mc = 0.024$ Å is called the Compton wavelength.

Example

A 1.332 MeV gamma photon from ^{60}Co is Compton scattered at an angle of 140°. Calculate the energy of the scattered photon and the Compton shift in wavelength. What is the momentum of the scattered photon?

Solution

The energy of the scattered photon is given by Eq. (7.12):

$$hv' = \frac{1.332 \text{ MeV}}{1 + (1.332/0.511)[1 - (-0.766)]} = 0.238 \text{ MeV}. \tag{7.15}$$

The Compton shift is given by Eq. (7.13) with $h/mc = 0.0242$ Å:

$$\Delta\lambda = (0.0242 \text{ Å})[1 - (-0.766)] = 0.0427 \text{ Å}. \tag{7.16}$$

The momentum of the scattered photon is

$$\frac{hv'}{c} = \frac{0.238 \text{ MeV} \times 1.6 \times 10^{-6} \text{ erg/MeV}}{3 \times 10^{10} \text{ cm/sec}} = 1.27 \times 10^{-17} \text{ g cm/sec}, \tag{7.17}$$

where we have used the fact that $1 \text{ erg} = 1 \text{ g cm}^2/\text{sec}^2$ (Appendix B).

We next examine some of the details of energy transfer in Compton scattering. The kinetic energy acquired by the secondary electron is given by

$$T = hv - hv'. \tag{7.18}$$

Substituting (7.12) for hv' and carrying out a few algebraic manipulations, one obtains (Problem 11)

$$T = hv \frac{1 - \cos\theta}{mc^2/hv + 1 - \cos\theta}. \tag{7.19}$$

The maximum kinetic energy, T_{max}, that a secondary electron can acquire occurs when $\theta = 180°$. In this case, Eq. (7.19) gives

$$T_{\text{max}} = \frac{2hv}{2 + mc^2/hv}. \tag{7.20}$$

When the photon energy becomes very large compared with mc^2, T_{max} approaches hv.

The recoil angle of the electron, φ in Fig. 7.4, is related to hv and θ. Using Eqs. (7.10) and (7.11) together with the trigonometric identity $\sin\varphi/\cos\varphi = \tan\varphi$, one obtains

$$\tan\varphi = \frac{hv' \sin\theta}{hv - hv' \cos\theta}. \tag{7.21}$$

Substituting from Eq. (7.12) for hv' gives

$$\tan\varphi = \frac{\sin\theta}{(1 + hv/mc^2)(1 - \cos\theta)}. \tag{7.22}$$

The trigonometric term in θ can be conveniently expressed in terms of the half-angle. Since $\sin\theta = 2 \sin\theta/2 \cos\theta/2$ and $1 - \cos\theta = 2 \sin^2\theta/2$, it reduces to

$$\frac{\sin\theta}{1 - \cos\theta} = \frac{2(\sin\theta/2)(\cos\theta/2)}{2\sin^2\theta/2} = \cot\frac{\theta}{2}. \tag{7.23}$$

Equation (7.22) can thus be written in the compact form

$$\cot\frac{\theta}{2} = \left(1 + \frac{hv}{mc^2}\right)\tan\varphi. \tag{7.24}$$

When θ is small, $\cot\theta/2$ is large and φ is near $90°$. In this case, the photon travels in the forward direction, imparting relatively little energy to the electron, which moves off nearly at right angles to the direction of the incident photon. As θ increases from $0°$ to $180°$, $\cot\theta/2$ decreases from ∞ to 0. Therefore, φ decreases from $90°$ to $0°$. The electron recoil angle φ in Fig. 7.4 is thus always confined to the forward direction ($0 \leq \varphi \leq 90°$), whereas the photon can be scattered in any direction.

Example
In the previous example a 1.332 MeV photon from ^{60}Co was scattered by an electron at an angle of 140°. Calculate the energy acquired by the recoil electron. What is the recoil angle of the electron? What is the maximum fraction of its energy that this photon could lose in a single Compton scattering?

Solution
Substitution into Eq. (7.19) gives the electron recoil energy,

$$T = 1.332 \frac{1 - (-0.766)}{0.511/1.332 + 1 - (-0.766)} = 1.094 \text{ MeV}. \tag{7.25}$$

Note from Eq. (7.15) that $T + h\nu' = 1.332$ MeV $= h\nu$, as it should. The angle of recoil of the electron can be found from Eq. (7.24). We have

$$\tan \varphi = \frac{\cot 140°/2}{1 + 1.332/0.511} = 0.101, \tag{7.26}$$

from which it follows that $\varphi = 5.76°$. This is a relatively hard collision in which the photon is backscattered, retaining only the fraction $0.238/1.332 = 0.179$ of its energy and knocking the electron in the forward direction. From Eq. (7.20),

$$T_{max} = \frac{2 \times 1.332}{2 + 0.511/1.332} = 1.118 \text{ MeV}. \tag{7.27}$$

The maximum fractional energy loss is $T_{max}/h\nu = 1.118/1.332 = 0.839$.

All of the details that we have worked out thus far for Compton scattering follow kinematically from the energy and momentum conservation requirements expressed by Eqs. (7.9)–(7.11). We have said nothing about how the photon and electron interact or about the *probability* that the photon will be scattered in the direction θ. The quantum-mechanical theory of Compton scattering, based on the specific photon–electron interaction, gives for the angular distribution of scattered photons the Klein–Nishina formula

$$\frac{d\sigma}{d\Omega} = \frac{e^4}{2m^2c^4} \left(\frac{\nu'}{\nu}\right)^2 \left(\frac{\nu}{\nu'} + \frac{\nu'}{\nu} - \sin^2 \theta\right) \text{ cm}^2. \tag{7.28}$$

Here $d\sigma/d\Omega$, called the differential scattering cross section, is the probability per incident photon/cm^2 that an electron will scatter the photon into a solid angle $d\Omega$ at angle θ. (Cross sections will be discussed in Section 7.8.) Equation (7.28) can be used with the other formulas above to obtain the energy spectrum of the Compton recoil electrons. This spectrum is shown in Fig. 7.5 for 1 MeV incident photons. The relative number of recoil electrons decreases from $T = 0$ until it begins to rise rapidly as T approaches $T_{max} = 0.796$ MeV, where the spectrum has its maximum value. The electron recoil spectra have similar shapes for other incident photon energies. The most probable collisions are those that transfer relatively large amounts of energy.

Of special importance for dosimetry is the average recoil energy, T_{avg}, of Compton electrons. For photons of a given energy, the average can be computed directly from the relative number distribution, such as that shown in Fig. 7.5. Table 7.1 gives values of T_{avg} for a range of photon energies. Also shown is the average fraction of the incident

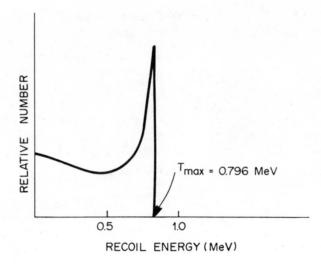

Figure 7.5. Relative number of Compton recoil electrons as a function of their energy for 1 MeV incident photons.

photon energy that is converted into kinetic energy of Compton electrons. This fraction steadily increases with photon energy.

Like the photoelectric effect, the Compton effect gave confirmation of the corpuscular nature of light. The discovery of quantum mechanics followed Compton's experiments by a few years. Modern quantum electrodynamics accounts very successfully for the dual wave–photon nature of electromagnetic radiation.

Table 7.1. Average Kinetic Energy, T_{avg}, of Compton Recoil Electrons
and Fraction of Incident Photon Energy, $h\nu$

Photon energy $h\nu$ (MeV)	Average recoil electron energy T_{avg} (MeV)	Average fraction of incident energy $T_{avg}/h\nu$
0.01	0.0002	0.0187
0.02	0.0007	0.0361
0.04	0.0027	0.0667
0.06	0.0056	0.0938
0.08	0.0094	0.117
0.10	0.0138	0.138
0.20	0.0432	0.216
0.40	0.124	0.310
0.60	0.221	0.368
0.80	0.327	0.409
1.00	0.440	0.440
2.00	1.06	0.531
4.00	2.43	0.607
6.00	3.86	0.644
8.00	5.34	0.667
10.0	6.84	0.684
20.0	14.5	0.727
40.0	30.4	0.760
60.0	46.6	0.776
80.0	62.9	0.787
100.0	79.4	0.794

7.5 PAIR PRODUCTION

A photon with an energy of at least twice the electron rest energy, $h\nu \geq 2mc^2$, can be converted into an electron–positron pair in the field of an atomic nucleus. Pair production can also occur in the field of an atomic electron, but the probability is considerably smaller and the threshold energy is $4mc^2$. (This process is often referred to as "triplet" production because of the presence of the recoiling atomic electron in addition to the pair.) When pair production occurs in a nuclear field, the massive nucleus recoils with negligible energy. Therefore, the photon energy $h\nu$ is converted into $2mc^2$ plus the kinetic energies T_+ and T_- of the partners:

$$h\nu = 2mc^2 + T_+ + T_-. \tag{7.29}$$

The distribution of the excess energy between the electron and positron is continuous; that is, the kinetic energy of either can vary from zero to a maximum of $h\nu - 2mc^2$. Furthermore, the energy spectra are almost the same for the two particles and depend on the atomic number of the nucleus. The threshold photon wavelength for pair production is 0.012 Å (Problem 20). Pair production becomes more likely with increasing photon energy, and the probability increases with atomic number approximately as Z^2.

The inverse process also occurs when an electron and positron annihilate to produce photons. A positron can annihilate in flight, although it is more likely first to slow down, attract an electron, and form positronium. Positronium is the bound system, analogous to the hydrogen atom, formed by an electron–positron pair orbiting about their mutual center of mass. Positronium exists for $\sim 10^{-10}$ sec before the electron and positron annihilate. Since the total momentum of positronium before decay is zero, at least two photons must be produced in order to conserve momentum. The most likely event is the creation of two 0.511 MeV photons going off in opposite directions. If the positron annihilates in flight, then the total photon energy will be $2mc^2$ plus its kinetic energy. Three photons are occasionally produced. The presence of 0.511 MeV annihilation photons around any positron source is always a potential radiation hazard.

7.6 PHOTONUCLEAR REACTIONS

Photons can be absorbed by an atomic nucleus and knock out a nucleon. This process is called photodisintegration. An example is gamma-ray capture by a $^{206}_{82}$Pb nucleus with emission of a neutron: $^{206}_{82}$Pb$(\gamma,n)^{205}_{82}$Pb. The photon must have enough energy to overcome the binding energy of the ejected nucleon, which is generally several MeV. Like the photoelectric effect, photodisintegration can occur only at photon energies above a threshold value. The kinetic energy of the ejected nucleon is equal to the photon energy minus the nucleon's binding energy.

The probability for photonuclear reactions is orders of magnitude smaller than the combined probabilities for the photoelectric effect, Compton effect, and pair production. However, unlike these processes, photonuclear reactions can produce neutrons, which often pose special radiation-protection problems. In addition, residual nuclei following photonuclear reactions are often radioactive. For these reasons, photonuclear reactions can be important around high-energy electron accelerators that produce energetic photons.

The thresholds for (γ,p) reactions are often higher than those for (γ,n) reactions because of the repulsive Coulomb barrier that a proton must overcome to escape from the nucleus (Fig. 3.1). Although the probability for either reaction is about the same in the lightest elements, the (γ,n) reaction is many times more probable than (γ,p) in heavy elements.

Other photonuclear reactions also take place. Two-nucleon knock-out reactions such as $(\gamma,2n)$ and (γ,np) occur, as well as (γ,α) reactions. Photon absorption can also induce fission in heavy nuclei.

Example
Compute the threshold energy for the (γ,n) photodisintegration of ^{206}Pb. The mass differences, Δ, are -23.79 MeV for ^{206}Pb, -23.77 MeV for ^{205}Pb, and 8.07 MeV for the neutron. What is the energy of a neutron produced by absorption of a 10 MeV photon?

Solution
The mass difference after the reaction is $-23.77 + 8.07 = -15.70$ MeV. The threshold energy needed to remove the neutron from ^{206}Pb is therefore $-15.70 - (-23.79) = 8.09$ MeV. Absorption of a 10 MeV photon would produce a neutron with kinetic energy $10 - 8.09 = 1.91$ MeV.

7.7 ATTENUATION, ABSORPTION, AND SCATTERING COEFFICIENTS

As pointed out at the beginning of this chapter, photon penetration in matter is governed statistically by the probability per unit distance traveled that a photon interacts by one physical process or another. This probability, denoted by μ, is called the linear attenuation coefficient and has the dimensions of inverse length (e.g., cm^{-1}). The coefficient μ depends on photon energy and on the material being traversed. The mass attenuation coefficient μ/ρ is obtained by dividing μ by the density ρ of the material. It is usually expressed in cm^2/g.

Monoenergetic photons are attenuated exponentially in a uniform target, as we now show. Figure 7.6 represents a narrow beam of N_0 monoenergetic photons incident normally on a slab. As the beam penetrates the absorber, some photons are scattered and some are absorbed. We let $N(x)$ represent the number of photons that reach a depth x without having interacted. The number that interact within the next small distance dx is proportional to N and to dx. Thus we may write

$$dN = -\mu N \, dx, \tag{7.30}$$

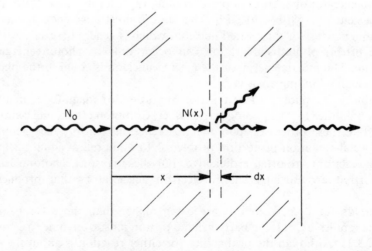

Figure 7.6. Pencil beam of N_0 monoenergetic photons incident on slab. The number of photons that reach a depth x without having an interaction is given by $N(x) = N_0 \, e^{-\mu x}$, where μ is the linear attenuation coefficient.

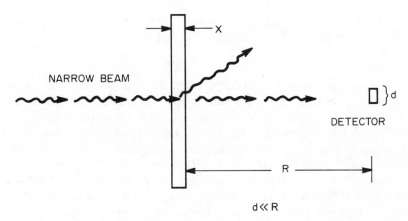

Figure 7.7. Illustration of "good" scattering geometry for measuring linear attenuation coefficient μ. Photons from a *narrow* beam that are absorbed or scattered by the absorber miss a small detector placed in beam line some distance away.

where the constant of proportionality μ is the linear attenuation coefficient. This equation is mathematically identical to Eq. (3.46) for radioactive decay. The solution, therefore, is the exponential function[†]

$$N(x) = N_0 e^{-\mu x}. \tag{7.31}$$

The linear attenuation coefficient can be measured by the experimental arrangement shown in Fig. 7.7. A narrow beam of monoenergetic photons is directed toward an absorbing slab of thickness x. A small detector of size d is placed at a distance $R \gg d$ behind the slab directly in the beam line. Under these conditions, referred to as "narrow-beam" or "good" scattering geometry, only photons that traverse the slab without interacting will be detected. One can measure the relative number of photons N/N_0 that reach the detector and use Eq. (7.31) to get the value of μ.

In practice, photon *intensity* is often measured, rather than *number*. The intensity I, which is also called the photon energy fluence rate, is defined as the energy per unit time that crosses a unit area at right angles to the beam (e.g., MeV cm^{-2} sec^{-1}). The number of photons per unit time crossing that area (e.g., cm^{-2} sec^{-1}) is called the fluence rate φ. For monoenergetic photons of energy $h\nu$, these two quantities are connected by the relation $I = \varphi h\nu$. By analogy with Eq. (7.31), we can write

$$I(x) = I_0 e^{-\mu x} \tag{7.32}$$

and

$$\varphi(x) = \varphi_0 e^{-\mu x}, \tag{7.33}$$

where I_0 and φ_0 are the intensity and fluence rate at zero depth. When the photon beam is not monoenergetic, these equations can be applied separately at each energy to calculate attenuation.

The magnitude of the linear attenuation coefficient depends on contributions from the

---

[†]Also, in analogy with Eq. (3.67), $1/\mu$ is the average distance traveled by a photon before interacting, and hence μ is also called the inverse mean free path.

photoelectric effect, Compton scattering, and pair production. (We neglect other interactions, which could be included, but are usually minor.) We write

$$\mu = \mu_{PE} + \mu_{CT} + \mu_{PP}, \tag{7.34}$$

where μ_{PE}, μ_{CT}, and μ_{PP} denote, respectively, the linear attenuation coefficients for the photoelectric effect, total Compton scattering, and pair production. The respective mass attenuation coefficients are μ_{PE}/ρ, μ_{CT}/ρ, and μ_{PP}/ρ for a material of density ρ.

In dosimetry we are interested in the energy deposited by secondary electrons in an absorber as a result of photon interactions. This energy is related to the linear attenuation coefficients in the following way. In Fig. 7.7, if a photon undergoes photoelectric absorption, then all of its energy is transferred to an electron, which in turn, unless the slab is thin compared with secondary electron ranges, deposits all of its energy in the slab. In this case, the attenuation coefficient μ_{PE} also describes energy deposition by secondary electrons in an absorber.

If the photon produces an electron–positron pair, the relationship between the energy deposited and the attenuation coefficient is different. In this case, an energy $2mc^2 = 1.022$ MeV is required to supply the rest energy of the pair. The fraction of the photon energy $h\nu$ that appears as the combined kinetic energies of the electron and positron is therefore $(h\nu - 1.022)/h\nu$, with $h\nu$ in MeV. The coefficient that governs energy deposition by secondary electrons as a result of pair production is, to a first approximation, $(h\nu - 1.022)\mu_{PP}/h\nu$. Subsequent bremsstrahlung and positron annihilation can convert some energy into photons that escape the absorber; such effects are not included.

When a photon is Compton scattered, the situation is still different. Only that part of the photon's energy that is transferred to the recoil electron is deposited locally in the absorber, the rest being carried away by the scattered photon. The total Compton linear attenuation coefficient is the sum of two parts,

$$\mu_{CT} = \mu_{CS} + \mu_{CA}. \tag{7.35}$$

The Compton *scattering* coefficient μ_{CS} describes the fraction of the photon energy that is scattered out of the narrow beam by the absorber in Fig. 7.7; the Compton *absorption* coefficient μ_{CA} describes the fraction deposited in the absorber (by the recoil electron).

For calculating energy deposition in a target irradiated by monoenergetic photons we define the linear *absorption* coefficient as

$$\mu_A = \mu_{PE} + \mu_{CA} + \left(\frac{h\nu - 1.022}{h\nu}\right)\mu_{PP}. \tag{7.36}$$

This factor, which is smaller than the attenuation coefficient μ, also has the dimensions of inverse length. The difference

$$\mu_S = \mu - \mu_A \tag{7.37}$$

between the attenuation and absorption coefficients is called the scattering coefficient. The mass absorption coefficient, μ_A/ρ, is also frequently used. For a monoenergetic photon beam of intensity I_0, the rate of energy deposition in an absorber of thickness x can be calculated by using μ_A in Eq. (7.32). The quantity

$$I_A(x) = I_0 e^{-\mu_A x} \tag{7.38}$$

gives the *total* photon intensity that is transmitted by the absorber. By comparison, Eq. (7.32) gives the intensity that is transmitted by photons that have no interaction, as would be registered by the small detector in Fig. 7.7. Since $\mu_A < \mu$, it follows that $I_A(x) > I(x)$.

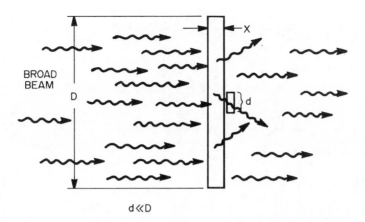

Figure 7.8. Illustration of "poor" scattering geometry for measuring the linear absorption coefficient μ_A. Detector placed near the absorber registers scattered as well as unscattered photons from a *broad* beam of monoenergetic photons.

It remains to discuss how μ_A can be measured. In Fig. 7.8 a parallel, *broad* beam of monoenergetic photons is incident on the absorber with the detector placed directly behind it. With this "broad-beam" or "poor" geometry, the detector measures scattered as well as unscattered radiation. The conditions for Eq. (7.38) can be approximately satisfied, and the detector reading will be close to I_A, thus giving a measure of μ_A.

We shall use Eq. (7.38) to calculate energy deposition from photon beams, although it is often only approximate. In practice, multiple Compton scattering, bremsstrahlung, and other factors can lead to inaccuracies. The resulting complicated patterns of energy deposition, especially in thick absorbers, can be described by buildup factors, which we treat in Chapter 13. Equation (7.38) is nevertheless useful in many practical applications, such as estimating dose in persons exposed to X- or gamma rays, filtration needed for medical X-ray units, and shielding requirements.

Figures 7.9 and 7.10 give the mass attenuation and mass absorption coefficients for five elements over a range of photon energies. The structure of these curves reflects the physical processes we have been discussing. At low photon energies the binding of the atomic electrons is important and the photoelectric effect is the dominant interaction. High-Z materials provide greater attenuation and absorption, which decrease rapidly with increasing photon energy. The coefficients for Pb and U rise abruptly when the photon energy is sufficient to eject a photoelectron from the K shell of the atom. The curves for the other elements show the same structure at lower energies. When the photon energy is several hundred keV or greater, the binding of the atomic electrons becomes relatively unimportant and the dominant interaction is Compton scattering. Since the elements (except hydrogen) contain about the same number of electrons per unit mass, there is not a large difference between the values of the mass attenuation and mass absorption coefficients for the different elements. Compton scattering continues to be important above the 1.022 MeV pair-production threshold until the latter process takes over as the more probable. Attenuation by pair production is enhanced by a large nuclear charge of the absorber.

Inspection of Figs. 7.9 and 7.10 shows that the coefficients for lead are larger than those for most common materials over the entire photon energy range. Per unit weight, lead is one of the most effective photon shielding materials there is. Because it is also rela-

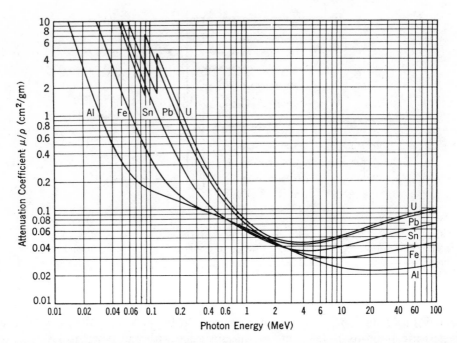

Figure 7.9. Mass attenuation coefficients for various elements. [Reprinted with permission from K.Z. Morgan and J.E. Turner, eds., *Principles of Radiation Protection*, p. 107, John Wiley, New York (1967). Copyright 1967 by John Wiley]

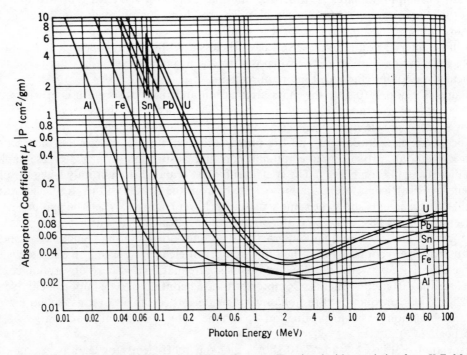

Figure 7.10. Mass absorption coefficients for various elements. [Reprinted with permission from K.Z. Morgan and J.E. Turner, eds., *Principles of Radiation Protection*, p. 106, John Wiley, New York (1967). Copyright 1967 by John Wiley]

tively inexpensive, it is almost universally used when photon shielding is required. It can be used to line the walls of X-ray rooms, incorporated into aprons worn by personnel around X-ray equipment, and fabricated into containers for gamma sources. Lead bricks afford a convenient and effective way to erect shielding around a gamma-ray source. The design of lead shielding is described in Chapter 13.

Figures 7.11 and 7.12 give the mass attenuation and mass absorption coefficients for a number of materials of special importance in radiation protection. The curves for air, H_2O, and tissue are almost the same, as would be expected because they consist of similar elements. Concrete is a better absorber in the photoelectric and pair-production energy domains. (Barytes concrete contains barite, or $BaSO_4$, as a high-Z material.) Lead glass is also an effective photon absorber. All of the materials have about the same properties over the Compton energy region.

It is instructive to see how the individual physical processes contribute to the attenuation and absorption coefficients as functions of photon energy. Figure 7.13 for water shows μ_{PE}, μ_{CS}, μ_{CA}, and μ_{PP}, which appear in Eqs. (7.34) and (7.35), as well as the coefficients μ and μ_A given by Eqs. (7.34) and (7.36). At the lowest energies (<15 keV), the photoelectric effect accounts for virtually all of the interaction, and so $\mu = \mu_A$. As the photon energy increases, μ_{PE} drops rapidly and goes below μ_{CS}. At 100 keV most of the attenuation in water is due to Compton scattering of the photons (μ_{CS}) with some Compton absorption (μ_{CA}) and very little photoelectric absorption (μ_{PE}). At still higher photon energies, the Compton collision spectrum becomes relatively harder (cf. last column in Table 7.1), and μ_{CA} exceeds μ_{CS}, although both decline beyond about 600 keV. The Compton coefficients continue to decrease at higher energies as pair production becomes the dominant process.

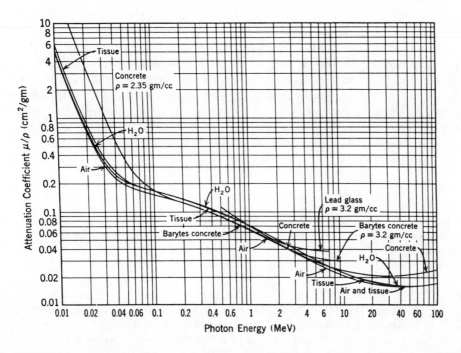

Figure 7.11. Mass attenuation coefficients for various materials. [Reprinted with permission from K.Z. Morgan and J.E. Turner, eds., *Principles of Radiation Protection*, p. 106, John Wiley, New York (1967). Copyright 1967 by John Wiley]

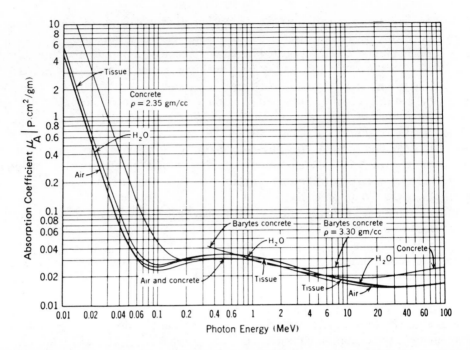

Figure 7.12. Mass absorption coefficients for various materials. [Reprinted with permission from K.Z. Morgan and J.E. Turner, eds., *Principles of Radiation Protection*, p. 108, John Wiley, New York (1967). Copyright 1967 by John Wiley]

Other coefficients for photon penetration are also used, depending on the information wanted. For example, the mass *energy-transfer* coefficient is defined like μ_A/ρ from Eq. (7.36), except that the fraction of energy emitted as fluorescent X-rays following photoelectron ejection is subtracted. This energy escapes from the immediate locale where the photoelectron is absorbed. For the mass *energy-absorption* coefficient, the fraction of the energy lost by secondary electrons as bremsstrahlung is also subtracted. Unfortunately, no uniform terminology is used to describe the various attenuation and absorption coefficients. In reading the literature, one should be careful to determine exactly what an author means by them.

Example
What thicknesses of concrete and of lead are needed to reduce the intensity of a narrow 500 keV photon beam to one-fourth its value? Compare the thicknesses in cm and in g/cm². Repeat for 1.5 MeV photons.

Solution
We use Eq. (7.32) with $I(x)/I_0 = 0.25$. The mass attenuation coefficients μ/ρ obtained from Figs. 7.9 and 7.11 are shown in Table 7.2. At 500 MeV, the linear attenuation coefficient for concrete is $\mu = (0.089 \text{ cm}^2/\text{g})(2.35 \text{ g/cm}^3) = 0.209 \text{ cm}^{-1}$ and that for Pb is $\mu = (0.14)(11.4) = 1.60 \text{ cm}^{-1}$. Using Eq. (7.32), we have for concrete

$$0.25 = e^{-0.209x}, \tag{7.39}$$

Table 7.2. Mass Attenuation Coefficients

$h\nu$	μ/ρ (cm^2/g)	
	Concrete $\rho = 2.35$ g/cm^3	Pb $\rho = 11.4$ g/cm^3
500 keV	0.089	0.14
1.5 MeV	0.052	0.051

giving $x = 6.63$ cm; for Pb,

$$0.25 = e^{-1.60x}, \tag{7.40}$$

giving $x = 0.866$ cm. The concrete shield is thicker by a factor of $6.63/0.866 = 7.65$. In g/cm^2, the concrete thickness is 6.63 cm $\times$ 2.35 g/cm^3 = 15.6 g/cm^2, while that for Pb is $0.866 \times 11.4 = 9.87$ g/cm^2. The concrete shield is more massive in thickness by a factor of $15.6/9.87 = 1.58$. Lead is a more efficient absorber than concrete for 500 keV photons on the basis of weight. Photoelectric absorption is important at this energy and the higher atomic number of lead is effective. The calculation can be repeated in exactly the same way for 1.5 MeV photons. Instead, we do it a little differently by using the mass attenuation coefficient directly, writing the exponent in Eq. (7.32) as $\mu x = (\mu/\rho)\rho x$. For 1.5 MeV photons incident on concrete,

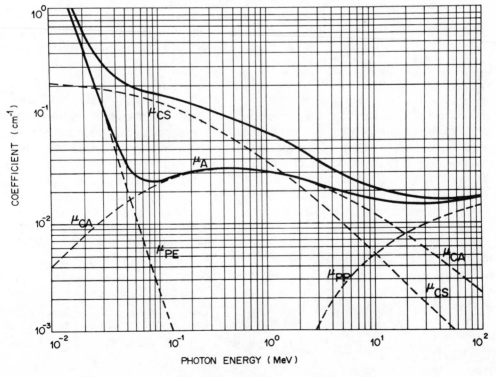

Figure 7.13. Linear attenuation and absorption coefficients for photons in water.

$$0.25 = e^{-0.052\rho x}, \tag{7.41}$$

giving $\rho x = 26.7$ g/cm^2 and $x = 11.4$ cm. For Pb,

$$0.25 = e^{-0.051\rho x}, \tag{7.42}$$

and so $\rho x = 27.2$ g/cm^2 and $x = 2.39$ cm. At this energy the Compton effect is the principal inter-action that attenuates the beam, and therefore all materials (except hydrogen) give comparable attenuation per g/cm^2.

Example
A beam of 1 MeV photons is incident on a 10 cm aluminum slab ($\rho = 2.70$ g/cm^3) at a rate of 10^3 photons/sec. What fraction of the photons is transmitted without being absorbed or scattered? What fraction of the photon energy is absorbed by the slab? How much energy is transmitted per second by the slab?

Solution
From Fig. 7.9, the mass attenuation coefficient is $\mu/\rho = 0.062$ cm^2/g; and so $\mu = 0.167$ cm^{-1}. The fraction of photons that do not interact is $e^{-\mu x} = e^{-0.167 \times 10} = 0.187$. To find the fraction of the energy absorbed by the slab, we must use the absorption coefficient. From Fig. 7.10, we have $\mu_A/\rho = 0.027$ cm^2/g; and so $\mu_A = 0.0729$ cm^{-1}. The fraction of the energy of the photon beam that is transmitted by the slab is $e^{-\mu_A x} = e^{-0.0729 \times 10} = 0.482$. This fraction includes both the energy of the 18.7% of the photons that do not interact and the energy of the photons that are Compton scattered. Therefore, the fraction of the incident photon energy that is absorbed in the slab is $1 - 0.482 = 0.518$. The photon beam has an incident energy rate of 10^3 MeV/sec. The rate of energy transmission by the slab is $0.482 \times 10^3 = 482$ MeV/sec.

7.8 CROSS SECTIONS

The linear attenuation coefficient, μ, gives the probability that a photon will interact per unit distance that it travels. If there are n atoms per unit volume in a uniform medium then one can express this probability by writing

$$\mu = n\sigma, \tag{7.43}$$

where σ, which has the dimensions of area, is called the atomic cross section. In a uni-form, parallel beam of photons, it represents the probability per unit fluence that an interaction occurs in a unit volume (of any shape) that contains n atoms. Although σ can be thought of as the cross-sectional area presented to a photon for interaction, it is usu-ally not equal to the actual physical size of the atom. Like μ, σ can vary over several orders of magnitude with changing photon energy. Since σ is usually a small number, it is often expressed in a unit called the barn, with 1 barn $= 10^{-24}$ cm^2. Generally, σ is referred to as the *microscopic* cross section and μ is called the *macroscopic* cross section.

In general, n in Eq. (7.43) can also represent the number of molecules, electrons, or nuclei per unit volume in an absorber. Then σ is called the molecular, electronic, or nuclear cross section. Individual atomic cross sections can simply be added to obtain molecular cross sections for photon interactions. We shall use nuclear cross sections for neutron interactions in the next chapter.

Atomic cross sections can be used in place of the linear attenuation and absorption coefficients to express interaction probabilities on a per atom basis. Combining Eqs. (7.34) and (7.43), for example, we may write

$$\sigma = \sigma_{PE} + \sigma_{CT} + \sigma_{PP}, \tag{7.44}$$

where the three quantities on the right-hand side are the individual atomic cross sections for the three principal photon interactions. Similarly, from Eq. (7.35), the total Compton cross section is

$$\sigma_{CT} = \sigma_{CS} + \sigma_{CA}, \tag{7.45}$$

where σ_{CS} and σ_{CA} are the scattering and absorption cross sections. Also, from (7.36), the absorption cross section is

$$\sigma_A = \sigma_{PE} + \sigma_{CA} + \left(\frac{h\nu - 1.022}{h\nu} \right) \sigma_{PP}. \tag{7.46}$$

Cross sections represent collision probabilities per unit fluence. The Klein–Nishina formula (7.28) is another example of a cross section, one that gives the probability that an incident photon will be scattered into a solid angle $d\Omega$ at angle θ. It is called a *differential cross section*, because it is differential in angle. It therefore gives the angular distribution of scattered photons. Integrating the differential cross section over all solid angles ($d\Omega = 2\pi \cos\theta\, d\theta$ with $0 \le \theta \le \pi$ rad) and multiplying by the atomic number Z (electrons per atom) gives the total atomic cross section for Compton scattering,

$$\sigma_{CT} = 2\pi Z \int_0^\pi \frac{d\sigma}{d\Omega} \cos\theta\, d\theta, \tag{7.47}$$

that enters Eq. (7.44). The Compton scattering cross section, which describes the fraction of the photon energy that is carried off by scattered photons, is given by weighting the differential scattering cross section by ν'/ν:

$$\sigma_{CS} = 2\pi Z \int_0^\pi \frac{\nu'}{\nu} \frac{d\sigma}{d\Omega} \cos\theta\, d\theta. \tag{7.48}$$

The Compton absorption cross section is the difference $\sigma_{CA} = \sigma_{CT} - \sigma_{CS}$.

Example
An 8 MeV photon is incident on water. The microscopic cross sections for total Compton scattering and pair production are, respectively, 0.599 barn and 0.119 barn per molecule. Calculate the mass attenuation coefficient.

Solution
Like the attenuation coefficients, cross sections for different interactions are additive [Eq. (7.44)]. Therefore, the total interaction cross section is $\sigma = 0.599 + 0.119 = 0.718$ barn $= 7.18 \times 10^{-25}$ cm^2 per molecule. ($\sigma_{PE} = 0$ for 8 MeV photons.) The number of water molecules/cm^3 is $n = 3.34 \times 10^{22}$ cm^{-3}. Equation (7.43) gives for the linear attenuation coefficient $\mu = 3.34 \times 10^{22}$ cm$^{-3} \times 7.18 \times 10^{-25}$ cm$^2 = 2.40 \times 10^{-2}$ cm^{-1}. The mass attenuation coefficient is $\mu/\rho = 2.40 \times 10^{-2}$ cm^{-1}/1 g cm$^{-3} = 2.40 \times 10^{-2}$ cm^2/g. (Cf. Fig. 7.13.)

7.9 PROBLEMS

1. The work function for lithium is 2.3 eV. (a) Calculate the maximum kinetic energy of photoelectrons produced by photons with energy of 12 eV. (b) What is the cutoff frequency? (c) What is the threshold wavelength?
2. The threshold wavelength for tungsten is 2700 Å. What is the maximum kinetic energy of photoelectrons produced by photons of wavelength 2200 Å?
3. What stopping potential is needed to turn back a photoelectron having a kinetic energy of 3.84×10^{-12} ergs?

4. The threshold wavelength for producing photoelectrons from sodium is 5650 Å. (a) What is the work function? (b) What is the maximum kinetic energy of photoelectrons produced by light with a wavelength of 4000 Å?

5. Light of wavelength 1900 Å is incident on a nickel surface (work function 4.9 eV). (a) Calculate the stopping potential. (b) What is the cutoff frequency for nickel? (c) What is the threshold wavelength?

6. (a) Make a sketch of the stopping potential versus light frequency for the photoelectric effect in silicon, which has a work function of 4.4 eV. (b) Write an equation for the stopping potential as a function of frequency. (c) What stopping potential is needed for light of frequency 1.32×10^{15} sec^{-1}?

7. (a) If the curve in the last problem is extended to values of ν below threshold, show that it intersects the vertical axis at a voltage $-\varphi_0/e$, where φ_0 is the work function and e is the electronic charge. (b) Are the curves for two metals, having different threshold values, ν_0, parallel?

8. In an experiment with potassium, a potential difference of 0.80 V is needed to stop the photoelectron current when light of wavelength 4140 Å is used. When the wavelength is changed to 3000 Å, the stopping potential is 1.94 V. Use these data to determine the value of Planck's constant.

9. From the information given in Problem 8, determine (a) the threshold wavelength, (b) the cutoff frequency, (c) the work function of potassium.

10. Derive Eq. (7.12) from Eqs. (7.9)–(7.11).

11. Show that Eq. (7.19) follows from (7.18) and (7.12).

12. A 1 MeV photon is Compton scattered at an angle of 55°. Calculate (a) the energy of the scattered photon, (b) the change in wavelength, (c) the angle of recoil of the electron, (d) the recoil energy of the electron.

13. (a) What is the maximum recoil energy that an electron can acquire from an 8 MeV photon? (b) At what angle of scatter will an 8 MeV photon lose 95% of its energy in a Compton scattering? (c) Sketch a curve showing the fraction of energy lost by an 8 MeV photon as a function of the angle of Compton scattering.

14. At what energy can a photon lose at most one-half of its energy in Compton scattering?

15. In a Compton scattering experiment a photon is observed to be scattered at an angle of 122° while the electron recoils at an angle of 17° with respect to the incident photon direction. (a) What is the incident photon energy? (b) What is the frequency of the scattered photon? (c) How much energy does the electron receive? (d) What is the recoil momentum of the electron?

16. Monochromatic X-rays of wavelength 0.5 Å and 0.1 Å are Compton scattered from a graphite target and the scattered photons are viewed at an angle of 60° with respect to the incident photon direction. (a) Calculate the Compton shift in each case. (b) Calculate the photon energy loss in each case.

17. Show that the fractional energy loss of a photon in Compton scattering is given by

$$\frac{T}{h\nu} = \frac{\Delta\lambda}{\lambda + \Delta\lambda}.$$

18. How much energy will 10^4 scattered photons deposit in the graphite target in Fig. 7.2? (Use Table 7.1.)

19. Use Eq. (7.12) to write the formula for the Klein–Nishina cross section (7.28) in terms of the photon scattering angle θ:

$$\frac{d\sigma}{d\Omega} = \frac{e^4}{2m^2c^4} \left[\frac{1}{1 + \dfrac{h\nu}{mc^2}(1 - \cos\theta)} \right]^2 (1 + \cos^2\theta)$$

$$\times \left\{ 1 + \frac{\left(\dfrac{h\nu}{mc^2}\right)^2 (1 - \cos\theta)^2}{(1 + \cos^2\theta)\left[1 + \dfrac{h\nu}{mc^2}(1 - \cos\theta)\right]} \right\} \text{ cm}^2.$$

20. Show that the threshold photon wavelength for producing an electron–positron pair is 0.012 Å.
21. A 4 MeV photon creates an electron–positron pair in the field of a nucleus. What is the total kinetic energy of the pair?
22. Calculate the threshold energy for the reaction $^{12}_{6}C(\gamma,n)^{11}_{6}C$. The mass differences are given in Appendix D.
23. Calculate the energy of the proton ejected in the $^{16}_{8}O(\gamma,p)^{15}_{7}N$ reaction with a 20 MeV photon. The mass differences are $\Delta = -4.737$ MeV for ^{16}O, $\Delta = 0.100$ MeV for ^{15}N, and $\Delta = 7.289$ MeV for the proton.
24. (a) What is the threshold energy for the $^{206}_{82}Pb(\gamma,p)^{205}_{81}Tl$ reaction, which competes with $^{206}Pb(\gamma,n)^{205}Pb$? The mass difference of ^{205}Tl is $\Delta = -23.81$ MeV and that of the proton is $\Delta = 7.289$ MeV. (b) Why is the (γ,p) threshold lower than that of (γ,n)? (c) Which process is more probable?
25. An experiment is carried out with monoenergetic photons in the "good" geometry shown in Fig. 7.7. The relative count rate of the detector is measured with different thicknesses x of tin used as absorber. The following data are measured:

x (cm)	0	0.50	1.0	1.5	2.0	3.0	5.0
Relative count rate	1.00	0.861	0.735	0.621	0.538	0.399	0.210

(a) What is the value of the linear attenuation coefficient? (b) What is the value of the mass attenuation coefficient? (c) What is the photon energy?
26. Show that Eq. (7.30) implies that μ is the probability per unit distance that a photon interacts.
27. A narrow beam of 10^4 photons/sec is normally incident on a 6 mm aluminum sheet. The beam consists of equal numbers of 200 keV photons and 2 MeV photons. (a) Calculate the number of photons/sec of each energy that are transmitted without interaction through the sheet. (b) How much energy is removed from the narrow beam per second by the sheet? (c) How much energy is absorbed in the sheet per second?
28. (a) Show that a 1.43 mm lead sheet has the same thickness in g/cm^2 as the aluminum sheet in the last problem. (b) Calculate the number of photons/sec of each energy that are transmitted without interaction when the Al sheet in Problem 27 is replaced by 1.43 mm of Pb. (c) Give a physical reason for any differences or similarities in the answers to Problems 27(a) and 28(b).
29. A narrow beam of 400 keV photons is incident normally on a 2 mm copper liner. (a) What fraction of the photons have an interaction in the liner? (b) What thickness of Cu is needed to reduce the fraction of photons that are transmitted without interaction to 10%? (c) If Al were used instead of Cu, what thickness would be needed in (b)? (d) How do the answers in (b) and (c) compare when expressed in g/cm^2? (e) If lead were used in (b), how would its thickness in g/cm^2 compare with those for Al and Cu?
30. The mass attenuation coefficient of Pb for 70 keV photons is 3.0 cm^2/g and the mass absorption coefficient is 2.9 cm^2/g (Figs. 7.9 and 7.10). (a) Why are these two values almost equal? (b) How many cm of Pb are needed to reduce the transmitted intensity of a 70 keV X-ray beam to 1% of its original value? (c) What percentage of the photons penetrate this thickness without interacting?
31. A 0.5 MeV gamma-ray beam is normally incident on an aluminum absorber of thickness 8 g/cm^2. The beam intensity is 8.24×10^{-25} erg/cm^2 sec. (a) Calculate the fraction of photons transmitted without interaction. (b) What is the total transmitted beam intensity? (c) What is the rate of energy absorption in the aluminum?
32. What fraction of the energy in a 10 keV X-ray beam is deposited in 5 mm of soft tissue?
33. A broad beam of 2 MeV photons is normally incident on a 20 cm concrete shield. (a) What fraction is transmitted without interaction? (b) Estimate the fraction of the beam intensity that is transmitted by the shield.
34. A dentist places the window of a 100 kVp (100 keV, peak voltage) X-ray machine near the face of a patient to obtain an X-ray of the teeth. Without filtration, considerable low-energy (assume 20 keV) X-rays are incident on the skin. (a) If the intervening tissue has a thickness of 5 mm, calculate the fraction of the 20 keV intensity absorbed in it. (b) What thickness of aluminum filter would reduce the 20 keV radiation exposure by a factor of 10? (c) Calculate the reduction in the intensity of 100 keV X-rays transmitted by the filter. (d) After adding the fil-

ter, the exposure time need not be increased to obtain the same quality of X-ray picture. Why not?

35. From Fig. 7.13 determine the energy at which the photoelectric effect and Compton scattering contribute equally to the attenuation coefficient of water.

36. The linear attenuation coefficient of copper for 800 keV photons is 0.58 cm^{-1}. Calculate the atomic cross section.

37. The mass attenuation coefficient for 1 MeV photons in carbon is 0.0633 cm^2/g. (a) Calculate the atomic cross section. (b) Estimate the total Compton cross section per electron.

38. The mass attenuation coefficients of Cu and Sn for 200 keV photons are, respectively, 0.15 cm^2/g and 0.31 cm^2/g. What are the macroscopic cross sections?

39. Calculate the microscopic cross sections in Problem 38 for Cu and Sn.

40. A bronze absorber, made of 9 parts of Cu and 1 part Sn by weight, is exposed to 200 keV X-rays (see Problem 38). Calculate the linear and mass attenuation coefficients of bronze for photons of this energy.

41. The atomic cross sections for 1 MeV photon interactions with carbon and hydrogen are, respectively, 1.27 barns and 0.209 barns. (a) Calculate the linear attenuation coefficient for paraffin. (Assume the composition CH_2 and density 0.89 g/cm^3.) (b) Calculate the mass attenuation coefficient.

42. What is the atomic cross section of Fe for 400 keV photons? What is the absorption cross section?

CHAPTER 8
NEUTRONS, FISSION, AND CRITICALITY

8.1 INTRODUCTION

The neutron was discovered by Chadwick in 1932. Nuclear fission, induced by capture of a slow neutron in ^{235}U, was discovered by Hahn and Strassman in 1939. In principle, the fact that several neutrons are emitted when fission takes place suggested that a self-sustaining chain reaction might be possible. Under Fermi's direction, the world's first man-made nuclear reactor went critical on December 2, 1942.[†] The neutron thus occupies a central position in the modern world of atoms and radiation.

In this chapter we describe the principal sources of neutrons, their interactions with matter, nuclear fission, and criticality. The most important neutron interactions from the standpoint of radiation protection will be stressed.

8.2 NEUTRON SOURCES

Nuclear reactors are the most copious sources of neutrons. The energy spectrum of neutrons from the fission of ^{235}U extends from a few keV to more than 10 MeV. The average energy is about 2 MeV. Research reactors often have ports through which neutron beams emerge into experimental areas outside the main reactor shielding. These neutrons are usually degraded in energy, having passed through parts of the reactor core and coolant as well as structural materials. Figure 8.1 shows an example of a research reactor.

Particle accelerators are used to generate neutron beams by means of a number of nuclear reactions. For example, accelerated deuterons that strike a tritium target produce neutrons via the $^3H(d,n)^4He$ reaction, i.e.,

$$_1^2H + {}_1^3H \rightarrow {}_2^4He + {}_0^1n. \tag{8.1}$$

To obtain monoenergetic neutrons with an accelerator, excited states of the product nucleus are undesirable. Therefore, light materials are commonly used as targets for a proton or deuteron beam. Table 8.1 lists some important reactions that are used to obtain monoenergetic neutrons. The first two are exothermic and can be used with ions of a few hundred keV energy in relatively inexpensive accelerators. For a given ion-beam energy, neutrons leave a thin target with energies that depend on the angle of exit with respect to the incident beam direction.

An alpha source, usually radium, polonium, or plutonium and a light metal, such as beryllium or boron, can be mixed together as powders and encapsulated to make a

---

[†] In 1972 the French Atomic Energy Commission found unexpectedly low assays of $^{235}U/^{238}U$ isotopic ratios in uranium ores from the Oklo deposit in Gabon, Africa. Close examination revealed that several sites in the Oklo mine were natural nuclear reactors in the distant past. As far as is known, the Oklo phenomenon, as it is called, is unique. For more information the reader is referred to the article by M. Maurette, Fossil nuclear reactors, *Ann. Rev. Nucl. Sci.* **26**, 319 (1976).

Figure 8.1. Oak Ridge Research Reactor, a swimming-pool-type reactor in which water serves as coolant, moderator, and shield. Glow around reactor is blue light emitted as Cerenkov radiation (Section 9.4.5) by electrons traveling faster than light in the water. (Photo courtesy Oak Ridge National Laboratory, operated by Martin Marietta Energy Systems, Inc., for the Department of Energy)

"radioactive" neutron source. Neutrons are emitted as a result of (α,n) reactions, such as the following:

$$\,^4_2\text{He} + \,^9_4\text{Be} \rightarrow \,^{12}_6\text{C} + \,^1_0\text{n}. \tag{8.2}$$

Light metals are used in order to minimize the Coulomb repulsion between the alpha particle and nucleus. The neutron intensity from such a source dies off with the half-life of the alpha emitter. Neutrons leave the source with a continuous energy spectrum, because the alpha particles slow down by different amounts before striking a nucleus. The neutron and the recoil nucleus [e.g., $\,^{12}_6\text{C}$ in (8.2)] share a total energy equal to the sum of the

Table 8.1. Reactions Used to Produce Monoenergetic Neutrons with Accelerated Protons (p) and Deuterons (d)

Reaction	Q value (MeV)
$\,^3\text{H}(d,n)^4\text{He}$	17.6
$\,^2\text{H}(d,n)^3\text{He}$	3.27
$\,^{12}\text{C}(d,n)^{13}\text{N}$	−0.281
$\,^3\text{H}(p,n)^3\text{He}$	−0.764
$\,^7\text{Li}(p,n)^7\text{Be}$	−1.65

Table 8.2. (α,n) Neutron Sources

Source	Average neutron energy (MeV)	Half-life
^{210}PoBe	4.2	138 days
^{210}PoB	2.5	138 days
^{226}RaBe	3.9	1602 yr
^{226}RaB	3.0	1602 yr
^{239}PuBe	4.5	24400 yr

Q value and the kinetic energy that the alpha particle has as it strikes the nucleus. Some common (α,n) sources are shown in Table 8.2.

Similarly, photoneutron sources, making use of (γ,n) reactions, are also available. Several examples are listed in Table 8.3. In contrast to (α,n) sources, which emit neutrons with a continuous energy spectrum, monoenergetic photoneutrons can be obtained by selecting a nuclide that emits a gamma ray of a single energy. Photoneutron sources decay in intensity with the half-life of the photon emitter. All the sources in Table 8.3 are monoenergetic except the last; ^{226}Ra emits gamma rays of several energies. It is important for radiation protection to remember that all photoneutron sources have gamma-ray backgrounds of >1000 photons per neutron.

Some very heavy nuclei fission spontaneously, emitting neutrons in the process. They can be encapsulated and used as neutron sources. Examples of some important spontaneous-fission sources are ^{254}Cf, ^{252}Cf, ^{244}Cm, ^{242}Cm, ^{238}Pu, and ^{232}U. In most cases the half-life for spontaneous fission is much greater than that for alpha decay. An exception is ^{254}Cf, which decays almost completely by spontaneous fission with a 60 day half-life.

8.3 CLASSIFICATION OF NEUTRONS

It is convenient to classify neutrons according to their energies. At the low end of the scale, neutrons can be in approximate thermal equilibrium with their surroundings. Their energies are then distributed according to the Maxwell–Boltzmann formula. The energy of a thermal neutron is sometimes given as 0.025 eV, which is the most probable energy in the distribution at room temperature (20°C). The average energy of thermal neutrons at room temperature is 0.038 eV. Thermal-neutron distributions do not necessarily have to correspond to room temperature. "Cold" neutrons, with lower "temperatures," are produced at some facilities, while others generate neutrons with energy distributions characteristic of temperatures considerably above 20°C. Thermal neutrons gain and lose

Table 8.3. (γ,n) Neutron Sources

Source	Neutron energy (MeV)	Half-life
^{24}NaBe	0.83	14.8 hr
^{24}NaD$_2$O	0.22	14.8 hr
^{116}InBe	0.30	54 min
^{124}SbBe	0.024	60 days
^{140}LaBe	0.62	40 hr
^{226}RaBe	0.7 (maximum)	1602 yr

only small amounts of energy through elastic scattering in matter. They diffuse about until captured by atomic nuclei.

Neutrons of higher energies, up to about 0.01 MeV or 0.1 MeV (the convention is not precise), are known variously as "slow," "intermediate," or "resonance" neutrons. "Fast" neutrons are those in the next-higher-energy classification, up to about 10 MeV or 20 MeV. "Relativistic" neutrons have still higher energies.

8.4 INTERACTIONS WITH MATTER

Like photons, neutrons are uncharged and hence can travel appreciable distances in matter without interacting. Under conditions of "good geometry" (cf. Section 7.7) a narrow beam of monoenergetic neutrons is also attenuated exponentially by matter. The interaction of neutrons with electrons, which is electromagnetic in nature,† is negligible. In passing through matter a neutron can collide with an atomic nucleus, which can scatter it elastically or inelastically. The scattering is elastic when the total kinetic energy is conserved; i.e., when the energy lost by the neutron is equal to the kinetic energy of the recoil nucleus. When the scattering is inelastic, the nucleus absorbs some energy internally and is left in an excited state. The neutron can also be captured, or absorbed, by a nucleus, leading to a reaction, such as (n,p), (n,2n), (n,α), or (n,γ). The reaction changes the atomic mass number and/or atomic number of the struck nucleus.

Typically, a fast neutron will lose energy in matter by a series of (mostly) elastic scattering events. This slowing-down process is called neutron moderation. As the neutron energy decreases, scattering continues, but the probability of capture by a nucleus generally increases. If a neutron reaches thermal energies, it will move about randomly by elastic scattering until absorbed by a nucleus.

Cross sections for the interactions of neutrons with atomic nuclei vary widely and usually are complicated functions of neutron energy. Figure 8.2 shows the total cross sections for neutron interactions with hydrogen and carbon as functions of energy. Because the hydrogen nucleus (a proton) has no excited states, only elastic scattering and neutron capture are possible. The total hydrogen cross section shown in Fig. 8.2 is the sum of the cross sections for these two processes. The capture cross section for hydrogen is comparatively small, reaching a value of only 0.33 barns (1 barn = 10^{-24} cm^2) at thermal energies, where it is largest. Thermal-neutron capture is an important interaction in hydrogenous materials.

In contrast, the carbon cross section in Fig. 8.2 shows considerable structure, especially in the region 1–10 MeV. The nucleus possesses discrete excited states, which can enhance or depress the elastic and inelastic scattering cross sections at certain values of the neutron energy (cf. Fig. 3.2).

8.5 ELASTIC SCATTERING

As mentioned in the last section, elastic scattering is the most important process for slowing down neutrons; the contribution by inelastic scattering is usually small in comparison. We treat elastic scattering here.

---

†Although the neutron is electrically neutral, it has spin, a magnetic moment, and a nonzero *distribution* of electric charge within it. These properties, coupled with the charge and spin of the electron, give rise to electromagnetic forces between neutrons and electrons. These forces, however, are extremely weak. In contrast, the neutron interacts with protons and neutrons at close range by means of the strong, or nuclear, force.

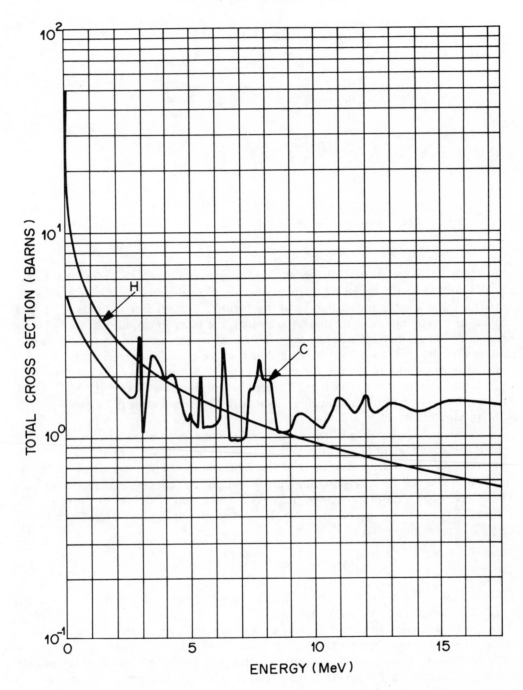

Figure 8.2. Total cross sections for neutrons with hydrogen and carbon as functions of energy.

The maximum energy that a neutron of mass M and kinetic energy E can transfer to a nucleus of mass m in a single (head-on) elastic collision is given by Eq. (4.4):

$$Q_{max} = \frac{4mME}{(M + m)^2}. \tag{8.3}$$

Table 8.4. Maximum Fraction of Energy Lost, Q_{max}/E from Eq. (8.3), by Neutron in Single Elastic Collision with Various Nuclei

Nucleus	Q_{max}/E
$^{1}_{1}H$	1.000
$^{2}_{1}H$	0.889
$^{4}_{2}He$	0.640
$^{9}_{4}Be$	0.360
$^{12}_{6}C$	0.284
$^{16}_{8}O$	0.221
$^{56}_{26}Fe$	0.069
$^{118}_{50}Sn$	0.033
$^{238}_{92}U$	0.017

Setting $M = 1$, we can calculate the maximum fraction of a neutron's energy that can be lost in a collision with nuclei of different atomic-mass numbers m. Some results are shown in Table 8.4 for nuclei that span the periodic system. For ordinary hydrogen, because the proton and neutron masses are equal, the neutron can lose all of its kinetic energy in a head-on, billiard-ball-like collision. As the nuclear mass increases, one can see how the efficiency of a material per collision for moderating neutrons grows progressively worse. As a rule of thumb, the average energy lost per collision is approximately one-half the maximum. This rule is exact when the scattering is isotropic in the center-of-mass system of the neutron and nucleus, as it is for hydrogen at neutron energies up to about 10 MeV.

An interesting consequence of the equality of the masses in n–p scattering is that the particles separate at right angles after collision, when the collision is nonrelativistic. Figure 8.3(a) represents a neutron of mass M and momentum MV approaching a stationary nucleus of mass m. After collision, in Figure 8.3(b), the particles have momenta mv' and MV'. The conservation of momentum requires that the sum of the vectors, $mv' + MV'$, be equal to the initial momentum vector MV, as shown in Figure 8.3(c). Since kinetic energy is conserved, we have

$$\frac{1}{2} MV^2 = \frac{1}{2} mv'^2 + \frac{1}{2} MV'^2. \tag{8.4}$$

If $M = m$, then $V^2 = v'^2 + V'^2$, which implies the Pythagorean theorem for the triangle in (c). Therefore, v' and V' are at right angles.

The elastic scattering of neutrons plays an important role in neutron energy measurements. As discussed in the next chapter, under suitable conditions the recoil energies of nuclei in a proportional-counter gas under neutron bombardment can be measured. The nuclear recoil energy and angle are directly related to the neutron energy. For example, as illustrated in Fig. 8.4, when a neutron of energy E_n strikes a proton, which recoils with energy E_p' at an angle θ with respect to the incident-neutron direction, then the conservation of energy and momentum requires that

$$E_p' = E_n \cos^2 \theta. \tag{8.5}$$

Thus, if E_p' and θ can be measured individually for a number of incident neutrons, one obtains the incident neutron spectrum directly. (The proton-recoil neutron spectrometer is discussed in Section 9.5.2.) More often, only the energies of the recoil nuclei in the gas

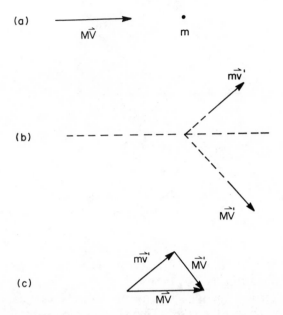

Figure 8.3. Momenta of colliding particles (a) before and (b) after collision. (c) Representation of momentum conservation.

(e.g., ^{3}He, ^{4}He, or ^{1}H and ^{12}C from CH$_4$) are determined, and the neutron energy spectrum must be unfolded from its statistical relationship to the recoil-energy spectra. The unfolding is further complicated by the fact that the recoil tracks do not always lie wholly within the chamber gas (wall effects).

Because of the abundance of hydrogen in soft tissue, n–p scattering is often the dominating mechanism whereby neutrons deliver dose to tissue. As we shall see in Section 10.9, over 85% of the "first-collision" dose in soft tissue (composed of H, C, O, and N) arises from n–p scattering for neutron energies below 10 MeV.

8.6 REACTIONS

In this section we describe several neutron reactions that are important in various aspects of neutron detection and radiation protection. The way in which the reactions are used for detection will be described in Chapter 9.

^{1}H(n, γ)^{2}H

We have already mentioned the capture of thermal neutrons by hydrogen. This reaction is an example of radiative capture; i.e., neutron absorption followed by the immediate emission of a gamma photon. Explicitly,

$$^1_0n + \,^1_1H \rightarrow \,^2_1H + \,^0_0\gamma. \tag{8.6}$$

Since the thermal neutron has negligible energy by comparison, the gamma photon has the energy $Q = 2.22$ MeV released by the reaction, which represents the binding energy of the deuteron.† When tissue is exposed to thermal neutrons, the reaction (8.6) provides a

---

†The energetics were worked out in Chapter 3 [Eq. (3.8)].

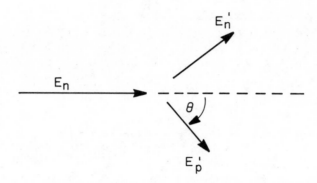

Figure 8.4. Schematic representation of elastic scattering of a neutron by a proton. The initial neutron energy E_n is given in terms of the proton recoil energy E'_p and angle θ by Eq. (8.5).

source of gamma rays that delivers dose to the tissue. The capture cross section for the reaction (8.6) for thermal neutrons is 0.33 barns.

Capture cross sections for low-energy neutrons generally decrease as the reciprocal of the velocity as the neutron energy increases. This phenomenon is often called the "$1/v$ law." Thus, if the capture cross section σ_0 is known for a given velocity v_0 (or energy E_0), then the cross section at some other velocity v (or energy E), can be estimated from the relations

$$\frac{\sigma}{\sigma_0} = \frac{v_0}{v} = \sqrt{\frac{E_0}{E}}. \tag{8.7}$$

These expressions can generally be used for neutrons of energies up to 100 eV or 1 keV, depending on the absorbing nucleus.

Example
The capture cross section for the reaction (8.6) for thermal neutrons is 0.33 barns. Estimate the cross section for neutrons of energy 10 eV.

Solution
The thermal-neutron energy is usually assumed to be the most probable value, $E_0 = 0.025$ eV. Applying Eq. (8.7) with $\sigma_0 = 0.33$ barns and $E = 10$ eV, we find for the capture cross section at 10 eV,

$$\sigma = \sigma_0 \sqrt{\frac{E_0}{E}} = 0.33 \sqrt{\frac{0.025}{10}} = 0.0165 \text{ barns.} \tag{8.8}$$

$^3\text{He}(n,p)^3\text{H}$
The cross section for thermal-neutron capture is 5330 barns and the energy $Q = 765$ keV is released by the reaction following thermal-neutron capture. Some proportional counters used for fast-neutron monitoring contain a little added ^3He for calibration purposes. A polyethylene sleeve slipped over the tube thermalizes incident neutrons by the time they enter the counter gas. The pulse-height spectrum then shows a peak, which identifies the channel number in which a pulse height of 765 keV is registered. With the energy per channel thus established, the polyethylene sleeve can be removed and the instrument used for fast-neutron monitoring. Other neutron devices use ^3He as the proportional-counter gas. They measure a continuum of pulse heights due to the recoil ^3He nuclei from elastic scattering. In addition, when a neutron with kinetic energy T is captured by a ^3He nucleus, an energy of $T + 765$ keV is released (see Fig. 9.32).

^{6}Li(n,t)^{4}He

This reaction, which produces a ^{3}H nucleus, or triton (t), and has a Q value of 4.78 MeV, is also used for thermal-neutron detection. The cross section is 940 barns, and the isotope ^{6}Li is 7.40% abundant. Neutron-sensitive LiI scintillators can be made, and Li can also be added to other scintillators to register neutrons. Lithium enriched in the isotope ^{6}Li is available.

^{10}B(n, α)^{7}Li

For this reaction, $\sigma = 3840$ barns for thermal neutrons. The isotope ^{10}B is 19.8% abundant. In 93% of the reactions the ^{7}Li nucleus is left in an excited state and emits a 0.48 MeV gamma ray. The total kinetic energy shared by the alpha particle and ^{7}Li recoil nucleus is then $Q = 2.31$ MeV. The other 7% of the reactions go to the ground state of the ^{7}Li nucleus with $Q = 2.79$ MeV. BF$_3$ is a gas that can be used directly in a neutron counter. Boron is also employed as a liner inside the tubes of proportional counters for neutron detection. It is also used as a neutron shielding material. Boron enriched in the isotope ^{10}B is available. Additional information is given in Section 9.5.2.

^{14}N(n,p)^{14}C

Since nitrogen is a major constituent of tissue, this reaction, like neutron capture by hydrogen, contributes to neutron dose. The cross section for thermal neutrons is 1.70 barns and the Q value is 0.626 MeV. Since their ranges in tissue are small, the energies of the proton and ^{14}C nucleus are deposited locally at the site where the neutron was absorbed. This reaction makes less than a 1% relative contribution to dose for neutron energies less than about 10 MeV. At higher energies the contribution is of the order of 1% or slightly greater.

^{23}Na(n, γ)^{24}Na

Absorption of a neutron by ^{23}Na gives rise to the radioactive isotope ^{24}Na. The latter has a half-life of 15.0 hr and emits two gamma rays, having energies of 2.75 MeV and 1.37 MeV, per disintegration. The thermal-neutron capture cross section is 0.534 barns. Since ^{23}Na is a normal constituent of blood, activation of blood sodium can be used as a dosimetric tool when persons are exposed to relatively high doses of neutrons, e.g., in a criticality accident.

^{32}S(n,p)^{32}P

For this reaction to occur, the neutron must have an energy of at least 2.7 MeV. It is an example of but one of many threshold reactions used for neutron detection. As described in Section 9.5.2, the simultaneous activation of foils made from a series of nuclides with different thresholds provides a means of estimating neutron spectra. The existence of ^{32}S in human hair has also been used to help estimate high-energy (>2.7 MeV) neutron doses to persons exposed in criticality accidents. The product ^{32}P, a pure beta emitter with a maximum energy of 1.17 MeV and a half-life of 14.3 days, is easily counted.

^{113}Cd(n, γ)^{114}Cd

Because of the large, 21,000 barn, thermal-neutron capture cross section of ^{113}Cd, cadmium is used as a neutron shield and as a reactor control-rod material. The relative abundance of the ^{113}Cd isotope is 12.3%. The absorption cross section of ^{113}Cd for neutrons is large from thermal energies up to ~0.5 eV, where it drops off sharply. A method for measuring the ratio of thermal to fast neutrons consists of comparing the induced activities in two identical foils (e.g., indium), one bare and the other covered with a cadmium shield. The latter absorbs essentially all neutrons with energies below the so-called cadmium cutoff of ~0.5 eV.

^{115}In$(n, \gamma)^{116m}$In

The cross section for thermal-neutron capture by ^{115}In (95.7% abundant) is 157 barns and the metastable ^{116m}In decays with a half-life of 54.3 min. The induced activity in indium foils worn by persons suspected having been exposed to neutrons can be checked as a quick-sort method following a criticality accident. In practical cases the method is sensitive enough to permit detection with an ionization chamber as well as a GM or scintillation survey instrument. The degree of foil activity depends so strongly on the orientation of the exposed person, the neutron-energy spectrum, and other factors that it does not provide a useful basis for even a crude estimate of dose. However, the fact that an exposure occurred (or did not) can thus be established.

^{197}Au$(n, \gamma)^{198}$Au

This isotope, which is 100% abundant, has a thermal-neutron capture cross section of 98.8 barns. Although not as sensitive as indium, its longer half-life of 2.70 days permits monitoring at longer times after exposure.

^{235}U(n,f)

Fission (f) is discussed in Section 8.8. Because of the large release of energy (~200 MeV), the fission process provides a distinct signature for detecting thermal neutrons, even in high backgrounds of other types of radiation (cf. Section 9.5.1).

8.7 NEUTRON ACTIVATION

The time dependence of the activity induced by neutron capture can be described quantitatively. If a sample containing N_T target atoms with cross section σ cm^2 is exposed to monoenergetic neutrons with a fluence rate expressed as neutrons cm^{-2} sec^{-1}, then the production rate of daughter atoms from neutron absorption is $\varphi\sigma N_T$ sec^{-1}. If the number of daughter atoms in the sample is N and the decay constant is λ, then the rate of loss of daughter atoms from the sample is λN. Thus, the rate of change dN/dt sec^{-1} in the number of daughter atoms present at any time while the sample is being bombarded is given by

$$\frac{dN}{dt} = \varphi\sigma N_T - \lambda N. \tag{8.9}$$

To solve this equation, we assume that the fluence rate is constant and that the original number of target atoms is not significantly depleted, so that N_T is also constant. Without the term $\varphi\sigma N_T$, which is then constant, Eq. (8.9) would be the same linear homogeneous equation as Eq. (3.47). Therefore, we try a solution to (8.9) in the same form as (3.52) for the homogeneous equation plus an added constant. Substituting $N = a + be^{-\lambda t}$ into Eq. (8.9) gives

$$-b\lambda e^{-\lambda t} = \varphi\sigma N_T - a\lambda - b\lambda e^{-\lambda t}. \tag{8.10}$$

The exponential terms on both sides cancel, and one finds that $a = \varphi\sigma N_T/\lambda$. Thus, the general solution is

$$N = \frac{\varphi\sigma N_T}{\lambda} + be^{-\lambda t}. \tag{8.11}$$

The constant b depends on the initial conditions. Specifying that no daughter atoms are present when the neutron irradiation begins (i.e., $N = 0$ when $t = 0$), we find from (8.11) that $b = -\varphi\sigma N_T/\lambda$. Thus we obtain the final expression

$$\lambda N = \varphi\sigma N_T(1 - e^{-\lambda t}). \tag{8.12}$$

The left-hand side expresses the activity of the daughter as a function of the time t. The quantity $\varphi\sigma N_T$ is called the saturation activity because it represents the maximum activity obtainable when the sample is irradiated for a long time ($t \to \infty$). A sketch of the function (8.12) is shown in Fig. 8.5.

When the neutrons are not monoenergetic, the terms in Eq. (8.9) can be replaced by integrals of the energy-dependent cross section σ and the neutron energy spectrum. Alternatively, Eq. (8.12) can be used as is, provided the proper average cross section is used for σ.

Example

A 3 g sample of ^{32}S is irradiated with fast neutrons having a constant fluence rate of 155 cm^{-2} sec^{-1}. The cross section for the reaction $^{32}S(n,p)^{32}P$ is 0.200 barns and the half-life of ^{32}P is $T = 14.3$ days. What is the maximum ^{32}P activity that can be induced? How many days are needed for the level of the activity to reach three quarters of the maximum?

Solution

The total number of target atoms is $N_T = \frac{3}{32} \times 6.02 \times 10^{23} = 5.64 \times 10^{22}$. The maximum (saturation) activity is $\varphi\sigma N_T = (155 \text{ cm}^{-2}\text{sec}^{-1})(0.2 \times 10^{-24} \text{ cm}^2)(5.64 \times 10^{22}) = 1.75 \text{ sec}^{-1}$. Expressed in curies, the saturation activity is $1.75/(3.7 \times 10^{10}) = 4.73 \times 10^{-11}$ Ci. The time t needed to reach three quarters of this value can be found from Eq. (8.12) by writing $\frac{3}{4} = 1 - e^{-\lambda t}$. Then $e^{-\lambda t} = \frac{1}{4}$ and $t = 2T = 28.6$ days. Note that the buildup toward saturation activity is analogous to the approach to secular equilibrium by the daughter of a long-lived parent (Sect. 3.12).

Example

Estimate the fraction of the ^{32}S atoms that would be consumed in the last example in 28.6 days.

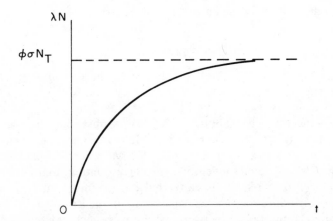

Figure 8.5. Buildup of induced activity λN, as given by Eq. (8.12), during neutron irradiation at constant fluence rate.

Solution

The rate at which ^{32}S atoms are used up is $\varphi \sigma N_T = 1.75 \text{ sec}^{-1}$. Since $t = 28.6$ days $= 2.47 \times 10^6$ sec, the number of ^{32}S atoms lost is $1.75 \times 2.47 \times 10^6 = 4.32 \times 10^6$. The fraction of ^{32}S atoms consumed, therefore, is $4.32 \times 10^6/(5.64 \times 10^{22}) = 7.66 \times 10^{-17}$, a negligible amount. Note that fractional burnup does not depend on the sample size N_T. The problem can also be worked by writing for the desired fraction $\varphi \sigma N_T t/N_T = \varphi \sigma t = 155 \times 0.2 \times 10^{-24} \times 2.47 \times 10^6 = 7.66 \times 10^{-17}$. The assumption of constant N_T for the validity of Eq. (8.12) is therefore warranted. The actual fraction of ^{32}S atoms consumed during a long time t is, of course, less than $\varphi \sigma t$.

8.8 FISSION

As described in Section 3.2, the binding energy per nucleon for heavy elements decreases as the atomic mass number increases (Fig. 3.3). Thus, when heavy nuclei are split into smaller pieces, energy is released. Alpha decay is an example of one such process that is spontaneous. With the discovery of nuclear fission, another process was realized, in which the splitting was much more dramatic and the energy release almost two orders of magnitude greater. Nuclear fission can be induced in certain nuclei as a result of absorbing a neutron. With ^{235}U, ^{239}Pu, and ^{233}U, absorption of a thermal neutron can set up vibrations in the nucleus which cause it to become so distended that it splits apart under the mutual electrostatic repulsion of its parts. The thermal-neutron fission cross sections for these isotopes are, respectively, 580, 747, and 525 barns. A greater activation energy is required to cause other nuclei to fission. An example is ^{238}U, which requires a neutron with a kinetic energy in excess of 1 MeV to fission. Cross sections for such "fast-fission" reactions are much smaller than those for thermal fission. The fast-fission cross section for ^{238}U, for instance, is 0.29 barns. Also, fission does not always result when a neutron is absorbed by a fissionable nucleus. ^{235}U fissions only 85% of the time after thermal-neutron absorption.

Nuclei with an odd number of nucleons fission more readily following neutron absorption than do nuclei with an even number of nucleons. This fact is related to the greater binding energy per nucleon found in even–even nuclei, as mentioned in Section 3.2. The ^{235}U nucleus is even–odd in terms of its proton and neutron numbers. Addition of a neutron transforms it into an even–even nucleus with a larger energy release than that following neutron absorption by ^{238}U.

Fissionable nuclei break up in a number of different ways. The ^{235}U nucleus splits in some 40 or so modes following the absorption of a thermal neutron. One typical example is the following:

$$\begin{smallmatrix}1\\0\end{smallmatrix}\text{n} + \begin{smallmatrix}235\\92\end{smallmatrix}\text{U} \rightarrow \begin{smallmatrix}147\\57\end{smallmatrix}\text{La} + \begin{smallmatrix}87\\35\end{smallmatrix}\text{Br} + 2\begin{smallmatrix}1\\0\end{smallmatrix}\text{n}. \tag{8.13}$$

An average energy of about 195 MeV is released in the fission process, distributed as shown in Table 8.5. The major share of the energy (162 MeV) is carried away by the charged fission fragments, such as the La and Br fragments in (8.13). Fission neutrons and gamma rays account for another 12 MeV. Subsequent fission-product decay accounts for 10 MeV and neutrinos carry off 11 MeV. In a new reactor, in which there are no fission products, the energy output is about 175 MeV per fission. In an older reactor, with a significant fission-product inventory, the corresponding figure is around 185 MeV. The energy produced in a reactor is converted mostly into heat from the stopping of charged particles, including the recoil nuclei struck by neutrons and the secondary electrons produced by gamma rays. Neutrinos escape with negligible energy loss.

As exemplified by (8.13), nuclear fission produces asymmetric masses with high probability. The mass distribution of fission fragments from ^{235}U is thus bimodal.

Table 8.5. Average Distribution of Energy Among Products Released by Fission of ^{235}U

Kinetic energy of charged fission fragments	162 MeV
Fission neutrons	6
Fission gamma rays	6
Subsequent beta decay	5
Subsequent gamma decay	5
Neutrinos	11
Total	195 MeV

All fission fragments are radioactive and most decay through several steps to stable daughters. The decay of the collective fission-product activity following the fission of a number of atoms at $t = 0$ is given by

$$A \cong 10^{-16} t^{-1.2} \text{ curies/fission,} \tag{8.14}$$

where t is in days. This expression can be used for estimating residual fission-product activity between about 10 sec and 1000 hr.

The average number of neutrons produced per fission of ^{235}U is 2.5. This number must exceed unity in order for a chain reaction to be possible. Some 99.36% of the fission neutrons are emitted promptly (in $\sim 10^{-14}$ sec) from the fission fragments, while the other (delayed) neutrons are emitted later (up to ~ 1 min or more). The delayed neutrons play an important role in the ease of control of a nuclear reactor, as discussed in the next section.

8.9 CRITICALITY

An assembly of fissionable material is said to be critical when, on the average, exactly one of the several neutrons emitted in the fission process causes another nucleus to fission. The power output of the assembly is then constant. The other fission neutrons are either absorbed without fission or else escape from the system. Criticality thus depends upon geometrical factors as well as the distribution and kinds of the material present. If an average of more than one fission neutron produces fission of another nucleus, then the assembly is said to be supercritical, and the power output increases. If less than one fission occurs per fission neutron produced, the unit is subcritical.

Criticality is determined by the extent of neutron multiplication as successive generations are produced. If N_i thermal neutrons are present in a system, their absorption will result in a certain number N_{i+1} of next-generation thermal neutrons. The effective multiplication factor is defined as

$$k_{\text{eff}} = \frac{N_{i+1}}{N_i}. \tag{8.15}$$

The system is critical if $k_{\text{eff}} = 1$, exactly; supercritical if $k_{\text{eff}} > 1$; and subcritical if $k_{\text{eff}} < 1$.

It is useful to discuss k_{eff} independently of the size and shape of an assembly. Therefore, we introduce the infinite multiplication factor, k_∞, for a system that is infinite in extent. For a finite system one can then write

$$k_{\text{eff}} = L k_\infty, \tag{8.16}$$

where L is the probability that a neutron will not escape. The value of k_∞ will depend on several factors, as we now describe for uranium fuel.

Of N_i total thermal neutrons present in the ith generation in an infinite system, generally only a fraction f, called the thermal utilization factor, will be absorbed in the fissionable fuel (viz., ^{235}U and ^{238}U). The rest will be absorbed by other kinds of atoms (moderator and impurities). If η represents the average number of fission neutrons produced per thermal-neutron capture in the uranium fuel, then the disappearance of the N_i thermal neutrons will result in the production of $N_i f \eta$ fission neutrons that belong to the next generation. Some of these neutrons will produce fast fission in ^{238}U before they have a chance to become thermalized. The fast fission factor ϵ is defined as the ratio of the total number of fission neutrons and the number produced by thermal fission. Then the absorption of the original N_i thermal neutrons results in the production of a total of $N_i f \eta \epsilon$ fission neutrons. Not all of these will become thermalized, because they may undergo radiative capture in the fuel (^{238}U) and moderator. Many materials have resonances in the (n, γ) cross section at energies of several hundred eV and downward. Letting p represent the resonance escape probability (i.e., the probability that a fast neutron will slow down to thermal without radiative capture) we obtain for the total number N_{i+1} of thermal neutrons in the next generation

$$N_{i+1} = N_i f \eta \epsilon p. \tag{8.17}$$

From the definition (8.15), it follows that the infinite-system multiplication factor is given by

$$k_\infty = f \eta \epsilon p. \tag{8.18}$$

The right-hand side of Eq. (8.18) is called the four-factor formula, describing the multiplication factor for an infinitely large system. The factors f, ϵ, and p depend on the composition and enrichment of the fuel and its physical distribution in the moderator. For a pure uranium-metal system, $f = 1$. The thermal utilization factor can be small if competition for thermal absorption by other materials is great. For pure natural uranium, the fast-fission factor has the value $\epsilon = 1.3$; for a homogeneous distribution of fuel and moderator, $\epsilon \sim 1$. Generally, but depending on the particular circumstances, p ranges from ~ 0.7 to ~ 1 for enriched systems. For pure ^{235}U, $p = 1$. For natural uranium metal, $p = 0$; and so such a system — even of infinite extent — will not be critical. The fourth factor, η, depends only on the fuel. The isotope ^{235}U emits an average of 2.5 neutrons per fission. However, since thermal capture by ^{235}U results in fission only 85% of the time, it follows that, for pure ^{235}U, $\eta = 0.85 \times 2.5 = 2.1$. For other enrichments, $\eta < 2.1$.

Example
Given the thermal-neutron fission cross section of ^{235}U, 580 barns, and the thermal-neutron absorption cross section for ^{238}U, 2.8 barns, find the value of η for natural uranium, which consists of 0.72% ^{235}U and 99.28% ^{238}U.

Solution
By definition, η is the average number of fission neutrons produced per thermal neutron absorbed in the uranium fuel (^{235}U plus ^{238}U), which is the only element present. From the given fission cross section it follows that the thermal-neutron absorption cross section for ^{235}U, which fissions 85% of the time, is $580/0.85 = 682$ barns. The fraction of thermal-neutron absorption events in the natural uranium fuel that result in fission is therefore $0.0072 \times 580/(0.0072 \times 682 + 0.9928 \times 2.8) = 0.543$. Since, on the average, 2.5 neutrons are emitted when ^{235}U fissions, it follows that $\eta = 0.543 \times 2.5 = 1.36$.

Example
Compute η for solid uranium metal that is 90% enriched in ^{235}U.

Solution

Similar to the last problem, we have here

$$\eta = \frac{0.90 \times 580}{0.90 \times 682 + 0.10 \times 2.8} \times 2.5 = 2.13. \tag{8.19}$$

Note that the largest possible value for η (that for pure ^{235}U) is $\frac{580}{682} \times 2.5 = 0.85 \times 2.5 = 2.13$, as pointed out above.

In its basic form, a nuclear reactor is an assembly that consists of fuel (usually enriched uranium); a moderator, preferably of low atomic mass number, for thermalizing neutrons; control rods made of materials with a high thermal-neutron absorption cross section (e.g., cadmium or boron steel); and a coolant to remove the heat generated. With the control rods fully inserted in the reactor, the multiplication constant k is less than unity. As a rod is withdrawn, k increases; and when $k = 1$, the reactor becomes critical and produces power at a constant level. Further withdrawal of control rods makes $k > 1$ and causes a steady increase in the power level. When the desired power level is attained, the rods are partially reinserted to make $k = 1$, and the reactor operates at a steady level.

As mentioned in the last section, 99.36% of the fission neutrons from ^{235}U are prompt; i.e., they are emitted immediately in the fission process, while the other 0.64% are released at times of the order of seconds to over a minute after fission. Since a fission neutron can be thermalized in a fraction of a second, the existence of these delayed neutrons greatly facilitates the control of a uranium reactor. If $1 < k < 1.0064$, the reactor is said to be in a delayed critical condition, since the delayed neutrons are essential to maintaining the chain reaction. The rate of power increase is then sufficiently slow to allow control through mechanical means, such as the physical adjustment of control-rod positions. When $k > 1.0064$, the chain reaction can be maintained by the prompt neutrons alone, and the condition is called prompt critical. The rate of increase in the power level is then much faster than when delayed critical.

Whenever fissionable material is chemically processed, machined, transported, stored, or otherwise handled, care must be taken to prevent accidental criticality. Generally, procedures for avoiding criticality depend on limiting the total mass or concentration of fissionable material present and on the geometry in which it is contained. For example, an infinitely long, water-reflected cylinder of aqueous solution with a concentration of 75 g ^{235}U per liter is subcritical as long as its diameter is less than 6.3 in. Without water reflection the limiting diameter is 8.7 in. When using such "always safe" geometry, attention must be given to the possibility that two or more subcritical units could become critical in close proximity.

8.10 PROBLEMS

1. Calculate the threshold energy for the ^{3}H(p,n)^{3}He reaction in Table 8.1.
2. What is the maximum energy of the neutrons produced when 10 MeV protons strike a ^{7}Li target (Table 8.1)?
3. What is the maximum energy of a neutron emitted from a ^{210}PoBe source? For ^{9}Be, $\Delta = 11.351$ MeV.
4. Calculate the neutron energy from a ^{24}NaD$_2$O source (Table 8.3).
5. (a) What is the maximum energy that a 4 MeV neutron can transfer to a ^{10}B nucleus in an elastic collision?
   (b) Estimate the average energy transferred per collision.
6. (a) Estimate the average energy that a 2 MeV neutron transfers to a deuteron in a single collision.
   (b) What is the maximum possible energy transfer?
7. Make a rough estimate of the number of collisions that a neutron of 2 MeV initial energy makes with deuterium in order for its energy to be reduced to 1 eV.

8. Repeat Problem 7 for a 4 MeV neutron in carbon.

9. A parallel beam of neutrons incident on H_2O produces a 4 MeV recoil proton in the straight-ahead direction.
   (a) What is the maximum energy of the neutrons?
   (b) What is the maximum recoil energy of an oxygen nucleus?

10. (a) Estimate the average recoil energy of a carbon nucleus scattered elastically by 1 MeV neutrons.
    (b) What is the average recoil energy of a hydrogen nucleus?
    (c) Discuss the relative importance of these two reactions as a basis for producing biological effects in soft tissue exposed to 1 MeV neutrons.

11. Derive Eq. (8.5).

12. Why does the photon in the reaction (8.6) get the energy $Q = 2.22$ MeV, while the deuteron gets negligible energy?

13. Estimate the capture cross section for 19 eV neutrons by ^{235}U.

14. (a) Estimate the capture cross section of ^{10}B for 100 eV neutrons.
    (b) What is the capture probability per cm for a 100 eV neutron in pure ^{10}B?
    (c) Estimate the probability that a 100 eV neutron will penetrate a 1 cm ^{10}B shield and produce a fission in a 1 mm ^{239}Pu foil (density 18.5 g/cm^3) behind it. Neglect energy loss of the neutron due to elastic scattering.

15. The reaction $^{10}_5B(n, \alpha)^7_3Li$ leaves the ^7_3Li nucleus in the ground state 7% of the time. Otherwise, the reaction leaves the ^7_3Li nucleus in an excited state, from which it decays to the ground state by emission of a 0.48 MeV gamma ray. For ^{10}B, $\Delta = 12.052$ MeV, for ^7Li, $\Delta = 14.907$ MeV.
    (a) Calculate the Q value of the reaction in both cases.
    (b) Calculate the alpha-particle energy in both cases.

16. How much energy is released per minute by the $^{14}_7N(n,p)^{14}_6C$ reaction in a 10 g sample of soft tissue bombarded by 2×10^{10} thermal neutrons/(cm^2 sec)? Nitrogen atoms constitute 3% of the mass of soft tissue.

17. A metal sample to be analyzed for its cobalt content is exposed to thermal neutrons at a constant fluence rate of 7.20×10^{10} cm^{-2} sec^{-1} for 10 days. The thermal-neutron absorption cross section for ^{59}Co (100% abundant) to form ^{60}Co is 37 barns. If, after the irradiation, the sample shows a count rate of 23 counts/min from ^{60}Co, how many grams of cobalt are present?

18. What is the saturation activity of ^{24}Na that can be induced in a 400 g sample of NaCl with a constant thermal-neutron fluence rate of 5×10^{10} cm^{-2} sec^{-1}? The isotope ^{23}Na is 100% abundant and has a thermal-neutron capture cross section of 0.53 barns.

19. A sample containing 62 g of ^{31}P (100% abundant) is exposed to 2×10^{11} thermal neutrons cm^{-2} sec^{-1}. If the thermal-neutron capture cross section is 0.19 barns, how much irradiation time is required to make a 1 Ci source of ^{32}P?

20. Estimate the fission-product activity 48 hr following a criticality accident in which there were 5×10^{15} fissions.

21. If the exposure rate at a given location in Problem 20 is 5 R/hr 10 hr after the accident, what will it be there exactly 1 wk after the accident?

22. The "seven–ten" rule for early fallout from a nuclear explosion states that, for every sevenfold increase in time after the explosion, the exposure rate decreases by a factor of 10. Using this rule and the exposure rate at 1 hr as a reference value, estimate the relative exposure rates at 7, 7×7, and $7 \times 7 \times 7$ hr. Compare with ones obtained with the help of Eq. (8.14).

23. A reactor goes critical for the first time, operates at a power level of 50 W for 3 hr, and is then shut down. How many fissions occurred?

24. What is the fission-product inventory in curies for the reactor in the last problem 8 hr after shutdown?

25. Calculate η for uranium enriched to 3% in ^{235}U. The thermal-neutron fission cross section of ^{235}U is 580 barns and the thermal-neutron absorption cross section for ^{238}U is 2.8 barns.

26. If $k = 1.0012$, by what factor will the neutron population be increased after 10 generations?

27. How many generations are needed in the last problem to increase the power by a factor of 1000?

CHAPTER 9
METHODS OF RADIATION DETECTION

This chapter describes ways in which ionizing radiation can be detected and measured. Section 9.5 covers special methods applied to neutrons. In Chapter 10 we shall see how these techniques are applied in radiation dosimetry.

9.1 IONIZATION IN GASES

Ionization Current

Figure 9.1(a) illustrates a uniform, parallel beam of monoenergetic charged particles that steadily enter a gas chamber across an area A with energy E and come to rest in the chamber. A potential difference V applied across the parallel chamber plates P_1 and P_2 gives rise to a uniform electric field between them. As the particles slow down in the chamber, they ionize gas atoms by ejecting electrons and leaving positive ions behind. The ejected electrons can immediately produce additional ion pairs. If the electric field strength, which is proportional to V, is relatively weak, then only a few of the total ion pairs will drift apart under its influence, and a small current I will flow in the circuit. Most of the other ion pairs will recombine to form neutral gas atoms. As shown in Fig. 9.1(b), the current I can be increased by increasing V up to a value V_0, at which the field becomes strong enough to collect all of the ion pairs produced by the incident radiation and its secondary electrons. Thereafter, the current remains on a plateau at its saturation value I_0 when $V > V_0$.

Since it is readily measurable, it is important to see what information the saturation current gives about the radiation. The intensity, or energy fluence rate, of a beam of radiation is the energy that crosses a unit area normal to the beam per unit time. If the fluence rate is φ particles/(cm^2 sec), then the intensity $\mathfrak{I}$ of the radiation entering the chamber is given by $\mathfrak{I} = \varphi E$. If W denotes the average energy needed to produce an ion pair when a particle of initial energy E stops in the chamber, then the average number N of ion pairs produced by an incident particle and its secondary electrons is $N = E/W$. The average charge (either + or −) produced per particle is Ne, where e is the magnitude of the electronic charge. The saturation current I_0 in the circuit is equal to the product of Ne and φA, the total number of particles that enter the chamber per unit time. Therefore, we have

$$I_0 = Ne\varphi A = \frac{e\varphi AE}{W}.$$

(9.1)

It follows that

$$\mathfrak{I} = \varphi E = \frac{I_0 W}{eA},$$

(9.2)

showing that the beam intensity is proportional to the saturation current.

The important relationship (9.2) is of limited use, because it applies to a uniform, par-

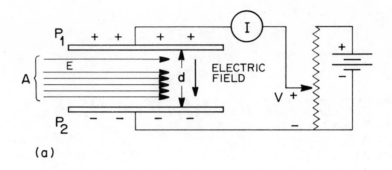

(a)

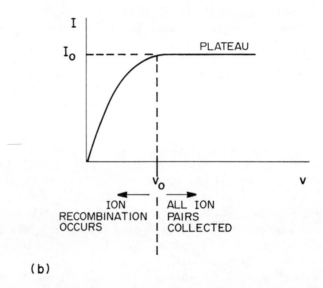

(b)

Figure 9.1. (a) Monoenergetic beam of particles stopping in parallel-plate ionization chamber with variable potential difference V applied across plates P_1 and P_2 (seen edge on). (b) Plot of current I vs. V.

allel beam of radiation. However, since the rate of total energy absorption in the chamber gas, $\dot{E}_{abs}$, is given by $\dot{E}_{abs} = \mathcal{I}A$, we may write in place of Eq. (9.2)

$$\dot{E}_{abs} = \frac{I_0 W}{e}. \tag{9.3}$$

Thus, the saturation current gives a direct measure of the rate of energy absorption in the gas. The relationship (9.3) holds independently of any particular condition on beam geometry, and is therefore of great practical utility. Fortunately, it is the energy absorbed in a biological system that is relevant for dosimetry; the radiation intensity itself is of secondary importance.

Example
Good electrometers measure currents as small as 10^{-16} amp. What is the corresponding rate of energy absorption in a parallel-plate ionization chamber containing a gas for which $W = 30$ eV/ip?

Solution
From Appendix B, 1 A = 1 C/sec. Equation (9.3) gives

$$\dot{E}_{abs} = \frac{(10^{-16} \text{ C/sec}) \times 30 \text{ eV/ip}}{1.6 \times 10^{-19} \text{ C/ip}} = 1.88 \times 10^4 \text{ eV/sec.} \qquad (9.4)$$

Ionization measurements are very sensitive. This average current would be produced, for example, by a single 18.8 keV beta particle stopping in the chamber per second.

W Values

Figure 9.2 shows W values for protons (H), alpha particles (He), and carbon and nitrogen ions of various energies in nitrogen gas, N_2. The values represent the average energy expended per ion pair when a particle of initial energy E and all of the secondary electrons it produces stop in the gas. The value for electrons, $W_\beta = 34.6$ eV/ip, shown by the horizontal line, is about the same as that for protons at energies $E > 10$ keV. W values for heavy ions, which are constant at high energies, increase with decreasing energy because a larger fraction of energy loss results in excitation rather than ionization of the gas. Elastic scattering of the ions by nuclei also causes a large increase at low energies.

The data in Fig. 9.2 indicate that W values for a given type of charged particle are approximately independent of its initial energy, unless that energy is small. This fact is of great practical significance, since it often enables absorbed energy to be inferred from measurement of the charge collected, independently of the energy spectrum of the inci-

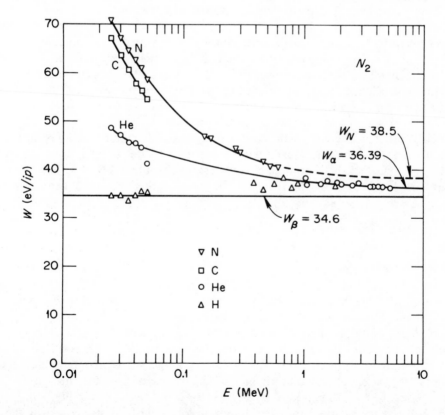

Figure 9.2. *W*-values for electrons, protons, alpha particles, carbon ions, and nitrogen ions in nitrogen gas as a function of particle energy *E*. The points represent experimental data, through which the curves are drawn. (Courtesy Oak Ridge National Laboratory, operated by Martin Marietta Energy Systems, Inc., for the Department of Energy)

Table 9.1. W Values, W_α and W_β, for Alpha and Beta Particles in Several Gases

Gas	W_α (eV/ip)	W_β (eV/ip)	W_α/W_β
He	43	42	1.02
H_2	36	36	1.00
O_2	33	31	1.06
CO_2	36	33	1.09
CH_4	29	27	1.07
C_2H_4	28	26	1.08
Air	36	34	1.06

dent particles. Alternatively, the rate of energy absorption can be inferred from measurement of the current.

W values for many polyatomic gases are in the range 25–35 eV/ip. Table 9.1 gives some values for alpha and beta particles in a number of gases.

Ionization Pulses

In addition to measuring absorbed energy, a parallel-plate ionization chamber operated in the plateau region [Fig. 9.1(b)] can also be used to count particles. When a charged particle enters the chamber, the potential difference across the plates momentarily drops slightly while the ions are being collected. After collection, the potential difference returns to its original value. The electrical pulse that occurs during ion collection can be amplified and recorded electronically to register the particle. Furthermore, if the particle stops in the chamber, then since the number of ion pairs is proportional to its original energy the size of each pulse can be used to determine the energy spectrum. While such measurements can, in principle, be carried out, pulse ionization chambers are of limited practical use because of the attendent electronic noise.

The noise problem is greatly reduced in a proportional counter. Such a counter utilizes a gas enclosed in a tube often made with a fine wire anode running along the axis of a conducting cylindrical-shell cathode, as shown schematically in Fig. 9.3. The electric field strength at a distance r from the center of the anode in this cylindrical geometry is given by

$$\mathcal{E}(r) = \frac{V}{r \ln(b/a)}, \tag{9.5}$$

where V is the potential difference between the central anode and the cylinder wall, b is the radius of the cylinder, and a is the radius of the anode wire. With this arrangement, very large field strengths are possible when a is small in the region near the anode, where r is also small. This fact is utilized as follows.

Consider the pulse produced by an alpha particle that stops in the counter gas. When the applied voltage is low, the tube operates like an ionization chamber. The number of ion pairs collected, or pulse height, is small if the voltage is low enough to allow recombination. As the potential difference is increased, the size of the pulse increases and then levels off over the plateau region, typically up to ~200 V, as shown in Fig. 9.4. When the potential difference is raised to a few hundred volts, the field strength near the anode increases to the point where electrons produced by the alpha particle and its secondary electrons acquire enough energy there to ionize additional gas atoms. Gas multiplication then occurs, and the number of ions collected in the pulse is proportional to the original

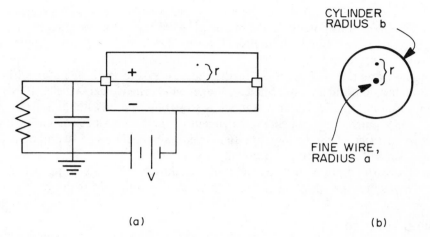

Figure 9.3. (a) Schematic side view and (b) end view of cylindrical proportional-counter tube. Variation of electric field strength with distance r from center of anode along cylindrical axis is given by Eq. (9.5).

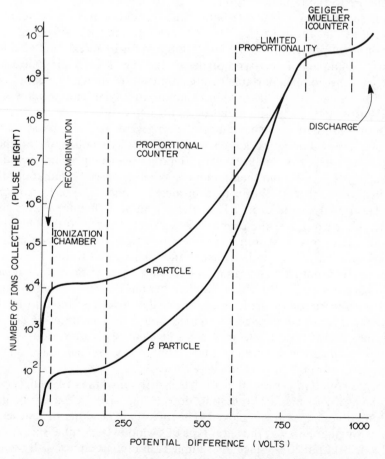

Figure 9.4. Regions of operation of gas-filled cylindrical ionization chamber operated in pulse mode.

number produced by the alpha particle and its secondaries. The tube operates as a proportional counter up to potential differences of ~700 V and can be used to measure the energy spectrum of individual alpha particles stopping in the gas. Gas multiplication factors of ~10^4 are typical.

When the potential difference is further increased the tube operates with limited proportionality, and then, at still higher voltage, enters the Geiger–Mueller (GM) region. In the latter mode, the field near the anode is so strong that any initial ionization of the gas results in a pulse, the size being independent of the number of initial ion pairs. With still further increase of the voltage, the field eventually becomes so strong that it ionizes gas atoms directly and the tube continually discharges.

Compared with an alpha particle, the pulse-height curve for a beta particle is similar, but smaller, as Fig. 9.4 shows. The two curves merge in the GM region.

Gas-Filled Detectors

Most ionization chambers for radiation monitoring are air-filled and unsealed, although sealed types that employ air or other gases are common. Used principally to monitor beta, gamma, and X-radiation, their sensitivity depends on the volume and pressure of the gas and on the associated electronic readout components. The chamber walls are usually air equivalent or tissue equivalent in terms of the secondary-electron spectra they produce in response to the radiation.

Ionization chambers are available both as active and as passive detectors. An active detector, such as that illustrated by Fig. 9.1, gives an immediate reading in a radiation field through direct processing of the ionization current in an external circuit coupled to the chamber. Examples of this type of device include the free-air ionization chamber (Sect. 10.3) and the popular portable beta-gamma survey meter, the cutie pie (Fig. 9.5). The latter instrument often has a thin, aluminized Mylar end window to allow the entrance of soft beta rays and alpha particles for detection.

Passive ionization detectors sometimes require several steps to assess radiation exposure. The familiar cylindrical pocket chamber is basically a condenser of known capacitance C. For use, it is first connected to a charger-reader at a potential difference V, giving it a charge $Q = CV$, and then removed. When exposed to radiation, the ion pairs produced in the chamber gas begin to neutralize the charge. The loss of charge ΔQ is directly proportional to the drop ΔV in the potential difference across the chamber: $\Delta Q = C\Delta V$. After exposure, the chamber is connected again to the charger-reader for evaluation of the exposure. Another type of passive pocket ionization chamber utilizes two quartz fibers that are given like charge and then viewed through an eyepiece. They separate from one another along a calibrated scale by a distance that depends on the amount of charge they still bear. After the chamber has been in a radiation field, observation of the new separation of the fibers on the scale shows the amount of exposure. Figure 9.6 shows a direct-reading pocket chamber. Still other condenser-type devices include the R chambers, which are made with air-equivalent walls of various thicknesses and compositions to measure accurately the output of X-ray machines and radioactive sources over a wide range of photon energies.

Proportional counters can be used to detect different kinds of radiation, and under suitable conditions, to measure radiation dose (Chap. 10). A variety of gas mixtures, pressures, and tube shapes are employed. The tubes may be either of a sealed or gas-flow type. As illustrated in Fig. 9.7, the latter type of "windowless" tube is useful for counting alpha and soft beta particles, since the sample is in direct contact with the counter gas.

Pulse-height discrimination with proportional counters affords an easy means for

Figure 9.5. Portable ionization-chamber survey meter (cutie pie). (Courtesy Victoreen, Inc.)

detecting one kind of radiation in the presence of another. For example, to count a combined alpha–beta source with an arrangement like that in Fig. 9.7(a) or (b), one sets the discriminator level so that only pulses above a certain size are registered. One then measures the count rate at different operating voltages of the tube, leaving the discriminator level set. The resulting count rate from the alpha–beta source will have the general characteristics shown in Fig. 9.8. At low voltages, only the most energetic alpha particles will produce pulses large enough to be counted. Increasing the potential difference causes the count rate to reach a plateau when essentially all of the alpha particles are being counted. With a further increase in voltage, increased gas multiplication enables pulses from the

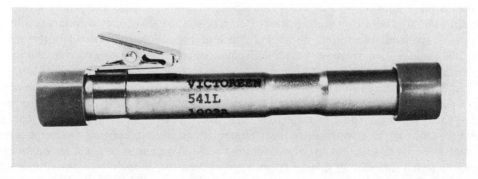

Figure 9.6. Condenser-type pocket ionization chamber. Exposure can be read directly through eyepiece in end. (Courtesy Victoreen, Inc.)

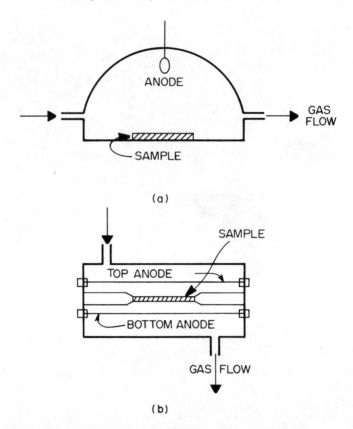

Figure 9.7. Diagram of (a) 2π and (b) 4π gas-flow proportional counters.

beta particles to surpass the discriminator level and be counted. At still higher voltages, a steeper combined alpha–beta plateau is reached. The use of proportional counters for neutron measurements is described in Sections 9.5.1 and 9.5.2. Gamma-ray discrimination is used to advantage in monitoring for neutrons in mixed gamma-neutron fields. The charged recoil nuclei from which the neutrons scatter generally produce large pulses compared to those from the Compton electrons and photoelectrons produced by the photons.

Figure 9.9 shows a portable Geiger–Mueller survey meter, which can also be used with a variety of probes, depending on the application. For beta–gamma surveys, cylindrical tubes, often equipped with a removable shield over a thin mica end window to permit detection of soft beta rays, can be used. Pancake-shaped GM tubes with large, thin mica windows are available for monitoring alpha and beta radiation. GM counters are frequently made to give an audible click with each count. They are very convenient and reliable radiation monitors.

Ideally, after the primary discharge in a GM tube, the positive ions from the counter gas drift to the cathode wall, where they are neutralized. Because of the high potential difference, however, some positive ions can strike the cathode with sufficient energy to release secondary electrons. Since these electrons can initiate another discharge, leading to multiple pulses, some means of quenching the discharge must be used. By one method, called external quenching, a large resistance between the anode and high-voltage supply reduces the potential difference after each pulse. This method has the disadvantage of making the tube slow ($\sim 10^{-3}$ sec) in returning to its original voltage. Internal quenching

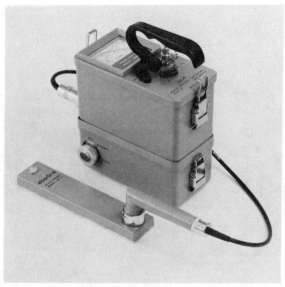

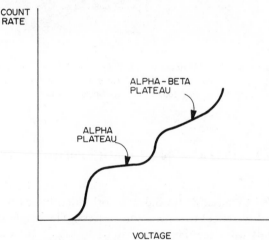

Figure 9.8. (Top) Portable gas-flow proportional counter with alpha probe. (Courtesy Eberline Instrument Corp.) (Bottom) Count rate vs. operating voltage for a proportional counter used with discriminator for counting mixed alpha–beta sources.

of a GM tube by addition of an appropriate gas is more common. The quenching gas is chosen with a lower ionization potential and a more complex molecular structure than the counter gas. When a positive ion of the counter gas collides with a molecule of the quenching gas, the latter, because of its lower ionization potential, can transfer an electron to the counter gas, thereby neutralizing it. Positive ions of the quenching gas, reaching the cathode wall, spend their energy in dissociating rather than producing secondary electrons. A number of organic molecules (e.g., ethyl alcohol) are suitable for internal quenching. Since the molecules are consumed by the dissociation process, organically quenched GM tubes have limited lifetimes ($\sim 10^9$ counts). Alternatively, the halogens chlorine and bromine are used for quenching. Although they dissociate, they later recombine. Halogen-quenched GM tubes are often preferred for extended use, although other factors limit their lifetimes.

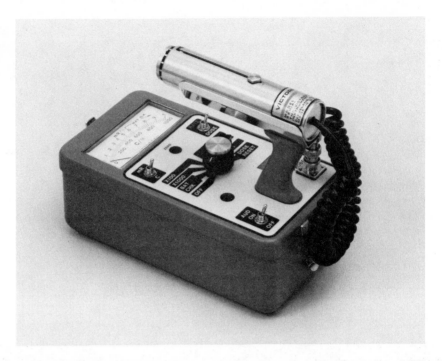

Figure 9.9. Portable survey meter with GM-tube probe. Instrument can be used with a variety of GM and scintillation probes to survey alpha, beta, gamma, and X-rays. Readout in counts per minute or mR/hour at ^{137}Cs energy. Switch on right allows integrated readout (counts or mR). Speaker can be turned on to give audible response. (Courtesy Victoreen, Inc.)

9.2 IONIZATION IN SEMICONDUCTORS

Band Theory of Solids

Solids are characterized by having allowed and forbidden bands of energy values which the most loosely bound (valence) electrons can possess. The origin of the band structure of energy levels in solids was described in Section 2.9 (Fig. 2.7). As shown schematically in Fig. 9.10, electrons can occupy continuum states with energies between 0 and E_0 in a valence band or with higher energies ($E > E_0 + E_G$) in a conduction band. There are no states with intermediate energies in a forbidden gap of width E_G.

The electrical properties of solids can be understood on the basis of the occupation of bands by electrons. If the valence band is full and the forbidden gap is about 5 eV or more wide, then the material is an insulator. No current will flow in response to an applied electric field. The ionic solid NaCl has full electronic shells and is an insulator. On the other hand, if there are vacancies in the valence band, then electrons can move freely under an applied field and the material is a conductor. The valence shell of solid Na, for example, is only half occupied by the single 3s electron per atom, and Na conducts electricity. Among the covalent solids, the width of the forbidden gap in diamond is 5.4 eV, making it an insulator. In Si the gap is 1.14 eV and in Ge, only 0.67 eV. Although the valence bands of these two metals are completely filled at low temperatures, making them insulators, a small fraction of their electrons are thermally excited into the conduction band at room temperatures ($kT \sim 0.025$ eV), giving them some conducting ability. Covalent solid insulators having an energy gap $E_G \sim 1$ eV are called intrinsic semiconductors.

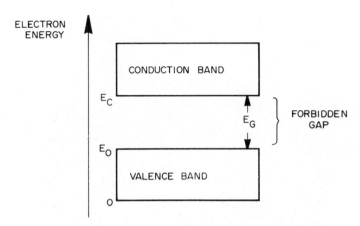

Figure 9.10. Band structure of solids. Electron states have continua of energy values in the valence and conduction bands. No states exist with energies in the range $E_G = E_C - E_O$ between the top of the valence band and the bottom of the conduction band.

These features can be analyzed quantitatively with the help of the electron-gas model mentioned in Section 2.9 in connection with metallic solids. In statistical mechanics it is shown that the average number of electrons per quantum state of energy E in the electron gas is given by

$$N = \frac{1}{e^{(E-E_F)/kT} + 1},\tag{9.6}$$

where k is the Boltzmann constant, T is the absolute temperature, and E_F denotes the Fermi energy. At absolute zero, all of the electronic states with energy E up to the Fermi energy are occupied with single electrons ($N = 1$) and no states with $E > E_F$ are occupied ($N = 0$). Since the Pauli exclusion principle prevents more than one electron from being in a given state, this configuration has the lowest energy possible, as expected at absolute zero. At temperatures $T > 0$ the Fermi energy is the energy at which the average number of electrons per quantum state is exactly $\frac{1}{2}$.

It is instructive to diagram the relative number of free electrons as a function of energy in various types of solids at different temperatures. Figure 9.11(a) shows the energy distribution of electrons in the conduction band of a conductor at a temperature above absolute zero ($T > 0$). Electrons occupy states with a thermal distribution of energies above E_C. The lower energy levels are filled, but unoccupied states are available for conduction near the top of the band. A diagram for the same conductor at $T = 0$ is shown in Fig. 9.11(b). All levels with $E < E_F$ are occupied and all with $E > E_F$ are unoccupied.

Figure 9.12(a) shows the electron energy distribution in an insulator with $T > 0$. The valence band is full and the forbidden-gap energy E_G (~5 eV) is so wide that the electrons cannot reach the conduction band at ordinary temperatures. Figure 9.12(b) shows a semiconductor in which $E_G \sim 1$ eV is considerably narrower than in the insulator. The tail of the thermal distribution [Eq. (9.6)] permits a relatively small number of electrons to have energies in the conduction band. In this case, the number of occupied states in the conduction band is equal to the number of vacant states in the valence band and E_F lies midway in the forbidden gap at an energy $E_0 + E_G/2$. At room temperature, the density of electrons in the conduction band is 1.5×10^{10} cm^{-3} in Si and 2.4×10^{13} cm^{-3} in Ge.

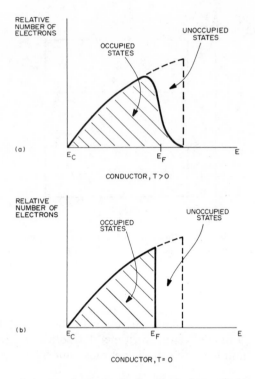

Figure 9.11. Relative number of electrons in the conduction band of a conductor at absolute temperatures (a) $T > 0$ and (b) $T = 0$.

Semiconductors

Figure 9.13 schematically represents the occupation of energy states for the semiconductor from Fig. 9.12(b) with $T > 0$. The relatively small number of electrons in the conduction band are denoted with minus signs and the equal number of positive ions they leave in the valence band with plus signs. The combination of two charges is called an electron–hole pair, roughly analogous to an ion pair in a gas. Under the influence of an applied electric field, electrons in the conduction band will move. In addition, electrons in the valence band move to fill the holes, leaving other holes in their place, which in turn are filled by other electrons, etc. This, in effect, causes the holes to migrate in the direction opposite to that of the electrons. The motions of both the conduction-band electrons and the valence-band holes contribute to the observed conductivity. The diagram in Fig. 9.13 represents an intrinsic (pure) semiconductor. Its inherent conductivity at room temperature is restricted by the small number of electron–hole pairs, which, in turn, is limited by the size of the gap compared with kT.

The conductivity of a semiconductor can be greatly enhanced by doping the crystal with atoms from a neighboring group in the periodic system. As an example, we consider the addition of a small amount of arsenic to germanium.† When a crystal is formed from the molten mixture, the arsenic impurity occupies a substitutional position in the germanium lattice, as indicated schematically in Fig. 9.14. (The As atom has a radius of 1.39 Å

---

†The reader is referred to the Periodic Table in the back of the book.

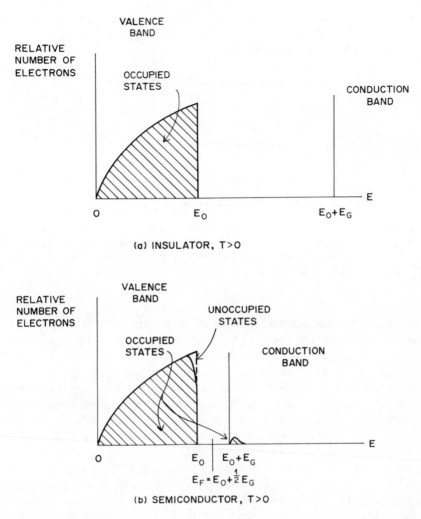

Figure 9.12. Relative number of electrons in valence and conduction bands of (a) an insulator and (b) a semiconductor with $T > 0$.

compared with 1.37 Å for Ge). Since As has five valence electrons, there is one electron left over after all of the eight covalent bonds have been formed with the neighboring Ge atoms. In Fig. 9.14, two short straight lines are used to represent a pair of electrons shared covalently by neighboring atoms and the loop represents, very schematically, the orbit of the extra electron contributed by As^+, which is in the crystal lattice. There is no state for the extra electron to occupy in the filled valence band. Since it is only very loosely bound to the As^+ ion (its orbit can extend over several tens of atomic diameters) this electron has a high probability of being thermally excited into the conduction band at room temperature. The conductivity of the doped semiconductor is thus greatly increased over its value as an intrinsic semiconductor. The amount of increase can be controlled by regulating the amount of arsenic added, which can be as little as a few parts per million. An impurity such as As that contributes extra electrons is called a donor and the resulting semiconductor is called n-type (negative).

Since little energy is needed to excite the extra electrons of an n-type semiconductor

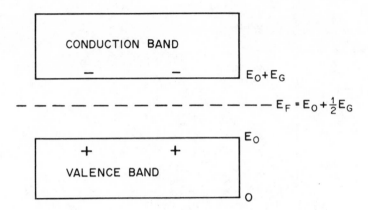

Figure 9.13. Occupation of energy states in an intrinsic semiconductor at room temperature. A relatively small number of electrons ($-$) are thermally excited into the conduction band, leaving an equal number of holes ($+$) in the valence band. Movement of both electrons and holes (in opposite directions) contributes to conductivity when an electric field is applied. The Fermi energy E_F lies at the middle of the forbidden gap.

into the conduction band, the energy levels of the donor impurity atoms must lie in the forbidden gap just below the bottom of the conduction band. The energy-level diagram for Ge doped with As is shown in Fig. 9.15. The donor states are found to lie 0.013 eV below the bottom of the conduction band, as compared with the total gap energy, $E_G = 0.67$ eV, for Ge. At absolute zero all of the donor states are occupied and no electrons are in the conduction band. The Fermi level lies between the donor levels and the bottom of the conduction band. As T is increased, thermally excited electrons enter the conduction band from the donor states, greatly increasing the conductivity. Antimony can also be used as a donor impurity in Ge or Si to make an n-type semiconductor.

Another type of semiconductor is formed when Ge or Si is doped with gallium or indium, which occur in the adjacent column to their left in the periodic system. In this case, the valence shell of the interposed impurity atom has one less electron than the number needed to form the regular covalent crystal. Thus, the doped crystal contains positively charged "holes," which can accept electrons. The dopant is then called an acceptor impurity and the resulting semiconductor, p-type (positive). Holes in the valence band move like positive charges as electrons from neighboring atoms fill them. Because of the ease with which valence-band electrons can move and leave holes with the impurity present, the effect of the acceptor impurity is to introduce electron energy levels in the forbidden gap slightly above the top of the valence band. Figure 9.16 shows the energy-level diagram for a p-type Ge semiconductor with Ga acceptor atoms added. The action of the p-type semiconductor is analogous to that of the n-type. At absolute zero all of the electrons are in the valence band; the Fermi level lying just above. When $T > 0$, thermally excited electrons occupy acceptor-level states, giving enhanced conductivity to the doped crystal.

Semiconductor Junctions

The usefulness of semiconductors as electronic circuit elements and for radiation measurements stems from the special properties created at a junction where n- and p-type semiconductors are brought into good thermodynamic contact. Figure 9.17 shows an electron energy-level diagram for an n–p junction. The two semiconductor types in contact form a single system with its own characteristic Fermi energy E_F. Because E_F lies

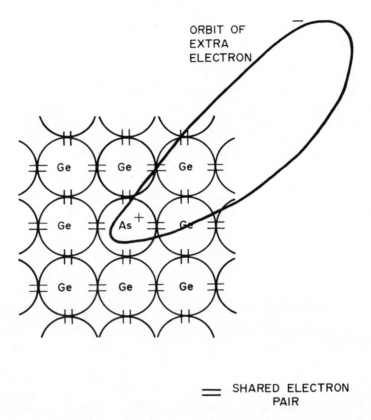

ORBIT OF
EXTRA
ELECTRON

═══ SHARED ELECTRON
PAIR

Figure 9.14. Addition of small quantity of pentavalent As to Ge crystal lattice provides very loosely bound "extra" electrons that have a high probability of being thermally excited into the conduction band at room temperatures. Arsenic is called a donor impurity and the resulting semiconductor, n-type.

just below the conduction band in the isolated n region and just above the valence band in the isolated p region, the bands must become deformed over the junction region, as shown in the figure. When the n- and p-type semiconductors are initially brought into contact, electrons flow from the donor impurity levels on the n side over to the lower-energy acceptor sites on the p side. This process accumulates negative charge on the p side of the junction region and leaves behind immobile positive charges on the n side in the form of ionized donor impurity atoms. The net effect at equilibrium is the separation of charge across the junction region (as indicated by the + and − symbols in Fig. 9.17) and the maintenance of the deformed bands.

The junction region over which the charge imbalance occurs is also called the depletion region, because any mobile charges initially there moved out when the two sides were joined. The depletion region acts, therefore, like a high-resistivity parallel-plate ionization chamber, making it feasible to use it for radiation detection. Ion pairs produced there will migrate out, their motion giving rise to an electrical signal. The performance of such a device is greatly improved by using a bias voltage to alleviate recombination and noise problems. The biased junction becomes a good rectifier, as described below.

Consider the n–p junction device in Fig. 9.18(a) with the negative side of an external bias voltage V applied to the n side. When compared with Fig. 9.17, it is seen that the applied voltage in this direction lowers the potential difference across the junction region

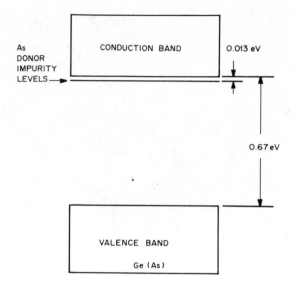

Figure 9.15. Energy-level diagram for Ge crystal containing As donor atoms (n-type semiconductor).

and causes a relatively large current *I* to flow in the circuit. Bias in this direction is called forward, and a typical current–voltage curve is shown at the right in Fig. 9.18(a). One obtains a relatively large current with a small bias voltage. When a reverse bias is applied in Fig. 9.18(b), comparison with Fig. 9.17 shows that the potential difference across the junction region increases. Therefore, a much smaller current flows—and in the opposite direction—under reverse bias, as illustrated on the right in Fig. 9.18(b). Note the vastly different voltage and current scales on the two curves in the figure. Such n–p junction devices are rectifiers, passing current readily in one direction but not the other.

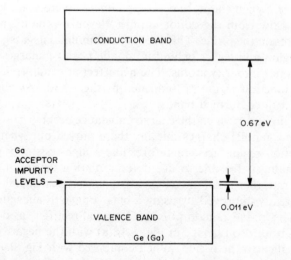

Figure 9.16. Energy-level diagram for Ge crystal containing Ga acceptor atoms (p-type semiconductor).

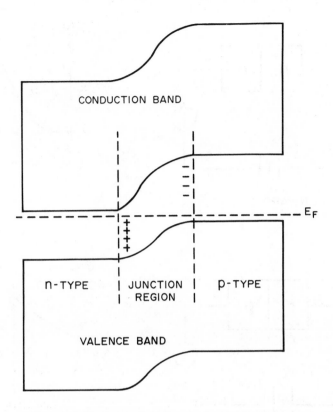

Figure 9.17. Energy-level diagram for n–p junction.

Radiation Measuring Devices

The reverse-biased n–p junction constitutes an attractive radiation detector. The depletion region, which is the active volume, has high resistivity, and ions produced there by radiation can be collected swiftly and efficiently. Like a gaseous ionization chamber, the number of ions is proportional to the energy deposited by a particle in the active volume. Therefore, the junction can be used as a spectrometer. The W values for Si and Ge are, respectively, 3.7 and 3.0 eV/ip. Unique among radiation detectors, the physical size of the depletion region can be varied by changing the bias voltage. For radiation measurements, especially with α and β rays, it is necessary to have the depletion region located near the surface of the device, so that the radiation can reach it with minimal energy loss in the intervening material. We describe several important diode-junction configurations for radiation applications.

Surface-barrier detectors can be made by starting with an n-type crystal and using one of several techniques to produce a high density of electron traps on the surface to act as the p side of the junction. One such barrier is made by etching the surface of an n-type Si crystal with acid, allowing a SiO_2 layer to form, and then evaporating a thin gold layer onto it for good electrical contact. Another type of surface barrier is made by evaporating aluminum onto the surface of a p-type crystal, forming an n-type contact. Figure 9.19 shows an arrangement for utilizing a surface-barrier detector.

Surface-barrier detectors are characterized by having very thin dead layers, e.g., ~0.05 mm. The oxide film and gold layer can also be made thin enough to allow α particles to

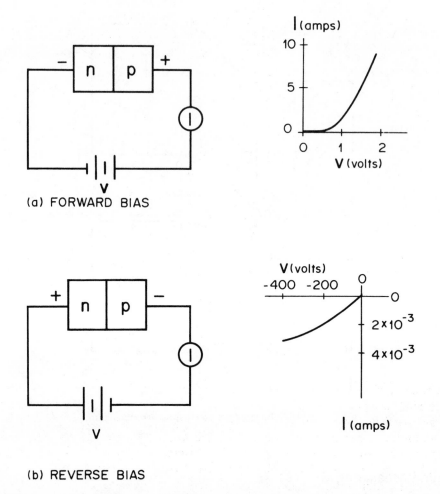

Figure 9.18. (a) Forward- and (b) reverse-biased n–p junctions and typical curves of current vs. voltage. Note the very different scales used for the two curves. Such an n–p junction is a good rectifier.

enter the sensitive volume. The detectors are used for spectroscopy with heavy charged particles, including fission fragments. They also can be employed for beta rays, although other semiconductor detectors are more common for this purpose.

Ion implantation is another method used to introduce impurities. Ions from an accelerator are directed at the semiconductor and the energy is varied to obtain the desired doped region. Very thin entrance layers result.

Diffused-junction detectors were used in the past. An n–p junction can be created by allowing the vapor of an n-type impurity (e.g., phosphorus) to diffuse into the surface of a p-type semiconductor.

The *particle identifier* utilizes two detectors in the configuration shown in Fig. 9.20. A heavy particle passes through a thin detector (1) and is stopped in a thick detector (2). The pulse height from 1 is proportional to the stopping power, $-dE/dx$, and that from 2 is proportional to the kinetic energy, $E = Mv^2/2$, with which the particle enters it, where M is the particle's mass and v is its velocity. The signals from 1 and 2 can be combined electronically in coincidence to form the product $E(-dE/dx)$. Since the ln term in the stopping-power formula [Eq. (4.7)] varies slowly with the particle energy E, it follows that

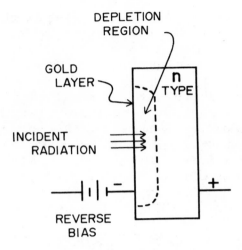

Figure 9.19. Example of surface-barrier detector.

$$E\left(-\frac{dE}{dx}\right) = kz^2M, \qquad (9.7)$$

approximately, where z is the particle's charge and k is a constant of proportionality. The quantity z^2M, which is thus determined by the measurement, is characteristic of a particular heavy particle, which can then be identified.

The major limitation of the detectors discussed so far is the restricted size of the depletion region that can be achieved. Thicknesses no greater than a few mm are obtained. While adequate for heavy charged particles and other short-range radiations, much greater active volumes are required for gamma-ray spectroscopy.

Lithium-drifted detectors, having large sensitive volumes ($\gtrsim 20$ mm), were in widespread use until the early 1980s. The large volumes can be obtained by drawing lithium

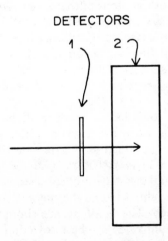

Figure 9.20. Particle identifer.

into a Ge or Si crystal under controlled conditions. With the lithium-drifted germanium [or Ge(Li)] detector, the lithium-ion mobility is so great that the crystals have to be maintained at low temperatures to preserve their properties. This is usually accomplished by keeping them at the temperature of liquid nitrogen ($-196°C$). Cooling of lithium-drifted silicon detectors is not necessary. Ge(Li) detectors have been widely used for gamma-ray spectroscopy, especially for photon energies of several hundred keV or more. They have extremely good energy resolution. Si(Li) detectors have been used for low-energy gamma and X-rays and for beta particles. Compared with Ge(Li) detectors, their lower atomic number and smaller sensitivity to temperature are helpful in these applications.

High-purity germanium detectors (HPGe). Today, high-purity germanium (HPGe) detectors have replaced most Ge(Li) detectors. The HPGe systems are operated at nearly liquid-nitrogen temperatures to reduce noise, but can be kept at room temperatures when not in use. Figure 9.21 shows two HPGe detector systems.

9.3 SCINTILLATION

General

Scintillation was the first method used to detect ionizing radiation (Roentgen having observed the fluorescence of a screen when he discovered X-rays). When radiation loses energy in a luminescent material, called a scintillator or phosphor, it causes electronic transitions to excited states in the material. The excited states decay by emitting photons, which can be observed and related quantitatively to the action of the radiation. If the decay of the excited state is rapid (10^{-8} or 10^{-9} sec) the process is called fluorescence; if it is slower, the process is called phosphorescence.

Scintillators employed for radiation detection are usually surrounded by reflecting surfaces to trap as much light as possible. The light is fed into a photomultiplier tube for generation of an electrical signal. There a photosensitive cathode converts a fraction of the photons into photoelectrons, which are accelerated through an electric field toward another electrode, called a dynode. In striking the dynode, each electron ejects a number of secondary electrons, giving rise to electron multiplication. These secondary electrons are then accelerated through a number of additional dynode stages (e.g., 10), achieving electron multiplication in the range 10^7-10^{10}. The magnitude of the final signal is proportional to the scintillator light output, which, under the right conditions, is proportional to the energy loss that produced the scintillation.

Since materials emit and absorb photons of the same wavelength, impurities are usually added to scintillators to trap energy at levels such that the wavelength of the emitted light will not fall into a self-absorption region. Furthermore, because many substances, especially organic compounds, emit fluorescent radiation in the ultraviolet range, impurities are also added as wavelength shifters. These lead to the emission of photons of visible light, for which glass is transparent and for which the most sensitive photomultiplier tubes are available.

Good scintillator materials should have a number of characteristics. They should efficiently convert the energy deposited by a charged particle or photon into detectable light. The efficiency of a scintillator is defined as the fraction of the energy deposited that is converted into visible light. The highest efficiency, about 13%, is obtained with sodium iodide. A good scintillator should also have a linear energy response; that is, the constant of proportionality between the light yield and the energy deposited should be independent of the particle or photon energy. The luminescence should be rapid, so that pulses are generated quickly and high count rates can be resolved. The scintillator should also be transparent to its own emitted light. Finally, it should have good optical quality for cou-

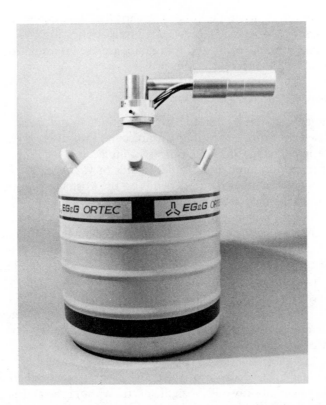

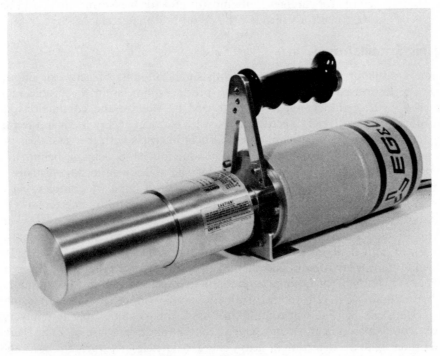

Figure 9.21. HPGe gamma-ray detector (top) mounted on dewar supplying liquid nitrogen. Portable HPGe spectrometer (bottom) can be detached from a large storage/fill dewar and used for a number of hours with its own reservoir of liquid nitrogen. (Courtesy Sanford Wagner, EG&G ORTEC)

pling to a light pipe or photomultiplier tube. The choice of a particular scintillation detector represents a balancing of these factors for a given application.

Two types of scintillators, organic and inorganic, are used in radiation detection. The luminescence mechanism is different in the two.

Organic Scintillators

Fluorescence in organic materials results from transitions in individual molecules. Incident radiation causes electronic excitations of molecules into discrete states, from which they decay by photon emission. Since the process is molecular, the same fluorescence can occur with the organic scintillator in the solid, liquid, or vapor state. Fluorescence in an inorganic scintillator, on the other hand, depends on the existence of a regular crystalline lattice, as described in the next section.

Organic scintillators are available in a variety of forms. Anthracene and stilbene are the most common organic crystalline scintillators, anthracene having the highest efficiency of any organic material. Organic scintillators can be polymerized into plastics. Liquid scintillators (e.g., xylene, toluene) are often used and are practical when large volumes are required. Radioactive samples can be dissolved or suspended in them for high-efficiency counting. Liquid scintillators are especially suited for measuring soft beta rays, such as those from ^{14}C or ^{3}H. High-Z elements (e.g., lead or tin) are sometimes added to organic scintillator materials to achieve greater photoelectric conversion, but usually at the cost of decreased efficiency.

Compared with inorganic scintillators, organic materials have much faster response, but generally yield less light. Because of their low-Z constituents, there are little or no photoelectric peaks in gamma-ray pulse-height spectra without the addition of high-Z elements. Organic scintillators are generally most useful for measuring alpha and beta rays and for detecting fast neutrons through the recoil protons produced.

Inorganic Scintillators

Inorganic scintillator crystals are made with small amounts of activator impurities to increase the fluorescence efficiency and to produce photons in the visible region. As shown in Fig. 9.22, the crystal is characterized by valence and conduction bands, as described in Section 9.2.1. The activator provides electron energy levels in the forbidden gap of the pure crystal. When a charged particle interacts with the crystal, it promotes electrons from the valence band into the conduction band, leaving behind positively charged holes. A hole can drift to an activator site and ionize it. An electron can then drop into the ionized site and form an excited neutral impurity complex, which then decays with the emission of a visible photon. Because the photon energies are less than the width of the forbidden gap, the crystal does not absorb them.

The alkali halides are good scintillators. In addition to its efficient light yield, sodium iodide doped with thallium [NaI(Tl)] is almost linear in its energy response. It can be machined into a variety of sizes and shapes. Disadvantages are that it is hygroscopic and somewhat fragile. NaI has become a standard scintillator material for gamma-ray spectroscopy. CsI(Na), CsI(Tl), and LiI(Eu) are samples of other inorganic scintillators. Silver-activated zinc sulfide is also commonly used. It is available only as a polycrystalline powder, from which thin films and screens can be made. The use of ZnS, therefore, is limited primarily to the detection of heavy charged particles. (Rutherford used ZnS detectors in his alpha-particle scattering experiments.) Glass scintillators are also widely used.

Figure 9.23 shows two examples of scintillator probes used for radiation detection. The probes contain both the scintillator material and the photomultiplier tube, ready for con-

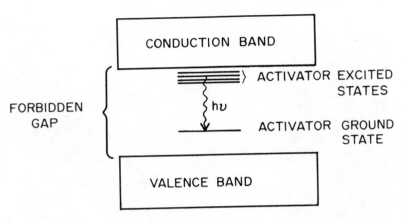

Figure 9.22. Energy-level diagram for activated crystal scintillator. Because the energy levels of the activator complex are in the forbidden gap, the crystal is transparent at the fluorescent photon energies $h\nu$.

nection to electronic amplifying and data processing equipment. The unit on the left contains a NaI(Tl) detector for gamma surveys; that on the right contains ZnS behind a Mylar window for monitoring alpha contamination.

Specialized scintillation devices have been designed for other specific purposes. One example is the phoswich (= phosphor sandwich) detector, which can be used to count beta particles or low-energy photons in the presence of high-energy photons. It consists of a thin NaI(Tl) crystal in front coupled to a larger scintillator of another material, often CsI(Tl), having a different fluorescence time. Signals that come from the photomultiplier tube can be distinguished electronically on the basis of the different decay times of the two phosphors to tell whether the light came only from the thin front crystal or from both crystals. In this way, the low-energy radiation can be counted in the presence of a high-energy gamma-ray background.

The use of several instruments for monitoring radiation emitted by isotopes inside the body is shown in Fig. 9.24. In this whole-body counter, a phoswich detector is shown in front of a HPGe detector, both placed above the chest to monitor low-energy photons, such as the X-rays and gamma rays emitted by the actinides Pu and Am. A large NaI(Tl) crystal below the table monitors photons in the range from about 200 keV to 3 MeV, such as those emitted by ^{60}Co, ^{137}Cs, and fission products.

Figure 9.25 shows a pulse-height spectrum measured with a 4 × 4 in. NaI(Tl) scintillator exposed to 662 keV gamma rays from ^{137}Cs. Several features should be noted. Only those photons that lose all of their energy in the crystal contribute to the total-energy peak, also called the photopeak. These include incident photons that produce a photoelectron directly and those that undergo one or more Compton scatterings and then produce a photoelectron. In the latter case, the light produced by the Compton recoil electrons and that produced by the final photoelectron add in producing a single scintillation pulse around 662 keV. (The light produced by Auger electrons and characteristic X-rays absorbed in the crystal is also included in the same pulse, these processes taking place rapidly). Other photons escape from the crystal after one or more Compton scatterings, and therefore do not deposit all of their energy in producing the scintillation. These events give rise to the continuous Compton distribution at lower pulse heights, as shown in the figure. The Compton edge at 478 keV is the maximum Compton-electron recoil energy, T_{max}, given by Eq. (7.20). The pulses that exceed T_{max} in magnitude come from two or more Compton recoil electrons produced by the photon before it escapes from the

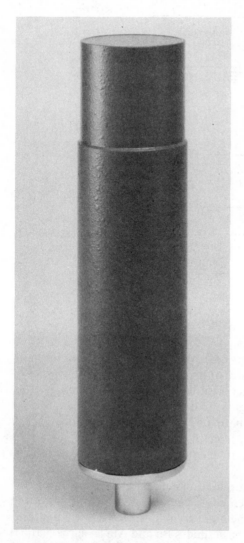
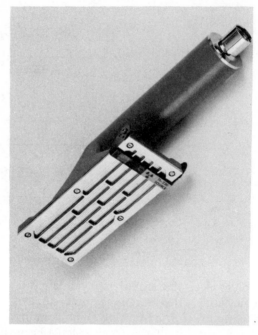

Figure 9.23. Examples of scintillator probes: left, NaI(Tl) for gamma surveys; right, ZnS for monitoring alpha contamination (courtesy Eberline Instrument Corp.). Scintillation probes can be recognized by their size and shape, because they include a photomultiplier tube.

crystal. The backscatter peak is caused by photons that are scattered into the scintillator from surrounding materials. The energy of a ^{137}Cs gamma ray that is scattered at 180° is $662 - T_{max} = 662 - 478 = 184$ keV. The backscattered radiation peaks at an energy slightly above this value, as shown.

The relative area under the total-energy peak and the Compton distribution in Fig. 9.25 depends on the size of the scintillator crystal. If the crystal is very large, then relatively few photons escape. Most of the pulses occur around 662 keV. If it is small, then only single interactions are likely and the Compton continuum is large. In fact, the ratio of the areas under the total-energy peak and the Compton distribution in a small detector is equal to the ratio of the photoelectric and Compton cross sections in the crystal material. Because of their small sizes, semiconductor detectors can have extensive Compton continua, which tend to obscure low-intensity photopeaks.

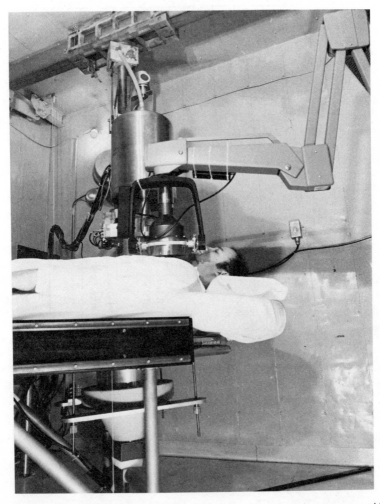

Figure 9.24. Whole-body counter, employing three detector systems to monitor photons over a wide energy range. See text. (Courtesy Carol D. Berger, Oak Ridge National Laboratory, operated by Martin Marietta Energy Systems, Inc., for the Department of Energy)

The occurrence of so-called escape peaks is accentuated when a detector is small, whether it be a scintillator or semiconductor. In NaI, for example, when a K-shell vacancy in iodine following photoelectric absorption is filled by an L-shell electron, a 28 keV characteristic X-ray is emitted. The X-ray will likely escape if the crystal is small, and a pulse size of $h\nu_0 - 28$ keV will be registered, where $h\nu_0$ is the energy of the incident photon. When the incident photons are monoenergetic, the pulse-height spectrum shows the escape peak, as illustrated in Figure 9.26. The relative size of the peaks at the two energies $h\nu_0$ and $h\nu_0 - 28$ keV depends on the physical dimensions of the NaI crystal. Germanium has an X-ray escape peak 11 keV below the photopeak. Another kind of escape peak can occur in the pulse-height spectra of monoenergetic high-energy photons of energy $h\nu_0$, which produce electron–positron pairs in the detector. The positron quickly stops and annihilates with an atomic electron, producing two 0.511 MeV photons. In small detectors one or both annihilation photons can escape without interacting, leading to escape peaks at the energies $h\nu_0 - 0.511$ and $h\nu_0 - 1.022$ MeV.

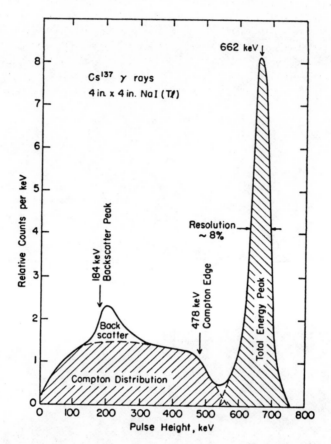

Figure 9.25. Pulse-height spectrum measured with 4×4 in. NaI (Tl) scintillator exposed to 662 keV gamma rays from ^{137}Cs. The resolution is about 8% of the peak energy. The maximum Compton-electron energy is 478 keV. [Reprinted with permission from R.D. Evans, "Gamma Rays," in *American Institute of Physics Handbook*, 3rd ed., p. 8–210, McGraw-Hill, New York (1972). Copyright 1972 by McGraw-Hill]

A comparison of the same gamma-ray pulse-height spectrum as measured with a NaI scintillation counter and with a Ge(Li) semiconductor detector is shown in Fig. 9.27. The superior energy resolution of the latter is clearly evident. The resolution of a detector depends on several factors, such as the efficiency of light or charge collection (ion recombination) and on electronic noise. It also depends on the statistical spread in the number of electrons that initiate the electrical pulse that is measured. This number, in turn, is governed by the magnitude of the average energy expended by the radiation to produce such an electron, as the following example illustrates.

Example
Monoenergetic 450 keV gamma rays are absorbed in a NaI(Tl) crystal having an efficiency of 12%. Seventy-five percent of the scintillation photons, which have an average energy of 2.8 eV, reach the cathode of a photomultiplier tube, which converts 20% of the incident photons into photoelectrons. Assume that variations in the pulse heights from different gamma photons are due entirely to statistical fluctuations in the number of visible photons per pulse that reach the cathode. (a) Calculate the average number of scintillation photons produced per absorbed gamma photon. (b) How many photoelectrons are produced, on the average, per gamma photon? (c) Estimate the relative error in the measured pulse height. (d) Interpret the relative error in terms of the energy resolution of this detector. (e) What is the average energy expended by the incident radiation to produce a

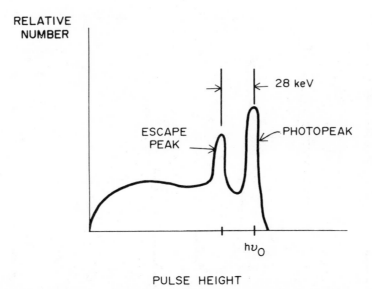

RELATIVE
NUMBER

ESCAPE
PEAK

28 keV

PHOTOPEAK

$h\nu_0$

PULSE HEIGHT

Figure 9.26. When characteristic X-rays from iodine escape from a NaI scintillator, an escape peak appears 28 keV below the photopeak.

photoelectron from the cathode of the photomultiplier tube (the "W value")? (f) What would be the relative error if the measurements were made with a Ge(Li) detector ($W = 3.0$ eV/ip)?

Solution

(a) The total energy of the visible light produced with 12% efficiency is 450 keV $\times$ 0.12 = 54.0 keV. The average number of scintillation photons is therefore 54,000/2.8 = 19,300. (b) The average number of photons that reach the photomultiplier cathode is 0.75 $\times$ 19,300 = 14,500, and so the average number of photoelectrons that produce a pulse is 0.20 $\times$ 14,500 = 2900. (c) By Eq. (E.7) from Appendix E, the relative error in this number is $1/\sqrt{2900} = 0.0186$, or 1.86%. (d) According to Table E.1, 68.3% of the pulse heights fall statistically within one standard deviation of the mean. Since 0.0186 $\times$ 450 = 8.37, we can express the resolution of the detector by saying that 68% of the area under the photopeak will lie within the interval 450 $\pm$ 8.4 keV. Resolution is also expressed in terms of the full width at half-maximum (FWHM) divided by the mean. With an assumed Gaussian shape, Eq. (E.16) gives $2.35/\sqrt{2900} = 0.0436$, and so FWHM = 19.6 keV. (Electron multiplication in the photomultiplier tube is so large that, by comparison, it introduces negligible statistical variation in the output signal.) (e) Since one 450 keV incident gamma photon produces an average of 2900 photoelectrons that initiate the signal, the "W value" for the scintillator is 450,000/2900 = 155 eV/photoelectron. (f) With the Ge(Li) detector, the number of ion pairs formed per incident photon would be 450,000/3.0 = 1.50×10^5. The relative error is $1/\sqrt{1.50 \times 10^5} = 0.00258 = 0.258\%$. Thus 68% of the pulses would lie in the interval 450 $\pm$ 1.16 keV. It follows that FWHM = 2.35 $\times$ 0.00258 $\times$ 450 = 2.73 keV. As mentioned above, although the energy resolution of a detector depends on other factors as well, the small value $W \sim 3$ eV/ip endows semiconductors with inherently better statistical resolution than other detectors. By comparison, $W \sim 30$ eV/ip for gases, and so the inherent resolution is poorer by a factor of $\sqrt{10}$. A W value of several hundred eV/photon is typical for a scintillator.

9.4 OTHER METHODS

Thermoluminescence

In the last section, in connection with Fig. 9.22, we described how ionizing radiation can produce electron–hole pairs in an inorganic crystal. These lead to the formation of

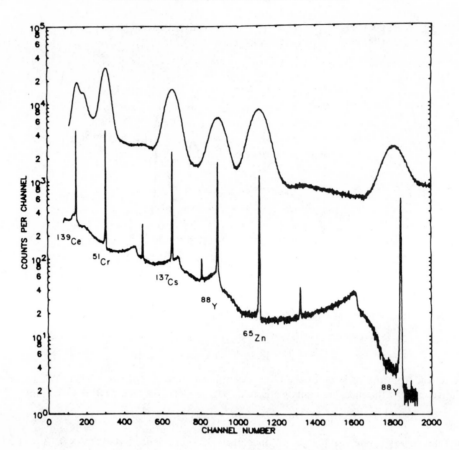

Figure 9.27. Comparison of gamma-ray spectra from a solution containing radionuclides as measured with a NaI scintillator and a Ge(Li) detector. [Reprinted with permission from *A Handbook of Radioactivity Measurements*, NCRP Report 58, p. 240, National Council on Radiation Protection and Measurements, Washington, D.C. (1978). Copyright 1978 National Council on Radiation Protection and Measurements]

excited states with energies that lie in the forbidden gap when particular added activator impurities are present. In a scintillation detector, it is desirable for the excited states to decay quickly to the ground states, so that prompt fluorescence results. In another class of inorganic crystals, called thermoluminescent dosimeters (TLDs), the crystal material and impurities are chosen so that the electrons and holes remain trapped at the activator sites at room temperature. Placed in a radiation field, a TLD crystal serves as an integrating detector, in which the number of trapped electrons and holes depends on its radiation exposure history.

After exposure, the TLD material is heated. As the temperature rises, trapped electrons and holes migrate and combine, with the accompanying emission of photons with energies of a few eV. Some of the photons enter a photomultiplier tube and produce an electronic signal. The sample is commonly processed in a TLD reader, which automatically heats the material, measures the light yield as a function of temperature, and records the information in the form of a glow curve, such as that shown in Fig. 9.28. Typically, several peaks occur as traps at different energy levels are emptied. The total light output or the area under the glow curve can be compared with that from calibrated TLDs to infer radiation dose. All traps can be emptied by heating to sufficiently high temperature, and the crystal reused.

A number of TLD materials are in use. Manganese-activated calcium sulfate, $CaSO_4$:Mn, is sensitive enough to measure doses of a few tens of μrad. Its traps are relatively shallow, however, and it has the disadvantage of "fading" significantly in 24 hr. $CaSO_4$:Dy is better. Another popular TLD crystal is LiF, which has inherent defects and impurities and needs no added activator. It exhibits negligible fading and is close to tissue in atomic composition. It can be used to measure gamma-ray doses in the range of about 0.01–1000 rad. The presence of ^{6}Li, which has a high thermal-neutron cross section, is utilized in detecting slow neutrons with LiF TLDs. Other TLD materials include CaF_2:Mn, CaF_2:Dy, and $Li_2B_4O_7$:Mn.

Particle Track Registration

A number of methods have been devised for directly observing the tracks of individual charged particles. Figure 4.1 shows an example of alpha- and beta-particle tracks in photographic film. In the cloud chamber, moisture from a supersaturated vapor condenses on the ions left in the wake of a passing charged particle, rendering the track visible. In the bubble chamber, tiny bubbles are formed as a superheated liquid starts to boil along a charged particle's track. Another device, the spark chamber, utilizes a potential difference between a stack of plates to cause a discharge along the ionized path of a charged particle that passes through the stack.

Track etching is possible in some organic polymers and in several types of glasses. A charged particle causes radiation damage along its path in the material. When treated chemically or electrochemically, the damaged sites are attacked preferentially and made visible, either with a microscope or to the unaided eye. Track etching is feasible only for particles of high LET. The technique is widely used in neutron dosimetry (e.g., CR-39 detectors). Although neutral particles do not produce a trail of ions, the tracks of the charged recoil particles they produce can be registered by techniques discussed here.

Chemical Dosimeters

Radiation produces chemical changes. One of the most widely studied chemical detection systems is the Fricke dosimeter, in which ferrous ions in a sulfate solution are oxidized by the action of radiation. As in all aqueous chemical dosimeters, radiation interacts with

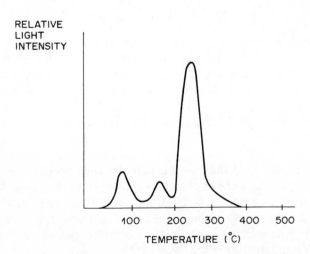

Figure 9.28. Typical TLD glow curve.

water to produce free radicals (e.g., H and OH), which are highly reactive. The OH radical, for example, can oxidize the ferrous ion directly: $Fe^{2+} + OH \rightarrow Fe^{3+} + OH^-$. After irradiation, aqueous chemical dosimeters can be analyzed by titration or light absorption. The useful range of the Fricke dosimeter is from about 4×10^3 to 4×10^4 rad. The dose measurements are accurate and absolute. The aqueous system approximates soft tissue.

Other chemical-dosimetry systems are based on ceric sulfate, oxalic acid, or a combination of ferrous sulfate and cupric sulfate. Doses of the order of 10 rad can be measured chemically with some chlorinated hydrocarbons, such as chloroform. Higher doses result in visible color changes in some systems.

Calorimetry

The energy imparted to matter from radiation is usually efficiently converted into heat. (Radiation energy can also be expended in nuclear transformations and chemical changes.) If the absorber is thermally insulated, as in a calorimeter, then the temperature rise can be used to infer absorbed dose absolutely. However, a relatively large amount of radiation is required for calorimetric measurements. An absorbed energy of 4.18×10^7 erg/g ($= 4.18 \times 10^5$ rad) in water raises the temperature only $1°C$ (Problem 39). Because they are relatively insensitive, calorimetric methods in dosimetry have been employed primarily for high-intensity radiation beams, such as those used for radiotherapy. Calorimetric methods are also utilized for the absolute calibration of source strength.

Cerenkov Detectors

When a charged particle travels in a medium faster than light, it emits visible electromagnetic radiation, analogous to the shock wave produced in air at supersonic velocities. The speed of light in a medium with index of refraction n is given by c/n, where c is the speed of light in a vacuum. Letting $v = \beta c$ represent the speed of the particle, we can express the condition for the emission of Cerenkov radiation as $\beta c > c/n$, or

$$\beta n > 1. \tag{9.8}$$

The light is emitted preferentially in the direction the particle is traveling and is confined to a cone with vertex angle given by $\cos \theta = 1/\beta n$. It follows from (9.8) that the threshold kinetic energy for emission of Cerenkov light by a particle of rest mass M is given (Problem 40) by

$$T = Mc^2(\sqrt{1 + 1/(n^2 - 1)} - 1). \tag{9.9}$$

The familiar "blue glow" seen from a swimming-pool reactor (e.g., Fig. 8.1) is Cerenkov radiation, emitted by energetic beta particles traveling faster than light in the water.

Cerenkov detectors are employed to observe high-energy particles. The emitted radiation can also be used to measure high-energy beta-particle activity in aqueous samples.

9.5 NEUTRON DETECTION

Slow Neutrons

Neutrons are detected through the charged particles they produce in nuclear reactions, both inelastic and elastic. In some applications, pulses from the charged particles are registered simply to infer the presence of neutrons. In other situations, the neutron energy spectrum is sought, and the pulses must be further analyzed. For slow neutrons (kinetic energies $T \lesssim 0.5$ eV) detection is usually the only requirement. For intermediate (0.5 eV $\lesssim T \lesssim 0.1$ MeV) and fast ($T \gtrsim 0.1$ MeV) neutrons, spectral measurements are frequently needed. We discuss slow-neutron detection methods first.

Table 9.2 lists the three most important nuclear reactions for slow-neutron detection. The reaction-product kinetic energies and cross sections are given for capture of thermal neutrons (energy = 0.025 eV). Since the incident kinetic energy of a thermal neutron is negligible, the sum of the kinetic energies of the reaction products is equal to the Q value itself. Given Q, equations analogous to Eqs. (3.18) and (3.19) can be applied to calculate the discrete energies of the two products, leading to the values given in Table 9.2. We shall describe slow-neutron detection by means of these reactions and then briefly discuss detection by fission reactions and foil activation.

^{10}B(n, α)

One of the most widely used slow-neutron detectors is a proportional counter using boron trifluoride (BF$_3$) gas. For increased sensitivity, the boron is usually highly enriched in ^{10}B above its 19.8% natural isotopic abundance. If the dimensions of the tube are large compared with the ranges of the reaction products, then pulse heights at the Q values of 2.31 MeV and 2.79 MeV should be observed with areas in the ratio 96:4, as shown in Fig. 9.29(a). With most practical sizes, however, a significant number of Li nuclei and alpha particles enter the wall of the tube, and energy lost there is not registered. Since the two reaction products are produced "back-to-back" to conserve momentum, when one strikes the wall the other is directed away from it. This wall effect introduces continua to the left of the peaks. As sketched in Fig. 9.29(b), one continuum takes off from the peak at 2.31 MeV and is approximately flat down to 1.47 MeV. (A similar continuum occurs below the small peak at 2.79 MeV.) Over this interval, the total energy (1.47 MeV) of the alpha particle is absorbed in the gas while only part of the energy of the Li nucleus is absorbed there, the rest going into the wall. Below 1.47 MeV, the spectrum again drops and is approximately flat down to the energy 0.84 MeV of the Li recoil. Pulses occur here when the Li nucleus stops in the gas and the alpha particle enters the wall.

The BF$_3$ proportional counter can discriminate against gamma rays, which are usually present with neutrons and produce secondary electrons that ionize the gas. Compared with the neutron reaction products, electrons produced by the photons are sparsely ionizing and give much smaller pulses. As indicated in Fig. 9.29(a), amplitude discrimination can be used to eliminate these counts as well as electronic noise if the gamma flux is not too large. In intense gamma fields, however, the pileup of multiple pulses from photons can become a problem.

In other counter designs, a boron compound is used to line the interior walls of the

Table 9.2. Reactions Used for Neutron Detection
(Numerical Data Apply to Thermal-Neutron Capture)

Reaction			Q value (MeV)	Product kinetic energies (MeV)	Cross section (barn)
$^{10}_{5}$B $+ ^{1}_{0}$n $\rightarrow$	$^{7}_{3}$Li* $+ ^{4}_{2}\alpha$	(96%)	2.31	$T_{Li} = 0.84$ $T_{\alpha} = 1.47$	3840
	$^{7}_{3}$Li $+ ^{4}_{2}\alpha$	(4%)	2.79	$T_{Li} = 1.01$ $T_{\alpha} = 1.78$	
$^{6}_{3}$Li $+ ^{1}_{0}$n $\rightarrow ^{3}_{1}$H $+ ^{4}_{2}\alpha$			4.78	$T_{H} = 2.73$ $T_{\alpha} = 2.05$	940
$^{3}_{2}$He $+ ^{1}_{0}$n $\rightarrow ^{3}_{1}$H $+ ^{1}_{1}$p			0.765	$T_{H} = 0.191$ $T_{p} = 0.574$	5330

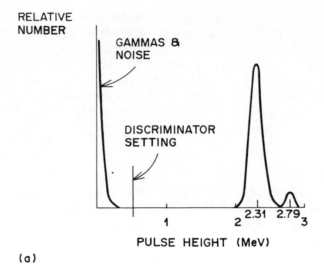

(a)

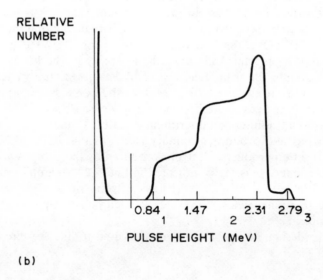

(b)

Figure 9.29. (a) Idealized pulse-height spectrum from large BF_3 tube in which reaction products are completely absorbed in the gas. (b) Spectrum from tube showing wall effects.

tube, in which another gas, more suitable for proportional counting than BF_3, is used. Boron-loaded scintillators (e.g., ZnS) are also employed for slow-neutron detection.

$^6Li(n,\alpha)$

As shown in Table 9.2, this reaction proceeds directly to the ground states of the products. Compared with $^{10}B(n,\alpha)$, it has a higher Q value (potentially better gamma-ray discrimination), but lower cross section (less sensitivity). The isotope ^6Li is 7.42% abundant in nature, but lithium enriched in ^6Li is available.

Lithium scintillators are frequently used for slow-neutron detection. Analogous to

NaI(Tl), crystals of LiI(Eu) can be employed. They can be made large compared with the ranges of the reaction products, so that the pulse-height spectra are free of wall effects. However, the scintillation efficiency is then comparable for electrons and heavy charged particles, and so gamma-ray discrimination is much poorer than with BF_3 gas.

Lithium compounds can be mixed with ZnS to make small detectors. Because secondary electrons produced by gamma rays easily escape, gamma-ray discrimination with such devices is good.

$^3He(n,p)$

This reaction has the highest cross section of the three in Table 9.2. Like the BF_3 tube, the ^3He proportional counter exhibits wall effects. However, ^3He is a better counter gas and can be operated at higher pressures with better detection efficiency. Because of the low value of Q, though, gamma discrimination is worse.

(n,f)

Slow-neutron-induced fission of ^{233}U, ^{235}U, or ^{239}Pu is utilized in fission counters. The Q value of ~200 MeV for each is large. About 165 MeV of this energy is converted directly into kinetic energy of the heavy fission fragments. Fission pulses are extremely large, enabling slow neutron counting to be done at low levels, even in a high background. Most commonly, the fissile material is coated on the inner surface of an ionization chamber. A disadvantage of fissionable materials is that they are alpha emitters, and one must sometimes contend with the pileup of alpha-particle pulses.

Activation Foils

Slow neutrons captured by nuclei induce radioactivity in a number of elements, which can be made into foils for neutron detection. The amount of induced activity will depend on a number of factors—the element chosen, the mass of the foil, the neutron energy spectrum, the capture cross section, and the time of irradiation. Examples of thermal-neutron activation-foil materials include Mn, Co, Cu, Ag, In, Dy, and Au.

Intermediate and Fast Neutrons

Nuclear reactions are also important for measurements with intermediate and fast neutrons. In addition, neutrons at these speeds can, by elastic scattering, transfer detectable kinetic energies to nuclei, especially hydrogen. Elastic recoil energies are negligible for slow neutrons. Detector systems can be conveniently discussed in four groups—those based on neutron moderation, nuclear reactions, elastic scattering alone, and foil activation.

Neutron Moderation

Two principal systems in this category have been developed; the long counter and moderating spheres enclosing a small thermal-neutron detector. A cross section of the cylindrical long counter is shown in Fig. 9.30. This detector can be constructed to give nearly the same response from a neutron of any energy from about 10 keV to 5 MeV. The long counter contains a BF_3 tube surrounded by an inner paraffin moderator, as shown. The instrument is sensitive to neutrons incident from the right. Those from other directions are either reflected or thermalized by the outer paraffin jacket and then absorbed in the B_2O_3 layer. Neutrons that enter from the right are slowed down in the inner paraffin moderator, high-energy neutrons reaching greater depths on the average than low-energy ones. With this arrangement, the probability that a moderated neutron will enter the BF_3

tube and be registered does not depend strongly on the initial energy with which it entered the counter. Holes on the front face make it easier for neutrons with energies <1 MeV to penetrate past the surface, from which they might otherwise be reflected. The long counter does not measure neutron spectra.

Neutron spectral information can be inferred by the use of polyethylene moderating spheres (Bonner spheres) of different diameters enclosing small lithium iodide scintillators at their centers. A series of five or more spheres, ranging in diameter from 2 to 12 in., is typically used. The different sizes provide varying degrees of moderation for neutrons of different energies. The response of each sphere is calibrated for monoenergetic neutrons from thermal energy to 10 MeV or more. The spheres are then exposed in an unknown neutron field and the count rates measured. An unfolding procedure is used to infer information about the neutron spectrum from knowledge of the calibration curves and the measured count rates. Because the unfolding procedure does not yield very precise results, this method is not widely used for spectral measurements. With a relatively large sphere, it is found that the response as a function of neutron energy is similar to the dose equivalent per neutron. It therefore serves as a neutron rem meter in many applied health physics operations. Such an instrument is shown in Fig. 9.31.

Nuclear Reactions

The $^6\text{Li}(n,\alpha)$ and $^3\text{He}(n,p)$ reactions are the only ones of major importance for neutron spectrometry. Ideally, an incident neutron of energy T that undergoes a reaction causes a detector to register a peak at an energy $Q + T$. In practice, many times another peak also occurs at an energy Q due to neutrons that have been slowed by multiple scattering in building walls and shielding around the detector. Slow-neutron cross sections can be orders of magnitude larger than at higher energies. The additional peak at Q is sometimes called the epithermal peak.

Crystals of LiI(Eu) are used in neutron spectroscopy. However, the nonlinearity of their response with the energy of the reaction products (tritons and alpha particles) is a serious handicap. Lithium-glass scintillators are also in use, principally as fast responding detectors in neutron time-of-flight measurements. In another type of neutron spectrometer, a thin LiF sheet is placed between two semiconductor diodes. At relatively low neu-

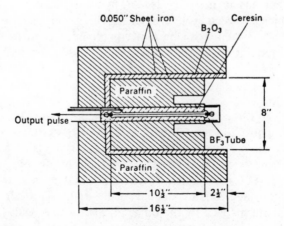

Figure 9.30. The long counter of Hanson and McKibben. [Reprinted with permission from A.O. Hanson and M.L. McKibben, "A neutron detector having uniform sensitivity from 10 keV to 5 MeV," *Phys. Rev.* **72**, 673 (1947). Copyright 1947 by the American Physics Society.]

Figure 9.31. BF$_3$ counter in center of 9 in. diameter cadmium-loaded polyethylene sphere is used to measure neutron dose equivalent. BF$_3$ tube allows excellent gamma rejection. (Courtesy Eberline Instrument Corp.)

tron energies T, the recoil products will tend to be ejected back-to-back, giving coincidence counts in both semiconductors with a total pulse height of $Q + T$, from which T can be ascertained.

In the ^{3}He(n,p) proportional counter, monoenergetic neutrons of energy T produce a peak at energy $T + 0.765$ MeV, as illustrated in Fig. 9.32. The epithermal peak is also shown at the energy $Q = 0.765$ MeV. In addition to these peaks from the reaction products, one finds a continuum of pulse heights from recoil ^{3}He nuclei that elastically scatter incident neutrons. It follows from Eq. (8.3) that the maximum kinetic energy that a neutron of mass $m = 1$ and kinetic energy T can transfer to a helium nucleus of mass $M = 3$ is

$$T_{max} = \frac{4mM}{(m + M)^2} T = \frac{4 \times 1 \times 3}{(1 + 3)^2} T = \frac{3}{4} T. \tag{9.10}$$

The elastic continuum in Fig. 9.32 thus extends up to the energy $0.75T$. Wall effects can be reduced in the ^{3}He proportional counter by using gas pressures of several atmospheres and also by adding a heavier gas (e.g., Kr).

Elastic Scattering

A number of instruments are based on elastic scattering alone, especially from hydrogen. As discussed in Section 8.5, a neutron can lose all of its kinetic energy T in a single head-on collision with a proton. Also, since n–p scattering is isotropic in the center-of-mass system for neutron laboratory energies up to ~10 MeV, the average energy imparted to protons by neutrons in this energy range is $T/2$.

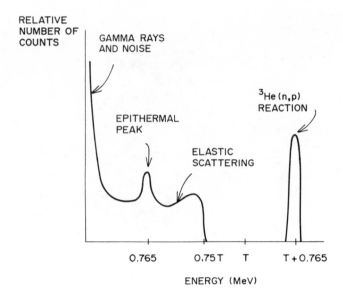

Figure 9.32. Pulse-height spectrum from ^{3}He proportional counter for monoenergetic neutrons of energy T.

Organic proton-recoil scintillators are available for neutron spectrometry in a variety of crystal, plastic, and liquid materials. The full proton recoil energies can be caught in these scintillators. Complications in the use of proton-recoil scintillators include non-linearity of response, multiple neutron scattering, and competing nuclear reactions. For applications in mixed fields, the gamma response can, in principle, be separated electronically from the neutron response on the basis of quicker scintillation.

Proportional counters have been designed with hydrocarbon gases, such as CH_4. These have inherently lower detection efficiencies than solid-state devices, but offer the potential for better gamma discrimination. Wall effects can be important. Proportional counters have also been constructed with polyethylene or other hydrogenous material surrounding the tube. One such device, based on the Bragg–Gray principle, will be discussed in Section 10.6.

A proton-recoil telescope, illustrated in Fig. 9.33, can be used to accurately measure the spectrum of neutrons in a collimated beam. At an angle θ, the energy T_p of a recoil proton from a thin target struck by a neutron of incident energy T is, by Eq. (8.5),

$$T_p = T \cos^2 \theta. \tag{9.11}$$

The $E(-dE/dx)$ coincidence particle identifier (Fig. 9.20) can be used to reduce background, eliminate competing events, and measure T_p.

Neutron spectra can also be inferred from the observed range distribution of recoil protons in nuclear track emulsions. Neutrons with at least several hundred keV of energy are needed to produce protons with recognizable tracks.

Threshold Foil Activation

Like low-energy neutrons, intermediate and fast neutrons can be detected by the radioactivity they induce in various elements. With many nuclides, a threshold energy exists for the required nuclear reaction. Some examples are given in Table 9.3. When foils of several nuclides are simultaneously exposed to a neutron field, differences in the induced

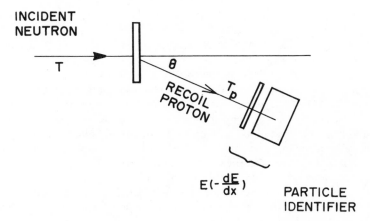

INCIDENT
NEUTRON

T

θ

RECOIL
PROTON

T_p

$E(-\frac{dE}{dx})$

PARTICLE
IDENTIFIER

Figure 9.33. Arrangement of recoil-proton telescope for measuring energy spectrum of a neutron beam.

activity between them can be used to obtain information about the neutron energy spectrum as well as the fluence. For example, if an exposed aluminum foil shows induced activity from ^{27}Mg and a simultaneously exposed cobalt foil shows no induced activity from ^{56}Mn, then one can infer that neutrons with energies 3.8 MeV $< T <$ 5.2 MeV were present. To obtain accurate spectral data from threshold-detector systems, one must take into account such factors as the masses of the particular isotopes in the foils, their neutron cross sections as functions of energy, the exposure history of the foils, and the half-lives of the induced radioisotopes.

9.6 PROBLEMS

1. How many electrons are collected per second in an ionization chamber when the current is 5×10^{-14} A? What is the rate of energy absorption if $W = 29.9$ eV/ip?
2. How many ion pairs does a 5.6 MeV alpha particle produce in N_2?
3. Why do the W values for heavy charged particles increase at low energies (Fig. 9.2)?
4. A beam of alpha particles produced a current of 10^{-14} in a parallel-plate ionization chamber for 8 sec. The chamber contained air at STP. (a) How many ion pairs were produced? (b) How much energy did the beam deposit in the chamber? (c) If the chamber volume was 240 cm^3, what was the energy absorbed per unit mass in the chamber gas (1 rad absorbed dose = 100 erg/g)?
5. A 10 cm^2 beam of charged particles is totally absorbed in an ionization chamber, producing a

Table 9.3. Reactions for Threshold Activation Detectors of Neutrons

Reaction	Threshold (MeV)
^{115}In(n,n')^{115m}In	0.5
^{58}Ni(n,p)^{58}Co	1.9
^{27}Al(n,p)^{27}Mg	3.8
^{56}Fe(n,p)^{56}Mn	4.9
^{59}Co(n,α)^{56}Mn	5.2
^{24}Mg(n,p)^{24}Na	6.0
^{197}Au(n,2n)^{196}Au	8.6
^{19}F(n,2n)^{18}F	11.6

saturation current of 10^{-6} A. If $W = 30$ eV/ip, what is the average beam intensity in units of eV sec^{-1} cm^{-2}?

6. A 5 MeV alpha-particle beam of cross-sectional area 2 cm^2 is stopped completely in an ionization chamber, producing a current of 10 μA under voltage saturation conditions. (a) If $W = 32$ eV/ip, what is the intensity of the beam? (b) What is the fluence rate?

7. A thin radioactive source placed in an ionization chamber emits 10^6 alpha particles per second with energy 3.81 MeV. The particles are completely stopped in the gas, for which $W = 36$ eV/ip. Calculate (a) the average number of ion pairs produced per second, (b) the current that flows under saturation conditions, (c) the amount of charge collected in 1 hr.

8. What is the ratio of the pulse heights from a 1 MeV proton ($W = 30$ eV/ip) and a 1 MeV carbon nucleus ($W = 40$ eV/ip) absorbed in a proportional counter?

9. Why is a GM counter not useful for determining the absorbed energy in a gas?

10. Assume that the W values for protons and carbon-recoil nuclei are both 30 eV/ip in C_2H_4 gas. What is the maximum number of ion pairs that can be produced by a 3 MeV neutron interacting elastically with (a) H or (b) C?

11. An alpha-particle source is fabricated into a thin foil. Placed first in a 2π gas-flow proportional counter, it shows only a single pulse height and registers 7080 counts/min (background negligible). The source is next placed in 4π geometry in an air ionization chamber operated under saturation conditions, where it produces a current of 5.56×10^{-12} A. Assume that the foil stops the recoil nuclei following alpha decay but absorbs a negligible amount of energy from the alpha particles. (a) What is the activity of the source in curies? (b) What is the alpha-particle energy? (c) Assume that the atomic mass number of the daughter nucleus is 206 and calculate its recoil energy.

12. Show that, at high energies, where the average number of electrons per quantum state is small, the quantum-mechanical distribution Eq. (9.6) approaches the classical Boltzmann distribution, $N = \exp(-E/kT)$.

13. Figure 9.12(b) shows the relative number of electrons at energies E in an intrinsic semiconductor when $T > 0$. Make such a sketch for the As-doped Ge semiconductor shown in Fig. 9.15.

14. What type of semiconductor results when Ge is doped with (a) Sb or (b) In?

15. Make a sketch like that in Fig. 9.14 for Ge doped with Ga (p-type semiconductor).

16. What sequence of events produces the 11 keV escape peak seen when pulse heights from monoenergetic photons are measured with a germanium detector?

17. Use the nonrelativistic stopping-power formula and show that Eq. (9.7) holds.

18. What are the relative magnitudes of the responses of an $E(-dE/dx)$ particle identifier to a proton, an alpha particle, and a stripped carbon nucleus?

19. The W value for silicon is 3.6 eV/ip. Calculate the mean number of ion pairs produced by a 300 keV beta particle absorbed in Si. What is the standard deviation, based on Poisson statistics (Appendix E)?

20. The W value for Ge is 3.0 eV/ip at 77 K. Based on Poisson statistics, calculate the expected energy resolution (FWHM) of a Ge(Li) detector for 0.662 MeV photons from ^{137}Cs. Assume that charge collection is complete and that electronic noise is negligible.

21. The resolution (FWHM/$\bar{n}$) of a certain Ge(Li) spectrometer ($W = 3.0$ eV/ip) for 662 keV ^{137}Cs gamma rays is found to be 0.223%. Assume that variations in pulse height are due entirely to statistical fluctuations in the number of ions produced by an absorbed photon. What is the resolution for the 1.332 MeV gamma ray from ^{60}Co?

22. Calculate the number of ion pairs produced by a 4 MeV alpha particle in (a) an ionization chamber filled with air ($W = 36$ eV/ip) and (b) a silicon surface-barrier detector ($W = 3.6$ eV/ip).

23. Using Poisson statistics, calculate the energy resolution (percent) of both detectors for the 4 MeV alpha particle in the last problem. How would the resolution of the two compare if the alpha-particle energy were 6 MeV?

24. A silicon semiconductor detector has a dead layer of 1 μm followed by a depletion region 250 μm in depth. (a) Are these detector dimensions suitable for alpha-particle spectroscopy up to 4 MeV? (b) For beta-particle spectroscopy up to 500 keV?

25. If 2.1 MeV of absorbed alpha-particle energy produces 41,100 scintillation photons of average wavelength 4800 Å in a scintillator, calculate its efficiency.

26. A 600 keV photon is absorbed in a NaI(Tl) crystal having an efficiency of 11.2%. The average wavelength of the scintillation photons produced is 5340 Å, and 21% of them produce a signal at the cathode of the photomultiplier tube. (a) Calculate the average energy from the incident radiation that produces one photoelectron at the cathode (the "W value"). (b) What is the energy resolution (FWHM) of the detector for this radiation? (c) What would be the resolution for 800 keV photons?

27. Figure 9.25 was obtained for a 4×4 in. NaI(Tl) crystal scintillator. Sketch the observed spectrum if measurements were made with the same source, but with a NaI crystal that was (a) very large, (b) very small. (c) In (b), interpret the relative areas under the photopeak and the Compton continuum.

28. The resolution (FWHM/$\bar{n}$) of a scintillator is 8.2% for 0.662 MeV gamma rays from ^{137}Cs. What is its resolution for 1.17 MeV gamma rays from ^{60}Co?

29. A 1.17 MeV gamma ray is Compton scattered once at an angle of 48° in a scintillator and again at an angle of 112° before escaping. (a) What (average) pulse height is registered? (b) If the photon were scattered once at 48° and then photoelectrically absorbed, what pulse height would be registered?

30. The mass attenuation coefficient of NaI (density = 3.67 g/cm³) for 500 keV photons is 0.90 cm²/g. What percentage of normally incident photons interact in a crystal 4 cm thick?

31. Calculate the energy of the backscatter peak for 3 MeV gamma photons.

32. A long-lived radioactive sample gives a mean count rate of 93 counts/min. What is the relative error when the rate is determined by counting for (a) 5 min, (b) 24 hr, (c) 1 wk (Appendix E)?

33. A 30 keV beta particle is absorbed in a scintillator having an efficiency of 8%. The average scintillation photon energy is 3.4 eV. If 21% of the photons are counted, what is the relative error in the measured beta-particle energy? What would be the relative error in a semiconductor ($W = 3$ eV/ip)?

34. According to Poisson statistics, how many counts are needed to have a relative error of less than 1%?

35. A sample of a long-lived radionuclide gives 939 counts in 3 min. How long must the sample be counted to determine the count rate to within ±3% with 95% confidence?

36. The count rate in the last problem is 5.22 counts/sec. What is the probability that exactly 26 counts would be observed in 5 sec? Is the use of Poisson statistics justified?

37. What are the pros and cons of using NaI scintillators vs. HPGe detectors for gamma-ray spectroscopy?

38. The following numbers N_i of ions are collected in registering the pulses from 10 identical particles: 1028, 1031, 1019, 1044, 1061, 1024, 1039, 1041, 1042, and 1051. (a) Calculate the variance. (b) What variance is expected if the N_i obey Poisson statistics? (c) Estimate the resolution of the counter, based on this small sample, by means of Eq. (E.16), Appendix E.

39. Calculate the temperature rise in a calorimetric water dosimeter that absorbs 10^5 erg/g from a radiation beam (= an absorbed dose of 1000 rad, a lethal dose if given acutely over the whole body). What absorbed dose is required to raise the temperature 1°C?

40. Prove Eq. (9.9).

41. (a) Calculate the threshold kinetic energy for an electron to produce Cerenkov radiation in water (index of refraction = 1.33). (b) What is the threshold energy for a proton?

42. An electron enters a water shield (index of refraction 1.33) with a speed $v = 0.90c$, where c is the speed of light in a vacuum. Assuming that 0.1% of its energy loss is due to Cerenkov radiation as long as this is possible, calculate the number of photons emitted if their average wavelength is 4200 Å.

43. Given the value $Q = 4.78$ MeV for the ^{6}Li(n,α)^{3}H reaction in Table 9.2, calculate T_H and T_α.

44. Sketch the pulse-height spectrum from a ^{3}He proportional counter exposed to thermal neutrons. Include wall effects at the appropriate energies.

45. (a) What is the maximum energy that a 3 MeV neutron can transfer to a ^{3}He nucleus by elastic

scattering? (b) What is the maximum energy that can be absorbed in a ^{3}He proportional counter from a 3 MeV neutron?

46. Sketch the pulse-height spectrum from a thin layered boron-lined proportional tube for thermal neutrons. Repeat for a layer thicker than the ranges of the reaction products.

47. Calculate the threshold energy for the reaction ^{24}Mg(n,p)^{24}Na shown in Table 9.3. The ^{24}Na nucleus is produced in an excited state from which it emits a 1.369 MeV gamma photon and goes to its ground state. (See Appendix D)

48. At what neutron energy is the cross section for the ^{3}He(n,p) reaction equal to the thermal-neutron cross section for the ^{6}Li(n,α) reaction?

49. In the proton-recoil telescope (Fig. 9.33), what is the energy of the scattered neutron if $\theta = 27°$ and $T_p = 1.18$ MeV?

CHAPTER 10
RADIATION DOSIMETRY

10.1 INTRODUCTION

Radiation dosimetry is the branch of science that attempts to quantitatively relate specific measurements made in a radiation field to chemical and/or biological changes that the radiation would produce in a target. Dosimetry is essential for quantifying the incidence of various biological changes as a function of the amount of radiation received (dose–effect relationships), for comparing different experiments, for monitoring the radiation exposures of individuals, and for surveillance of the environment. In this chapter we describe the principal concepts upon which radiation dosimetry is based and present methods for their practical utilization.

When radiation interacts with a target it produces excited and ionized atoms and molecules as well as large numbers of secondary electrons. The secondary electrons can produce additional ionizations and excitations until, finally, the energies of all electrons fall below the threshold necessary for exciting the medium. As we shall see in detail in the next chapter, the initial electronic transitions, which produce chemically active species, are completed in very short times ($\leq 10^{-15}$ sec) in local regions within the path traversed by a charged particle. These changes, which require the direct absorption of energy from the incident radiation by the target, represent the initial physical perturbations from which subsequent chemical and biological effects evolve. It is natural, therefore, to consider measurements of ionization and energy absorption as the basis for radiation dosimetry.

10.2 QUANTITIES AND UNITS

Exposure

Exposure is defined for gamma and X-rays in terms of the amount of ionization they produce in air. The unit of exposure is called the roentgen (R) and was introduced at the Radiological Congress in Stockholm in 1928 (Chap. 1). It was originally defined as that amount of gamma or X-radiation that produces in air 1 esu of charge of either sign per 0.001293 g of air. (This mass of air occupies 1 cm^3 at standard temperature and pressure.) The charge involved in the definition of the roentgen includes both the ions produced directly by the incident photons as well as ions produced by all secondary electrons. Since 1962, exposure has been defined by the ICRU as the quotient $\Delta Q/\Delta m$, where ΔQ is the sum of all charges of one sign produced in air when all the electrons liberated by photons in a mass Δm of air are completely stopped in air. The unit roentgen is now defined as

$$1 \text{ R} = 2.58 \times 10^{-4} \text{ C/kg.} \tag{10.1}$$

The concept of exposure applies only to electromagnetic radiation; the charge and mass used in its definition, as well as in the definition of the roentgen, refer only to air.

Example
Show that 1 esu/cm^3 in air at STP is equivalent to the definition (10.1) of 1 R of exposure.

Solution
Since the density of air at STP is 0.001293 g/cm^3 and 1 esu = 3.34×10^{-10} C (Appendix B), we have

$$\frac{1 \text{ esu}}{\text{cm}^3} = \frac{3.34 \times 10^{-10} \text{ C}}{0.001293 \text{ g} \times 10^{-3} \text{ kg/g}} = 2.58 \times 10^{-4} \frac{\text{C}}{\text{kg}}. \tag{10.2}$$

As discussed in Section 9.1, the average energy W needed by a charged particle and its secondaries to produce an ion pair in a gas is independent of the charged-particle energy except at low velocities, where the curves in Fig. 9.2 rise significantly. This fact enables one to relate the amount of ionization produced in a gas by radiation to the energy absorbed there [cf. Eq. (9.3)]. Therefore, we can find the energy absorbed per unit mass of air per roentgen of exposure. Since the electronic charge is 1.60×10^{-19} C, Eq. (10.1) gives for the number of ion pairs produced per kg of air for an exposure of 1 R:

$$1 \text{ R} = \frac{2.58 \times 10^{-4} \text{ C}}{\text{kg}} \times \frac{1}{1.60 \times 10^{-19} \text{ C/ip}} = 1.61 \times 10^{15} \frac{\text{ip}}{\text{kg}}. \tag{10.3}$$

Since $W = 34$ eV/ip for electrons in air, we find

$$1 \text{ R} = 1.61 \times 10^{15} \frac{\text{ip}}{\text{kg}} \times 34 \frac{\text{eV}}{\text{ip}} \times 1.6 \times 10^{-12} \frac{\text{erg}}{\text{eV}} = 8.76 \times 10^{4} \frac{\text{erg}}{\text{kg}} = 87.6 \frac{\text{erg}}{\text{g}}. \tag{10.4}$$

Thus, when the exposure is 1 R, the energy absorbed in air from the photon beam is 87.6 erg/g.

Absorbed Dose

The concept of exposure and the definition of the roentgen provide a practical, measurable standard for electromagnetic radiation in air. However, additional concepts are needed to apply to all kinds of radiation and to other materials, particularly tissue. Calculations show that a radiation exposure of 1 R would produce about 95 erg/g of absorbed energy in tissue. This unit is called the "rep" ("roentgen-equivalent-physical") and was formally used in radiation-protection work as a measure of the change produced in living tissue by radiation. The rep is no longer employed. Today, the primary physical quantity used in dosimetry is the absorbed dose. Whereas exposure applies only to electromagnetic radiation in air, absorbed dose is defined as the energy absorbed per unit mass from any kind of ionizing radiation in any kind of matter. The traditional unit of absorbed dose is the rad, defined as 100 erg/g. The newer, SI unit, 1 J/kg, is called the gray (Gy):

$$1 \text{ Gy} \equiv \frac{1 \text{ J}}{\text{kg}} = \frac{10^{7} \text{ erg}}{10^{3} \text{ g}} = 10^{4} \frac{\text{erg}}{\text{g}} = 100 \text{ rad}. \tag{10.5}$$

The absorbed dose is often referred to simply as the dose.

Dose Equivalent

It has long been recognized that the absorbed dose needed to achieve a given level of biological damage (e.g., 50% cell killing) is often different for different kinds of radiation. As discussed in the next chapter, radiation with a high LET (Sect. 6.3) is generally more damaging to a biological system per unit dose than radiation with a low LET (for example, cf. Fig. 11.11).

To allow for the different biological effectiveness of different kinds of radiation, the ICRP, NCRP, and ICRU (Chap. 1) have introduced the concept of dose equivalent for radiation-protection purposes. The dose equivalent H is defined as the product of the absorbed dose D and a dimensionless quality factor Q, which depends on LET as shown in Table 10.1:

$$H = QD. \tag{10.6}$$

In principle, other multiplicative modifying factors can be included along with Q to allow for additional considerations (e.g., dose fractionation), but these are not ordinarily used. For incident charged particles, the LET values in Table 10.1 refer directly to the LET of the radiation in water, expressed in keV per micron of travel. For incident neutrons, gamma rays, and other uncharged radiations, the LET refers to that which the secondary charged particles they generate would have in water. When the absorbed dose is expressed in rad, the unit of dose equivalent is called the rem ("roentgen-equivalent-man"). With the dose expressed in Gy, the SI dose-equivalent unit is called the sievert (Sv). It follows that 1 Sv = 100 rem.

In routine work involving radiation, the exposure of individuals is limited to a given level of dose equivalent, independent of the type of radiation involved. Also, the dose equivalents to an individual from different kinds of radiations are added to obtain his total dose equivalent. Thus, the dose equivalent is the quantity that must be determined for administrative and legal purposes in the monitoring of allowable radiation exposures.

Example

What is the absorbed dose in air when the exposure is 1 R? What would be the absorbed dose in tissue for an exposure of 1 R?

Solution

As we see from Eq. (10.4), 1 R = 87.6 erg/g of air. Therefore, 1 R = 0.876 rad. It follows from the discussion of the rep above that the absorbed dose in tissue would be 0.95 rad, if the exposure (in air) were 1 R.

Example

A worker receives a whole-body dose of 10 mrad from 2 MeV neutrons. Estimate the dose equivalent.

Table 10.1. Dependence of Quality Factor Q on LET of Radiation as Recommended by ICRP, ICRU, and NCRP

LET (keV/μm in water)	Q
3.5 or less	1
3.5–7.0	1–2
7.0–23	2–5
23–53	5–10
53–175	10–20
Gamma rays, X-rays, electrons, positrons of any LET	1

Solution

Most of the absorbed dose is due to the elastic scattering of the neutrons by the hydrogen in tissue (cf. Table 10.5). To make a rough estimate of the quality factor, we first find Q for a 1 MeV proton – the average recoil energy for 2 MeV neutrons. From Table 4.3 we see that the stopping power for a 1 MeV proton in water is 270 MeV/cm = 27.0 keV/μm. Since the stopping power and LET are the same (Sect. 6.3), we see from Table 10.1 that an estimate of $Q \sim 6$ should be reasonable for the recoil protons. The recoil O, C, and N nuclei have considerably higher LET values, but do not contribute as much to the dose as H. (LET is proportional to the square of a particle's charge.) Without going into more detail, we take the overall quality factor, $Q \sim 12$, to be twice that for the recoil protons alone. Therefore, the estimated dose equivalent is $H \sim 12 \times 10 = 120$ mrem. [The value $Q = 10$ is obtained from detailed calculations (cf. Table 10.4).]

10.3 MEASUREMENT OF EXPOSURE

Free-Air Ionization Chamber

Based on its definition, exposure can be measured operationally with the "free-air," or "standard," ionization chamber, sketched in Fig. 10.1. X-rays emerge from the target T of an X-ray tube and enter the free-air chamber through a circular aperature of area A, defining a right circular cone TBC of rays. Parallel plates Q and Q' in the chamber collect the ions produced in the volume of air between them with center P'.

The exposure in the volume DEFG in roentgens would be determined directly if the total ionization produced only by those ions that originate from X-ray interactions in the truncated conical volume DEFG could be collected and the resulting charge divided by the mass of air in DEFG. This mass is given by $M = \rho A' L$, where ρ is the density of air, A' is the cross-sectional area of the truncated cone at its midpoint P', and L, the thickness of the cone, is equal to the length of the collecting plates Q and Q'. Unfortunately, the plates collect all of the ions between them, not the particular set that is specified in the definition of the roentgen. Some electrons produced by X-ray interactions in DEFG escape this volume and produce ions that are not collected by the plates Q, Q'. Also, some ions from electrons originally produced outside DEFG are collected. Thus, only part of the ionization of an electron such as e_1 in Fig. 10.1 is collected, while ionization from an "outside" electron, such as e_2, is collected. When the distance from P to DG is sufficiently large (e.g., ~10 cm for 300 keV X-rays) electronic equilibrium will be real-

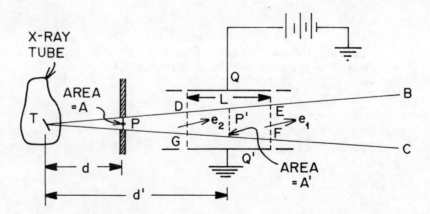

Figure 10.1. Schematic diagram of the "free-air" or "standard" ionization chamber.

ized; that is, there will be almost exact compensation between ionization lost from the volume DEFG by electrons, such as e_1, that escape and ionization gained from electrons, such as e_2, that enter. The distance from P to DG, however, should not be so large as to attenuate the beam significantly between P and P′. Under these conditions, when a charge q is collected, the exposure at P′ is given by

$$E_{P'} = \frac{q}{\rho A'L}. \tag{10.7}$$

In practice, one prefers to know the exposure E_P at P, the location where the entrance port is placed, rather than $E_{P'}$. By the inverse-square law, $E_P = (d'/d)^2 E_{P'}$. Since $A = (d/d')^2 A'$, Eq. (10.7) gives

$$E_P = \left(\frac{d'}{d}\right)^2 \frac{q}{\rho A'L} = \frac{q}{\rho AL}. \tag{10.8}$$

Example
The entrance port of a free-air ionization chamber has a diameter of 0.25 cm and the length of the collecting plates is 6 cm. Exposure to an X-ray beam produces a steady current of 2.6×10^{-10} A for 30 sec. The temperature is 26°C and the pressure is 750 mm Hg. Calculate the exposure rate and the exposure.

Solution
We can apply Eq. (10.8) to exposure rates as well as to exposure. The rate of charge collection is $\dot{q} = 2.6 \times 10^{-10}$ A $= 2.6 \times 10^{-10}$ C/sec. The density of the air under the stated conditions is $\rho = (0.00129)(273/299)(750/760) = 1.16 \times 10^{-3}$ g/cm³. The entrance-port area is $A = \pi(0.125)^2 = 4.91 \times 10^{-2}$ cm² and $L = 6$ cm. Equation (10.8) implies, for the exposure rate,

$$\dot{E}_P = \frac{\dot{q}}{\rho AL} = \frac{2.6 \times 10^{-10} \text{ C/sec}}{1.16 \times 10^{-3} \times 4.91 \times 10^{-2} \times 6 \text{ g}} \times \frac{1 \text{ R}}{2.58 \times 10^{-7} \text{ C/g}} = 2.95 \frac{\text{R}}{\text{sec}}. \tag{10.9}$$

The total exposure is 88.5 R.

Measurement of exposure with the free-air chamber requires some care and attention to details. For example, the collecting plates Q and Q′ in Fig. 10.1 must be recessed away from the active volume DEFG by a distance not less than the lateral range of electrons produced there. We have already mentioned minimum and maximum restrictions on the distance from P to DG. When the photon energy is increased, the minimum distance required for electronic equilibrium increases rapidly and the dimensions for a free-air chamber become excessively large for photons of high energy. For this and other reasons, the free-air ionization chamber and the roentgen are not used for photon energies above 3 MeV.

The Air-Wall Chamber

The free-air ionization chamber is not a practical instrument for measuring routine exposure. It is used chiefly as a primary laboratory standard. For routine use, chambers can be built with walls of a solid material, having photon response properties similar to those of air. Chambers of this type were discussed in Section 9.1.

Such an "air-wall" pocket chamber, built as a capacitor, is shown schematically in Fig. 10.2. A central anode, insulated from the rest of the chamber, is given an initial charge from a charger-reader device to which it is attached before wearing. When exposed to photons, the secondary electrons liberated in the walls and enclosed air tend to neutralize

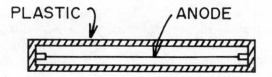

Figure 10.2. Air-wall pocket ionization chamber, having plastic wall with approximately the same response to photons as air.

the charge on the anode and lower the potential difference between it and the wall. The change in potential difference is directly proportional to the total ionization produced and hence to the exposure. Thus, after exposure to photons, measurement of the change in potential difference from its original value when the chamber was fully charged can be used to find the exposure. Direct-reading pocket ion chambers are available (Fig. 9.6).

Example

A pocket air-wall chamber has a volume of 2.5 cm³ and a capacitance of 7 $\mu\mu$F. Initially charged at 200 V, the reader showed a potential difference of 170 V after the chamber was worn. What exposure in roentgens may be inferred?

Solution

The charge lost is $\Delta Q = C \Delta V = 7 \times 10^{-12} \times (200 - 170) = 2.10 \times 10^{-10}$ C. The mass of air (we assume STP) is $M = 0.00129 \times 2.5 = 3.23 \times 10^{-3}$ g. It follows that the exposure is

$$\frac{2.10 \times 10^{-10} \text{ C}}{3.23 \times 10^{-3} \text{ g}} \times \frac{1 \text{ R}}{2.54 \times 10^{-7} \text{ C/g}} = 0.256 \text{ R}. \qquad (10.10)$$

Literally taken, the data given in this problem indicate only that the chamber was partially discharged. Charge loss could occur for reasons other than radiation (e.g., leakage from the central wire). Often two pocket ion chambers are worn simultaneously to improve reliability.

In practice, air-wall ionization chambers involve a number of compromises from an ideal instrument that measures exposure accurately. For example, if the wall is too thin, incident photons will produce insufficient ionization inside the chamber. If the wall is too thick, it will significantly attenuate the incident radiation. The optimal thickness is reached when, for a given photon field, the ionization in the chamber gas is a maximum. This value, called the equilibrium wall thickness, is equal to the range of the most energetic secondary electrons produced in the wall. In addition, a solid wall can be only approximately air equivalent. Air-wall chambers can be made with an almost energy-independent response from a few hundred keV to about 2 MeV — the energy range in which Compton scattering is the dominant photon interaction in air and low-Z wall materials.

10.4 MEASUREMENT OF ABSORBED DOSE

One of the primary goals of dosimetry is the determination of the absorbed dose in tissue exposed to radiation. The Bragg–Gray principle provides a means of relating ionization measurements in a gas to the absorbed dose in some convenient material from which a dosimeter can be fabricated. To obtain the tissue dose, either the material can be tissue equivalent or else the ratio of the absorbed dose in the material to that in tissue can be inferred from other information, such as calculations or calibration measurements.

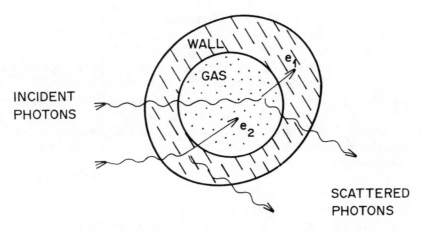

Figure 10.3. Gas in cavity enclosed by wall to illustrate Bragg–Gray principle.

Consider a gas in a walled enclosure irradiated by photons, as illustrated in Fig. 10.3. The photons lose energy in the gas by producing secondary electrons there, and the ratio of the energy deposited and the mass of the gas is the absorbed dose in the gas. This energy is proportional to the amount of ionization in the gas when electronic equilibrium exists between the wall and the gas. Then an electron, such as e_1 in Fig. 10.3, which is produced by a photon in the gas and enters the wall before losing all of its energy, is compensated by another electron, like e_2, which is produced by a photon in the wall and stops in the gas. When the walls and gas have the same atomic composition, then the energy spectra of such electrons will be the same irrespective of their origin, and a high degree of compensation can be realized. The situation is then analogous to the air-wall chamber discussed above. Electronic equilibrium requires that the wall thickness be at least as great as the maximum range of secondary charged particles. However, as with the air-wall chamber, the wall thickness should not be so great that the incident radiation is appreciably attenuated.

The Bragg–Gray principle states that, if a gas is enclosed by a wall of the same atomic composition and if the wall meets the above thickness conditions, then the energy absorbed per unit mass in the gas is equal to the number of ion pairs produced there times the W value divided by the mass m of the gas. Furthermore, the absorbed dose D_g in the gas is equal to the absorbed dose D_w in the wall. Denoting the number of ions in the gas by N_g, we write

$$D_w = D_g = \frac{N_g W}{m}. \tag{10.11}$$

When the wall and gas are of different atomic composition, the absorbed dose in the wall can still be obtained from the ionization in the gas. In this case, the cavity size and gas pressure must be small, so that secondary charged particles lose only a small fraction of their energy in the gas. The absorbed dose then scales as the ratio S_w/S_g of the mass stopping powers of the wall and gas:

$$D_w = \frac{D_g S_w}{S_g} = \frac{N_g W S_w}{m S_g}. \tag{10.12}$$

If neutrons, rather than photons, are incident, then in order to satisfy the Bragg–Gray principle the wall must be at least as thick as the maximum range of any secondary charged recoil particle that the neutrons produce in it.

As with the air-wall chamber for measuring exposure, condenser-type chambers that satisfy the Bragg–Gray conditions can be used to measure absorbed dose. Prior to exposure, the chamber is charged. The dose can then be inferred from the reduced potential difference across the instrument after it is exposed to radiation.

The determination of dose rate is usually made by measuring the current due to ionization in a chamber that satisfies the Bragg–Gray conditions. As the following example shows, this method is both sensitive and practical.

Example

A chamber satisfying the Bragg–Gray conditions contains 0.15 g of gas with a W value of 33 eV/ip. The ratio of the mass stopping power of the wall and the gas is 1.03. What is the current when the absorbed dose rate in the wall is 1 rad/hr?

Solution

We apply Eq. (10.12) to the dose rate, with $S_w/S_g = 1.03$. From the given conditions, $\dot{D}_w = 1$ rad/hr = (100 erg/g)/(3600 sec) = 0.0278 erg/g sec. The rate of ion-pair production in the gas is, from Eq. (10.12),

$$\dot{N}_g = \frac{\dot{D}_w m S_g}{W S_w} = \frac{0.0278 \text{ erg/g sec} \times 0.15 \text{ g}}{33 \text{ eV/ip} \times 1.6 \times 10^{-12} \text{ erg/eV} \times 1.03} = 7.67 \times 10^7 \frac{\text{ip}}{\text{sec}}. \qquad (10.13)$$

Since the electronic charge is 1.60×10^{-19} C, the current is $7.67 \times 10^7 \times 1.60 \times 10^{-19} = 1.23 \times 10^{-11}$ C/sec = 1.23×10^{-11} A. Simple electrometer circuits can be used to measure currents of 10^{-14} A, corresponding to dose rates less than 1 mrad/hr in this example.

10.5 MEASUREMENT OF X- AND GAMMA-RAY DOSE

Figure 10.4 shows the cross section of a spherical chamber of graphite that encloses CO_2 gas. The chamber satisfies the Bragg–Gray conditions for photons over a wide energy range, and so the dose D_C in the carbon wall can be obtained from the measured ionization in the CO_2 by means of Eq. (10.12). Since carbon is a major constituent of soft tissue, the wall dose approximates that in soft tissue D_t. Calculations show that, for

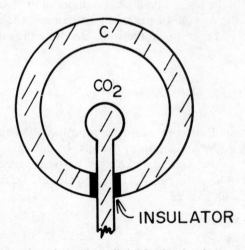

Figure 10.4. Cross section of graphite-walled CO_2 chamber for measuring photon dose.

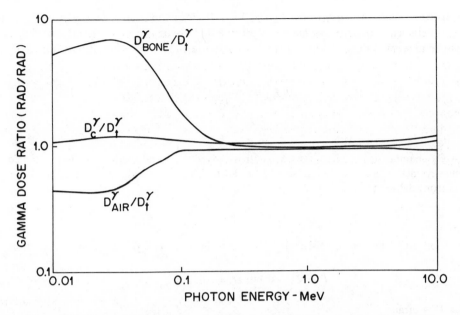

Figure 10.5. Ratio of absorbed doses in bone, air, and carbon to that in soft tissue, D_t.

photon energies between 0.2 MeV and 5 MeV, $D_t = 1.1 D_C$ to within 5%. Thus soft-tissue dose can be measured with an accuracy of 5% with the carbon chamber.

Generally, the dose in low-Z wall materials will approximate that in soft tissue over a wide range of photon energies. This fact leads to the widespread use of plastics and a number of other low-Z materials for gamma-dosimeter walls. The ratio of the absorbed dose in many materials relative to that in soft tissue has been calculated. Figure 10.5 shows several important examples. For photon energies from ~0.1 MeV to ~10 MeV, the ratios for all materials of low atomic number are near unity, because Compton scattering dominates. The curve for bone, in contrast to the other two, rises at low energies due to the larger cross section for photoelectric absorption in the heavier elements of bone (e.g., Ca and P).

10.6 NEUTRON DOSIMETRY

An ionization device, such as that shown in Fig. 10.4, used for measuring gamma-ray dose will show a reading when exposed to neutrons. The response is due to ionization produced in the gas by the charged recoil nuclei struck by neutrons in the walls and gas. However, the amount of ionization will not be proportional to the absorbed dose in tissue unless (1) the walls and gas are tissue equivalent and (2) the Bragg–Gray principle is satisfied for neutrons. As shown in Table 10.2, soft tissue consists chiefly of hydrogen, oxygen, carbon, and nitrogen, all having different cross sections as functions of neutron energy (cf. Fig. 8.2). The carbon wall of the chamber in Fig. 10.4 would respond quite differently from tissue to a field of neutrons of mixed energies, because the three other principal elements of tissue are lacking.

The C–CO_2 chamber in Fig. 10.4 and similar devices *can* be used for neutrons of a given energy if the chamber response has been calibrated experimentally as a function of neutron energy. Table 10.3 shows the relative response $P(E)$ of the C–CO_2 chamber to photons or to neutrons of a given energy for a fluence that delivers 1 rad to soft tissue. If $D_C^n(E)$ is absorbed dose in the carbon wall due to 1 tissue rad of neutrons of energy E and D_C^γ is the absorbed dose in the wall due to 1 tissue rad of photons, then, approximately,

$$P(E) = \frac{D_C^n(E)}{D_C^\gamma}. \tag{10.14}$$

Example

A C–CO_2 chamber exposed to 1 MeV neutrons gives the same reading as that obtained when gamma rays deliver an absorbed dose of 20 mrad to the carbon wall. What absorbed dose would the neutrons deliver to soft tissue?

Solution

From Table 10.3, the neutron tissue dose D_n would be given approximately by the relation

$$P(E)D_n = 0.149 \, D_n = 2.0 \text{ mrad}, \tag{10.15}$$

or $D_n = 134$ mrad.

Tissue-equivalent gases and plastics have been developed for constructing chambers to measure neutron dose directly. These materials are fabricated with the approximate relative atomic abundances shown in Table 10.2. In accordance with the proviso mentioned after Eq. (10.12), the wall of a tissue-equivalent neutron chamber must be at least as thick as the range of a proton having the maximum energy of the neutrons to be monitored.

More often than not, gamma rays are present when neutrons are. In monitoring mixed gamma–neutron radiation fields one generally needs to know the separate contributions that each type of radiation makes to the absorbed dose. One needs this information in order to assign the proper quality factor to the neutron part to obtain the dose equivalent. To this end, two chambers can be exposed — one C–CO_2 and one tissue equivalent — and doses determined by a difference method. The response R_T of the tissue-equivalent instrument provides the combined dose, $R_T = D_\gamma + D_n$. The reading R_C of the C–CO_2 chamber can be expressed as $R_C = D_\gamma + P(E)D_n$, where $P(E)$ is an appropriate average from Table 10.3 for the neutron field in question. The individual doses D_γ and D_n can be inferred from R_T and R_C.

Example

In an unknown gamma–neutron field, a tissue-equivalent ionization chamber registers 8.2 mrad/hr and a C–CO_2 chamber, 2.9 mrad/hr. What are the gamma and neutron dose rates?

Table 10.2. Principal Elements in Soft Tissue of Unit Density

Element	Atoms/cm^3
H	5.98×10^{22}
O	2.45×10^{22}
C	9.03×10^{21}
N	1.29×10^{21}

Table 10.3. Relative Response of C–CO_2 Chamber to Neutrons of Energy E and Photons [Eq. (10.14)]

Neutron energy (MeV)	$P(E)$
0.1	0.109
0.5	0.149
1.0	0.149
2.0	0.145
3.0	0.151
4.0	0.247
5.0	0.168
10.0	0.341
20.0	0.487

Solution

The instruments' responses can be written in terms of the dose rates as

$$\dot{R}_T = \dot{D}_\gamma + \dot{D}_n = 8.2 \tag{10.16}$$

and

$$\dot{R}_C = \dot{D}_\gamma + P(E)\dot{D}_n = 2.9. \tag{10.17}$$

Since we are not given any information about the neutron energy spectrum, we must assume some value of $P(E)$ in order to go further. We choose $P(E) \sim 0.15$, representative of neutrons in the lower MeV to keV range in Table 10.3. Subtracting both sides of Eq. (10.17) from (10.16) gives $\dot{D}_n = (8.2 - 2.9)/(1 - 0.15) = 6.24$ mrad/hr. It follows from (10.16) that $\dot{D}_\gamma = 1.96$ mrad/hr.

Very often, as the example illustrates, the neutron energy spectrum is not known and the difference method may not be accurate.

As mentioned in Section 9.5, the proportional counter provides a direct method of measuring neutron dose, and it has the advantage of excellent gamma discrimination. The pulse height produced by a charged recoil particle is proportional to the energy that the particle deposits in the gas. The Hurst fast-neutron proportional counter is shown in Fig. 10.6. To satisfy the Bragg–Gray principle, the polyethylene walls are made thicker than the range of a 20 MeV proton. The counter gas can be either ethylene (C_2H_4) or cyclopropane (C_3H_6), both having the same $H/C = 2$ ratio as the walls. A recoil proton or carbon nucleus from the wall or gas has high LET. Unless only a small portion of its path is in the gas it will deposit much more energy in the gas than a low-LET secondary electron produced by a gamma ray. Rejection of the small gamma pulses can be accomplished by electronic discrimination. Fast-neutron dose rates as low as 0.001 rad/hr can be measured in the presence of gamma fields with dose rates up to 100 rad/hr. In very intense fields signals from multiple gamma rays can "pile up" and give pulses comparable in size to those from neutrons.

The LET spectra of the recoil particles produced by neutrons (and hence neutron quality factors) depend on neutron energy. Table 10.4 gives the mean quality factors and fluence rates for monoenergetic neutrons that give a dose equivalent of 100 mrem in a 40 hr work week. The quality factors have been computed by averaging over the LET spectra of all charged recoil nuclei produced by the neutrons. For practical applications, using $\bar{Q} = 3$ for neutrons of energies less than 10 keV and $\bar{Q} = 10$ for higher energies will result in little error. Using $\bar{Q} = 10$ for all neutrons is acceptable, but may be overly conservative.

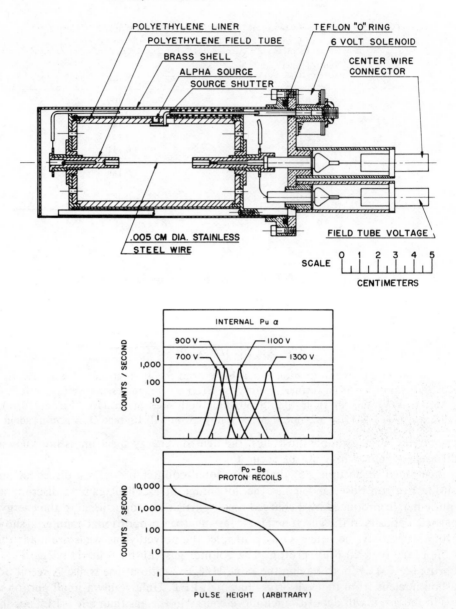

Figure 10.6. Hurst fast-neutron proportional counter. Internal alpha source in wall is used to provide pulses of known size for energy calibration. (Courtesy Oak Ridge National Laboratory, operated by Martin Marietta Energy Systems, Inc., for the Department of Energy)

Thus, in monitoring neutrons for radiation-protection purposes, one should generally know or estimate the neutron energy spectrum or LET spectrum (i.e., the LET spectrum of the recoil particles). Measurement of LET spectra is discussed in Section 10.8. Several methods of obtaining neutron energy spectra were described in Section 9.5. The neutron rem meter, shown in Fig. 9.31, was discussed previously.

Figure 10.7 shows an experimental setup for exposing anthropomorphic phantoms,

Table 10.4. Mean Quality Factors $\overline{Q}$ and Fluence Rates for Monoenergetic Neutrons
that Give a Maximum Dose-Equivalent Rate of 100 mrem in 40 hr†

Neutron energy (eV)	$\overline{Q}$	Fluence rate (cm^{-2} sec^{-1})
0.025 (thermal)	2	680
0.1	2	680
1.0	2	560
10.0	2	560
10^2	2	580
10^3	2	680
10^4	2.5	700
10^5	7.5	115
5×10^5	11	27
10^6	11	19
5×10^6	8	16
10^7	6.5	17
1.4×10^7	7.5	12
6×10^7	5.5	11
10^8	4	14
4×10^8	3.5	10

†From *Protection Against Neutron Radiation*, NCRP Report 38, National Council on Radiation Protection and Measurements, Washington, D.C. (1971).

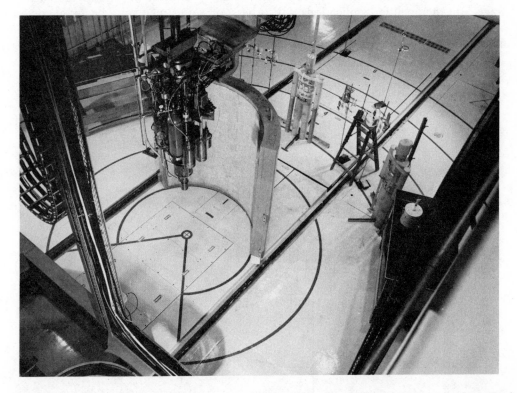

Figure 10.7. Anthropomorphic phantoms, wearing a variety of dosimeters in different positions, can be exposed to neutrons with known fluence and energy spectrum at the Health Physics Research Reactor. (Courtesy Oak Ridge National Laboratory, operated by Martin Marietta Energy Systems, Inc., for the Department of Energy)

wearing various types of dosimeters, to fission neutrons. A bare reactor is positioned above the circle, drawn on the floor, with an intervening shield placed between it and the phantoms, located 3 m away. In this Health Physics Research Reactor facility, the response of dosimeters to neutrons with a known energy spectrum and fluence can be studied.

Intermediate and fast neutrons incident on the body are subsequently moderated and can be backscattered at slow or epithermal energies through the surface they entered. Exposure to these neutrons can, therefore, be monitored by wearing a device, such as a TLD dosimeter enriched in ^{6}Li, that is sensitive to slow neutrons. Such a device is called an albedo-type neutron dosimeter. (For a medium A that contains a neutron source and an adjoining medium B that does not, the albedo is defined in reactor physics as the fraction of neutrons entering B that are reflected or scattered back into A).

10.7 DOSE MEASUREMENTS FOR CHARGED-PARTICLE BEAMS

For radiotherapy and for radiobiological experiments one needs to measure the dose or dose rate in a beam of charged particles. This is often accomplished by measuring the current from a thin-walled ionization chamber placed at different depths in a water target exposed to the beam, as illustrated in Fig. 10.8. The dose rate is proportional to the current. For monoenergetic particles of a given kind (e.g., protons) the resulting "depth-dose" curve has the reversed shape of the mass stopping-power curves in Fig. 4.3. The dose rate is a maximum in the region of the Bragg peak near the end of the particles' range. In therapeutic applications, absorbers or adjustments in beam energy are employed so that the beam stops at the location of a tumor or other tissue to be irradiated. In this way, the dose there (as well as LET) is largest, while the intervening tissue is

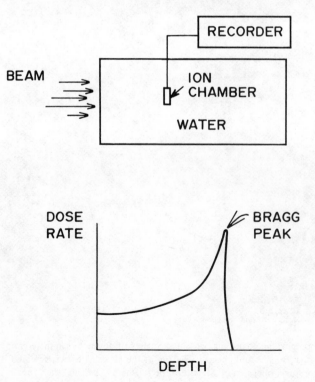

Figure 10.8. Measurement of dose or dose rate as a function of depth in water exposed to a charged-particle beam.

relatively spared. To further spare healthy tissue, a tumor can be irradiated from several directions.

If the charged particles are relatively low-energy protons (≤400 MeV), then essentially all of their energy loss is due to electronic collisions. The curve in Fig. 10.8 will then be similar in shape to that for the mass stopping power. Higher-energy protons undergo significant nuclear reactions, which attenuate the protons and deposit energy by nuclear processes. The depth–dose curve is then different from the mass stopping power. Other particles, such as charged pions, have strong nuclear interactions at all energies, and depth–dose patterns can be quite different.

10.8 DETERMINATION OF LET

To specify dose equivalent, one needs, in addition to the absorbed dose, the LET of incident charged particles or the LET of the charged recoil particles produced by incident neutral radiation (neutrons or gamma rays). As given in Table 10.1, the required quality factors are defined in terms of the LET in water, which, for radiation-protection purposes, is the same as the stopping power (cf. Sections 6.2 and 6.3). Stopping-power values of water for a number of charged particles are available (Figs. 4.3 and 5.1; Tables 4.3 and 5.1). These are used in many applications.

Radiation fields more often than not occur with a spectrum of LET values. H. H. Rossi and coworkers have developed methods for inferring LET spectra directly from measurements made with a proportional counter.† A spherically shaped counter (usually tissue equivalent) is used and a pulse-height spectrum measured in the radiation field. If the counter gas pressure is low, so that a charged particle from the wall does not lose a large fraction of its energy in traversing the gas, then the pulse size is equal to the product of the LET and the chord length. The distribution of isotropic chord lengths x in a sphere of radius R is given by the simple linear expression

$$P(x)\, dx = \frac{x}{2R^2}\, dx; \qquad (10.18)$$

that is, the probability that a given chord has a length between x and $x + dx$ is $P(x)\, dx$, this function giving unity when integrated from $x = 0$ to $2R$. Using analytic techniques, one can, in principle, unfold the LET spectrum from the measured pulse-height spectrum and the distribution $P(x)$ of track lengths through the gas. However, energy-loss straggling and other factors complicate the practical application of this method.

Precise LET determination presents a difficult technical problem. Usually, practical needs are satisfied by using rough estimates of the quality factor based on conservative assumptions.

10.9 DOSE CALCULATIONS

Absorbed dose, LET, and dose equivalent can frequently be obtained reliably by calculations. In this section we discuss several examples.

Alpha and Low-Energy Beta Emitters Distributed in Tissue

When a radionuclide is ingested or inhaled, it can become distributed in various parts of the body. It is then called an internal emitter. Usually a radionuclide entering the body

---

†Cf. *Microdosimetry*, ICRU Report 36, International Commission on Radiation Units and Measurements, Bethesda, MD (1983).

follows certain metabolic pathways and, as a chemical element, preferentially seeks specific body organs. For example, iodine concentrates in the thyroid; radium and strontium are bone seekers. In contrast, tritium (hydrogen) and cesium tend to distribute themselves throughout the whole body. If an internally deposited radionuclide emits particles that have a short range, then their energies will be absorbed in the tissue that contains them. One can then calculate the dose rate in the tissue from the activity concentration there. Such is the case when an alpha or low-energy beta emitter is embedded in tissue. If C denotes the average concentration, in Ci/cm^3, of the radionuclide in the tissue and E denotes the average alpha- or beta-particle energy, in MeV per disintegration, then the energy absorbed per unit volume of tissue is 3.7×10^{10} CE MeV/(cm^3 sec). If the density of the tissue in g/cm^3 is ρ, then the absorbed dose rate is

$$\dot{D} = 3.7 \times 10^{10} \, CE \, \frac{\text{MeV}}{\text{cm}^3 \, \text{sec}} \times 1.6 \times 10^{-6} \, \frac{\text{erg}}{\text{MeV}} \times \frac{1}{\rho \, \text{g/cm}^3} \times \frac{1}{100 \, \text{erg/(g rad)}} \quad (10.19)$$

or

$$\dot{D} = \frac{592CE}{\rho} \, \frac{\text{rad}}{\text{sec}}. \quad (10.20)$$

Note that this procedure gives the average dose rate in the tissue that contains the radionuclide. If the source is not uniformly distributed in the tissue, then the peak dose rate will be higher than that given by Eq. (10.20). The existence of "hot spots" for nonuniformly deposited internal emitters can complicate a meaningful organ-dose evaluation. Nonuniform deposition can occur, for example, when inhaled particulate matter becomes embedded in different regions of the lungs.

Example
What is the average dose rate in a 50 g sample of soft tissue that contains 3.25 μCi of ^{14}C?

Solution
Soft tissue has unit density, and so $\rho = 1$ g/cm^3 and $C = 3.25 \times 10^{-6}/50 = 6.50 \times 10^{-8}$ Ci/cm^3. The mean energy of ^{14}C beta particles is (Appendix D) $E = 0.045$ MeV. It follows from Eq. (10.20) that $\dot{D} = 592(6.50 \times 10^{-8})(0.045) = 1.73 \times 10^{-6}$ rad/sec.

Example
If the tissue sample in the last example is spherical in shape and the ^{14}C is distributed uniformly, make a rough estimate of the fraction of the beta-particle energy that escapes from the tissue.

Solution
We compare the range of a beta particle having the average energy $E = 0.045$ MeV with the radius of the tissue sphere. The sphere radius r is found by writing $50 = 4\pi r^3/3$, which gives $r = 2.29$ cm. From Table 5.1, the range of the beta particle is $R = 0.0036$ cm. Thus a beta particle of average energy emitted no closer than 0.0036 cm from the surface of the tissue sphere will be absorbed in the sphere. The fraction F of the tissue volume that lies at least this close to the surface can be calculated from the difference in the volumes of spheres with radii r and $r - R$. Alternatively, we can differentiate the expression for the volume, $V = 4\pi r^3/3$:

$$F = \frac{dV}{V} = \frac{3 \, dr}{r} = \frac{3 \times 0.0036}{2.29} = 4.72 \times 10^{-3}. \quad (10.21)$$

If we assume that one-half of the average beta-particle energy emitted in this outer layer is absorbed in the sphere and the other half escapes, then the fraction of the emitted beta-particle energy that escapes from the sphere is $F/2 = 2.4 \times 10^{-3}$, a very small amount.

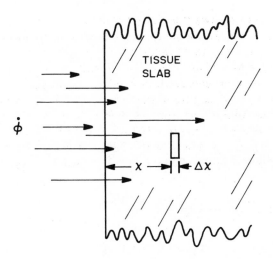

Figure 10.9. Uniform, parallel beam of charged particles normally incident on thick tissue slab. Fluence rate = $\dot{\varphi}$ cm^{-2} sec^{-1}.

Charged-Particle Beams

Figure 10.9 represents a uniform, parallel beam of monoenergetic charged particles of a given kind (e.g., protons) normally incident on a thick tissue slab with fluence rate $\dot{\varphi}$ cm^{-2} sec^{-1}. To calculate the dose rate at a given depth x in the slab, we consider a thin, disc-shaped volume element with thickness Δx in the x direction and area A normal to the beam. The rate of energy deposition in the volume element is $\dot{\varphi} A (-dE/dx) \Delta x$, where $-dE/dx$ is the (collisional) stopping power of the beam particles as they traverse the slab at depth x. [We ignore straggling (Chapter 6).] The dose rate $\dot{D}$ is obtained by dividing by the mass $\rho A \Delta x$ of the volume element, where ρ is the density of the tissue:

$$\dot{D} = \frac{\dot{\varphi} A (-dE/dx) \Delta x}{\rho A \Delta x} = \dot{\varphi} \left(-\frac{dE}{\rho \, dx} \right). \tag{10.22}$$

It follows that the dose per unit fluence at any depth is equal to the mass stopping power for the particles at that depth. If, for example, the mass stopping power is 3 MeV cm^2/g, then the dose per unit fluence can be expressed as 3 MeV/g. This analysis assumes that energy is deposited only by means of electronic collisions (stopping power). As discussed after Fig. 10.8, if significant nuclear interactions occur, e.g., as with high-energy protons, then accurate depth–dose curves cannot be calculated from Eq. (10.22). One can then resort to Monte Carlo calculations, in which the fates of individual incident and secondary particles are handled statistically on the basis of the cross sections for the various nuclear interactions that can occur.

Point Source of Gamma Rays

We next derive a simple formula for computing the dose rate in air from a point gamma source of strength C curies that emits an average total photon energy of E MeV per disintegration. The rate of energy release in the form of photons escaping from the source is

$$CE \, (\text{Ci MeV}) \times 3.7 \times 10^{10} \, \text{sec}^{-1} \, \text{Ci}^{-1} \times 1.6 \times 10^{-6} \, \text{erg/MeV} = 5.92 \times 10^4 CE \, \text{erg/sec.}$$

$$\tag{10.23}$$

Neglecting attenuation in air, we can write for the rate of energy flow per unit area (= intensity) through the surface of a sphere of radius r (cm) surrounding the source

$$I = \frac{5.92 \times 10^4 CE}{4\pi r^2} = \frac{4.71 \times 10^3 CE}{r^2} \quad \frac{\text{erg}}{\text{cm}^2\,\text{sec}}. \tag{10.24}$$

The rate of energy absorption in a volume element of air, having unit area and thickness dr at r (Fig. 10.10) is given by $I\mu_A\,dr$, where μ_A is the average energy absorption coefficient in cm^{-1} (Section 7.7) for the photons. If the density of the air is ρ g/cm^3, then the mass of air in the volume element in grams is $\rho\,dr$. The absorbed dose rate in the volume element is

$$\dot{D} = \frac{I\mu_A\,dr}{\rho\,dr} = \frac{4.71 \times 10^3 CE\mu_A}{r^2\rho} \quad \frac{\text{erg}}{\text{g sec}}. \tag{10.25}$$

At STP, $\rho = 0.001293$ g/cm^3, and so, converting to rad, we have

$$\dot{D} = \frac{3.64 \times 10^4 CE\mu_A}{r^2} \quad \frac{\text{rad}}{\text{sec}}. \tag{10.26}$$

This equation leads to a handy formula for gamma dose rate over the Compton range of photon energies, for which the energy absorption coefficient μ_A is approximately constant. As seen from Fig. 7.12, for photons of energies between about 60 keV and 2 MeV, the mass-absorption coefficient for air is, approximately, $\mu_A/\rho \cong 0.027$ cm^2/g, giving $\mu_A \cong 3.5 \times 10^{-5}$ cm^{-1}. Substitution of this value into Eq. (10.26) gives

$$\dot{D} \cong \frac{1.27CE}{r^2} \quad \frac{\text{rad}}{\text{sec}}. \tag{10.27}$$

At a distance of 1 m ($r = 100$ cm) from the source, we obtain for the hourly dose rate

$$\dot{D} \cong \frac{1.27CE}{10^4} \quad \frac{\text{rad}}{\text{sec}} \times \frac{3600\,\text{sec}}{\text{hr}} = 0.46CE \quad \frac{\text{rad}}{\text{hr}} \quad \text{at 1 m.} \tag{10.28}$$

This is a convenient formula for estimating the gamma dose rate in air at a distance of 1 m from a source of C curies where E is the photon energy in MeV emitted per disintegration. Since an absorbed dose of 0.876 rad in air corresponds to an exposure of 1 R [Eq. (10.4)], we can write for the exposure rate $\dot{\Gamma}$ in place of (10.28)

$$\dot{\Gamma} \cong 0.5CE \text{ R/hr} \quad \text{at 1 m.} \tag{10.29}$$

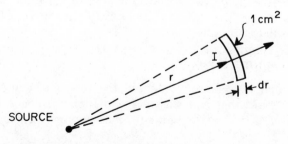

Figure 10.10. Rate of energy absorption by air thickness dr per cm^2 is $I\mu_A\,dr$, where μ_A is the energy absorption coefficient (cm^{-1}) (Section 7.7).

Also convenient is the expression for the exposure rate in R/hr at 1 ft (= 30.5 cm). From Eq. (10.27) we find

$$\dot{\Gamma} \cong \frac{1.27CE}{(30.5)^2} \frac{\text{rad}}{\text{sec}} \times 3600 \frac{\text{sec}}{\text{hr}} \times \frac{1}{0.876 \, \text{rad/R}}, \tag{10.30}$$

or

$$\dot{\Gamma} \cong 5.6CE \cong 6CE \, \frac{\text{R}}{\text{hr}} \quad \text{at 1 ft.} \tag{10.31}$$

Example
Estimate the exposure rate at a distance of 2 ft from a 100 mCi point source of ^{137}Cs.

Solution
The isotope emits a 0.662 MeV photon in 85% of the decays (Appendix D). The average photon energy emitter per disintegration is, therefore, $E = 0.85 \times 0.662 = 0.563$ MeV. The source strength is $C = 0.1$ Ci. Equation (10.31) gives the exposure rate at 1 ft. Taking into account the inverse-square dependence, we have, for the rate at 2 ft,

$$\dot{\Gamma} \cong \frac{6 \times 0.1 \times 0.563}{2^2} = 0.084 \, \text{R/hr.} \tag{10.32}$$

Example
What value of E should be used for ^{60}Co in Eqs. (10.29) and (10.31)?

Solution
^{60}Co emits two photons per disintegration, with energies 1.173 MeV and 1.332 MeV. Therefore, $E = 2.505$ MeV.

Neutrons

As discussed in Chapter 8, fast neutrons lose energy primarily by elastic scattering while slow and thermal neutrons have a high probability of being captured. The two principal capture reactions in tissue are ^{1}H(n,γ)^{2}H and ^{14}N(n,p)^{14}C. Slow neutrons are quickly thermalized by the body. The first capture reaction releases a 2.22 MeV gamma ray, which may deposit a fraction of its energy in escaping the body. In contrast, the nitrogen-capture reaction releases an energy of 0.626 MeV, which is deposited by the proton and recoil carbon nucleus in the immediate vicinity of the capture site. The resulting dose from exposure to thermal neutrons can be calculated, as the next example illustrates.

Example
Calculate the dose in a 150 g sample of soft tissue exposed to a fluence of 10^7 thermal neutrons/cm^2.

Solution
From Table 10.2, the density of nitrogen atoms in soft tissue is $N = 1.29 \times 10^{21}$ cm^{-3}, ^{14}N being over 99.6% abundant. The thermal-neutron capture cross section is $\sigma = 1.70 \times 10^{-24}$ cm^2 (Section 8.6). Each capture event by nitrogen results in the deposition of energy $E = 0.626$ MeV, which will be absorbed in the unit-density sample ($\rho = 1$ g/cm^3). The number of interactions per unit fluence per unit volume of the tissue is $N\sigma$. The dose from the fluence $\varphi = 10^7$ cm^{-2} is therefore

$$D = \frac{\varphi N\sigma E}{\rho} = \frac{10^7 \text{ cm}^{-2} \times 1.29 \times 10^{21} \text{ cm}^{-3} \times 1.70 \times 10^{-24} \text{ cm}^2 \times 0.626 \text{ MeV}}{1 \text{ g/cm}^3}$$

$$\times \frac{1.6 \times 10^{-6} \text{ erg}}{\text{MeV}} \times \frac{1}{100 \text{ erg/(g rad)}} = 2.20 \times 10^{-4} \text{ rad.} \tag{10.33}$$

Some additional dose would be deposited by the gamma rays produced by the ^{1}H(n,γ)^{2}H reaction, for which the cross section is 3.3×10^{-25} cm^2. However, in a tissue sample as small as 150 g, the contribution of this gamma-ray dose is negligible. It is not negligible in a large target, such as the whole body.

For fast neutrons, since the average energy transferred in a collision with a nucleus is about one-half the neutron energy (Section 8.5), the "first-collision" dose in tissue can be readily calculated. The first-collision dose is that delivered by neutrons that make only a single collision in the target. The first-collision dose closely approximates the actual dose when the mean free path of the neutrons is large compared with the dimensions of the target. A 5 MeV neutron, for example, has a macroscopic cross section in soft tissue of 0.051 cm^{-1}, and so its mean free path is $1/0.051 = 20$ cm. Thus, in a target the size of the body, a large fraction of 5 MeV neutrons will not make multiple collisions, and the first-collision dose can be used as a basis for approximating the actual dose. The first-collision dose is, of course, always a lower bound to the actual dose. Moreover, fast neutrons deposit most of their energy in tissue by means of collisions with hydrogen. Therefore, calculating the first-collision dose with tissue hydrogen often provides a simple, lower-bound estimate of fast-neutron dose.

Example
Calculate the first-collision dose to tissue hydrogen per unit fluence of 5 MeV neutrons.

Solution
The density of H atoms is $N = 5.98 \times 10^{22}$ cm^{-3} (Table 10.2) and the cross section for scattering 5 MeV neutrons is $\sigma = 1.61 \times 10^{-24}$ cm^2 (Fig. 8.2). The mean energy loss per collision, $\bar{E} = 2.5$ MeV, is one-half the incident neutron energy. The dose per unit neutron fluence from collisions with hydrogen is therefore (tissue density $\rho = 1$ g/cm^3)

$$D = \frac{N\sigma\bar{E}}{\rho} = \frac{5.98 \times 10^{22} \text{ cm}^{-3} \times 1.61 \times 10^{-24} \text{ cm}^2 \times 2.5 \text{ MeV}}{1 \text{ g/cm}^3} \times \frac{1.6 \times 10^{-6} \text{ erg/MeV}}{100 \text{ erg/(g rad)}}$$

$$= 3.85 \times 10^{-9} \text{ rad cm}^2. \tag{10.34}$$

Note that the units of "rad per neutron/cm^2" are rad cm^2.

Similar calculations of the first-collision doses due to collisions of 5 MeV neutrons with the O, C, and N nuclei in soft tissue give, respectively, contributions of 0.244×10^{-9}, 0.079×10^{-9}, and 0.024×10^{-9} rad cm^2, representing in total about an additional 10%. Detailed analysis shows that hydrogen recoils contribute approximately 85–95% of the first-collision soft-tissue dose for neutrons with energies between 10 keV and 10 MeV. Table 10.5 shows the analysis of first-collision neutron doses.

Detailed calculations of multiple neutron scattering and energy deposition in slabs and in anthropomorphic phantoms, containing soft tissue, bone, and lungs, have been carried out by Monte Carlo techniques. Computer programs are available, based on experimental cross-section data and theoretical algorithms, to transport individual neutrons through a target with the same statistical distribution of events that neutrons have in nature. Such Monte Carlo calculations can be made under general conditions of target

Table 10.5. Analysis of First-Collision Dose for Neutrons in Soft Tissue†

Neutron energy (MeV)	First-collision dose per unit neutron fluence for collisions with various elements (10^{-9} rad cm^2)				
	H	O	C	N	Total
0.01	0.091	0.002	0.001	0.000	0.094
0.02	0.172	0.004	0.001	0.001	0.178
0.03	0.244	0.005	0.002	0.001	0.252
0.05	0.369	0.008	0.003	0.001	0.381
0.07	0.472	0.012	0.004	0.001	0.489
0.10	0.603	0.017	0.006	0.002	0.628
0.20	0.914	0.034	0.012	0.003	0.963
0.30	1.14	0.052	0.016	0.003	1.21
0.50	1.47	0.122	0.023	0.004	1.62
0.70	1.73	0.089	0.029	0.005	1.85
1.0	2.06	0.390	0.036	0.007	2.49
2.0	2.78	0.156	0.047	0.012	3.00
3.0	3.26	0.205	0.045	0.018	3.53
5.0	3.88	0.244	0.079	0.024	4.23
7.0	4.22	0.485	0.094	0.032	4.83
10.0	4.48	0.595	0.157	0.046	5.28
14.0	4.62	1.10	0.259	0.077	6.06

†From "Measurement of Absorbed Dose of Neutrons and of Mixtures of Neutrons and Gamma Rays," *National Bureau of Standards Handbook 75*, Washington, D.C. (1961).

composition and geometry as well as incident neutron spectra and directions of incidence. Compilations of the results for a large number of neutrons then provide dose and LET distributions as functions of position, as well as any other desired information, to within the statistical fluctuations of the compilations. Using a larger number of neutron histories reduces the variance in the quantities calculated, but increases computer time.†

Figure 10.11 shows the results of Monte Carlo calculations carried out for 5 MeV neutrons incident normally on a 30 cm soft-tissue slab, approximating the thickness of the body. (The geometry is identical to that shown for the charged particles in Fig. 10.9.) The curve labeled E_T is the total dose, E_p is the dose due to H recoil nuclei (protons), E_γ is the dose from gamma rays from the ^{1}H(n, γ)^{2}H slow-neutron capture reaction, and E_H is the dose from the heavy (O,C,N) recoil nuclei. The total dose builds up somewhat in the first few cm of depth and then decreases as the beam becomes degraded in energy and neutrons are absorbed. The proton and heavy-recoil curves, E_p and E_H, show a similar pattern. As the neutrons penetrate, they are moderated and approach thermal energies. This is reflected in the rise of the gamma-dose curve, E_γ, which has a broad maximum over the region from about 6 cm to 14 cm. Note that the total dose decreases by an order of magnitude between the front and back of the slab.

The result of our calculation of the first-collision dose, $D = 3.85 \times 10^{-9}$ rad cm^2, due to proton recoils in the last example can be compared with the curve for E_p in Fig. 10.11. At the slab entrance, $E_p = 4.8 \times 10^{-9}$ rad cm^2 is greater than D, which, as we pointed out, is a lower bound for the actual dose from proton recoils. A number of neutrons are

---

†For a description of the basic Monte Carlo technique and its application to radiation dosimetry the reader is referred to the paper by J. E. Turner, H. A. Wright, and R. N. Hamm, A Monte Carlo primer for health physicists, *Health Phys.* **48**, 717–733 (1985).

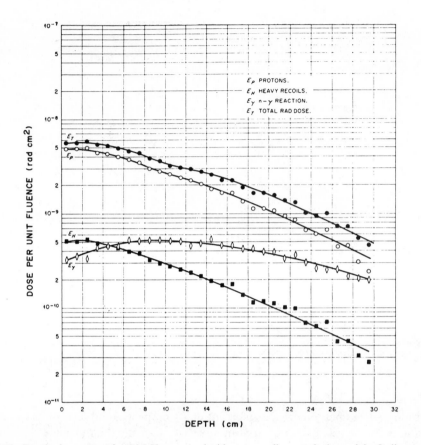

Figure 10.11. Depth–dose curves for 5 MeV neutrons incident normally on soft-tissue slab. Ordinate gives dose per unit fluence at different depths shown on the abscissa. [From "Protection Against Neutron Radiation up to 30 Million Electron Volts," *National Bureau of Standards Handbook 63*, p. 44, Washington, D.C. (1957)]

backscattered from within the slab to add to the first-collision dose deposited directly by the incident 5 MeV neutrons.

10.10 OTHER DOSIMETRIC CONCEPTS AND QUANTITIES

Kerma

A quantity related to dose for indirectly ionizing radiation (photons and neutrons) is the initial kinetic energy of all charged particles liberated by the radiation per unit mass. This quantity, which has the dimensions of absorbed dose, is called the *kerma* (Kinetic Energy Released per unit MAss). By definition, kerma includes energy that may subsequently appear as bremsstrahlung and it also includes Auger-electron energies. The absorbed dose generally builds up behind a surface irradiated by a beam of neutral particles to a depth comparable with the range of the secondary charged particles generated (cf. Fig. 10.11). The kerma, on the other hand, decreases steadily because of the attenuation of the primary radiation with increasing depth.

The first-collision "dose" calculated for neutrons in the last section is, more precisely stated, the first-collision "kerma." The two are identical as long as all of the initial kinetic energy of the recoil charged particles is absorbed locally in the target. Absorbed dose and kerma can be substantially different when bremsstrahlung escapes from the target. For additional information about kerma and other quantities discussed in this section, the reader is referred to ICRU Report 33.[†]

Microdosimetry

Absorbed dose is an averaged quantity and, as such, does not specifically reflect the stochastic, or statistical, nature of energy deposition by ionizing radiation in matter. Statistical aspects are especially important when one considers dose in small regions of an irradiated target, such as cell nuclei or other subcellular components. The subject of microdosimetry deals with these phenomena. Consider, for example, cell nuclei having a diameter ~5 μm. If the whole body receives a uniform dose of 100 mrad of low-LET radiation, then $\frac{2}{3}$ of the nuclei will have no ionizations at all and $\frac{1}{3}$ will receive an average dose of ~300 mrad. If, on the other hand, the whole body receives 100 mrad from fission neutrons, then 99.8% of the nuclei will receive no dose and 0.2% will have a dose of ~50 rad.[‡] The difference arises from the fact that the neutron dose is deposited by recoil nuclei, which have a short range. A proton having an energy of 500 keV has a range of 8×10^{-4} cm = 8 μm in soft tissue (Table 4.3), compared with a range of 0.174 cm for a 500 keV electron (Table 5.1). Both particles deposit the same energy. The proton range is comparable to the cell-nucleus diameter; the electron travels the equivalent of ~1740/5 = 350 nuclear diameters.

Specific Energy

When a particle or photon of radiation interacts in a volume of tissue, one refers to the interaction as an energy-deposition event. The energy deposited by the incident particle and all of the secondary electrons that interact in the volume is called the energy imparted, ϵ. Because of the statistical nature of radiation interaction, the energy imparted is a stochastic quantity. The specific energy (imparted) in a volume of mass m is defined as

$$z = \frac{\epsilon}{m}. \tag{10.35}$$

It has the dimensions of absorbed dose. When the volume is irradiated, it experiences a number of energy-deposition events, which are characterized by the distribution in the values of z that occur. The average absorbed dose in the volume from a number of events is the mean value of z. Studies of the distributions in z from different radiations in different-size small volumes of tissue are made in microdosimetry.

The example cited from the BEIR-III Report in the next-to-last paragraph can be conveniently described in terms of the distribution of specific energy z in the cell nuclei. For the low-LET radiation, $\frac{2}{3}$ of the nuclei have $z = 0$; in the other $\frac{1}{3}$, z varies widely with a mean value of ~300 mrad. For the fission neutrons, 99.8% of the nuclei have $z = 0$; in the other 0.2%, z varies by many orders of magnitude with a mean value of ~50 rad.

[†]*Radiation Quantities and Units*, ICRU Report 33, International Commission on Radiation Units and Measurements, Bethesda, MD (1980).

[‡]*The Effects of Populations of Exposure to Low Levels of Ionizing Radiation: 1980*, BEIR-III Report, p. 14. National Academy of Sciences, Washington, D.C. (1980).

10.11 PROBLEMS

1. (a) What is the average absorbed dose in a 40 cm^3 region of a body organ (density $= 0.93$ g/cm^3) which absorbs 3×10^5 MeV of energy from a radiation field?
   (b) If the energy is deposited by ionizing particles with an LET of 10 keV/μm in water, what is the dose equivalent?
   (c) Express the answers to (a) and (b) in both rads and rems as well as Gy and Sv.

2. A portion of the body receives 15 mrad from radiation with a quality factor $Q = 6$ and 22 mrad from radiation with $Q = 10$. (a) What is the total dose? (b) What is the total dose equivalent?

3. A beam of X-rays produces 4 esu of charge per second in 0.08 g of air. What is the exposure rate?

4. If all of the ion pairs are collected in the last example, what is the current?

5. A free-air ionization chamber operating under saturation conditions has a sensitive volume of 12 cm^3. Exposed to a beam of X-rays, it gives a reading of 5×10^{-6} mA. The temperature is 18°C and the pressure is 756 torr. What is the exposure rate?

6. A free-air ionization chamber with a sensitive volume of 14 cm^3 is exposed to gamma rays. A reading of 10^{-11} A is obtained under saturation conditions when the temperature is 20°C and the pressure is 762 torr.
   (a) Calculate the exposure rate.
   (b) How would the exposure rate be affected if the temperature dropped 2°C?

7. A current of 10^{-4} A is produced in a free-air chamber operating under saturation conditions. The sensitive volume, in which electronic equilibrium exists, contains 0.022 g of air. What is the exposure rate? (Assume $W = 34$ eV/ip.)

8. A pocket air-wall ionization chamber with a capacitance of 10^{-11} F contains 2.7 cm^3 of air. If it can be charged to 240 V, what is the maximum exposure that it can measure?

9. A pocket dosimeter with a volume of 2.2 cm^3 and capacitance of 8 pF was fully charged at a potential difference of 200 V. After being worn during a gamma-ray exposure, the potential difference was found to be 192 V. Assume STP conditions.
   (a) Calculate the exposure.
   (b) What charging voltage would have to be applied to the dosimeter if it is to read exposures of up to 3 R?

10. State the Bragg–Gray principle and the conditions for its validity.

11. What minimum wall thickness of carbon ($\rho = 2.62$ g/cm^3) is needed to satisfy the Bragg–Gray principle if the chamber pictured in Fig. 10.4 is to be used to measure absorbed dose from photons with energies up to 5 MeV? (Assume same mass stopping powers for carbon and water and use numerical data given in Chapter 5.)

12. An ionization chamber that satisfies the Bragg–Gray principle contains 0.12 g of CO_2 gas ($W = 33$ eV/ip). When exposed to a beam of gamma rays, a saturation current of 4.4×10^{-10} A is observed. What is the absorbed-dose rate in the gas?

13. An ionization chamber like that shown in Fig. 10.4 was exposed to gamma rays and 2.8×10^{12} ion pairs were produced per gram of CO_2 ($W = 33$ eV/ip). What was the absorbed dose in the carbon walls? The mass stopping powers of graphite and CO_2 for 1 MeV photons are, respectively, 1.648 and 1.680 MeV cm^2/g.

14. What disadvantages are there in using the C–CO_2 chamber to monitor neutrons?

15. In a laboratory test, a tissue-equivalent (TE) chamber and a C–CO_2 chamber both show correct readings of 1 rad in response to a gamma-ray exposure that produces 1 rad in tissue (i.e., when given a "tissue rad"). When exposed to a tissue rad of 0.5 MeV neutrons, the TE chamber reads 1 rad and the C–CO_2 chamber shows 0.149 rad. When the two chambers are used together to monitor an unknown mixed gamma-neutron field, the TE chamber reads 27 mrad/hr and the C–CO_2 chamber reads 21 mrad/hr. What are the individual gamma and neutron dose rates? (Assume that the neutrons have an energy of 0.5 MeV.)

16. Why should a neutron dosimeter contain a significant amount of hydrogen?

17. In a proportional counter such as the one shown in Fig. 10.6, why, on the average, would pulses produced by gamma photons be much smaller than pulses produced by fast neutrons?

18. If $W = 30$ eV/ip for a 2 MeV proton and $W = 40$ eV/ip for a 1 MeV carbon recoil nucleus in a proportional-counter gas, what is the ratio of the pulse heights produced by these two particles if they stop completely in the gas?

19. An air ion chamber having a volume of 2.5 cm^3 (STP) is placed at a certain depth in a water tank, as shown in Fig. 10.8. An electron beam incident on the tank produces a current of 0.004 μA in the chamber. What is the dose rate at that depth?

20. A 75 μA parallel beam of 4 MeV electrons passes normally through the flat surface of a sample of soft tissue in the shape of a disc. The diameter of the disc is 2 cm and its thickness is 0.5 cm. Calculate the average absorbed dose rate in the disc.

21. What is the average whole-body dose rate in a 22 g mouse that contains 5 μCi of ^{14}C distributed in its body?

22. A patient receives an injection of 3 mCi of ^{131}I, 30% of which goes to the thyroid, having a mass of 20 g. What is the average dose rate in the organ?

23. Tritium often gets into body water following an exposure and quickly becomes distributed uniformly throughout the body. What uniform concentration of ^{3}H, in Ci/g, would give a dose equivalent rate of 100 mrem/wk?

24. A soft-tissue disc with a radius of 0.5 cm and thickness of 1 mm is irradiated normally on its flat surface by a 6 μA beam of 100 MeV protons. Calculate the average dose rate in the sample.

25. An experiment is planned in which bean roots are to be placed in a tank of water at a depth of 2.2 cm and irradiated by a parallel beam of 10 MeV electrons incident on the surface of the water. What fluence rate would be needed to expose the roots at a dose rate of 1000 rad/min?

26. What is the exposure rate at a distance of 1 ft from a 20 mCi, unshielded, point source of ^{60}Co?

27. What is the activity of an unshielded point source of ^{60}Co if the exposure rate at 20 m is 6 R/min?

28. A worker accidently strayed into a room in which a small, bare vial containing 23 Ci of ^{131}I was being used to expose a sample. He remained in the room approximately 10 min, standing at a lab bench 5 m away from the source. Estimate the dose that the worker received.

29. A parallel beam of monoenergetic photons emerged from a source when the shielding was removed for a short time. The photon energy $h\nu$ and the total fluence φ of photons are known.
   (a) Write a formula from which one can calculate the absorbed dose in air in rad from $h\nu$, expressed in MeV, and φ, expressed in cm^{-2}.
   (b) Write a formula for calculating the exposure in R.

30. The thermal-neutron capture cross section for the ^{14}N(n,p)^{14}C reaction is 1.70 barn. Calculate (a) the Q value for the reaction and (b) the resulting dose in soft tissue per unit fluence of thermal neutrons.

31. A 100 cm^3 sample of water is exposed to 1500 thermal neutrons per cm^2 per sec. How many photons are emitted per second as a result of neutron capture by hydrogen? The cross section for the ^{1}H(n,γ)^{2}H reaction is 3.3×10^{-25} cm^2.

32. A uniform target with a volume of 5 L is exposed to 100 thermal neutrons per cm^2 per sec. It contains an unknown number of hydrogen atoms. While exposed to the thermal neutrons, it emits 1.11×10^4 photons/sec as the result of thermal-neutron capture by hydrogen (cross section = 0.33 barn). No other radiation is emitted. What is the density of H atoms in the target? Neglect attenuation of the neutrons and photons as they penetrate the target.

33. (a) Calculate the average recoil energies of a hydrogen nucleus and a carbon nucleus elastically scattered by 4 MeV neutrons.
   (b) What can one say about the relative contributions that these two processes make to absorbed dose and dose equivalent in soft tissue?

34. Using Table 10.5, plot the percentage of the first-collision tissue dose that is due to elastic scattering from hydrogen for neutrons with energies between 0.01 MeV and 14.0 MeV.

35. Calculate the first-collision dose rate per unit fluence for 14 MeV neutrons based on their interactions with tissue hydrogen alone. Compare the result with Table 10.5.

36. From Table 10.5, the total first-collision dose per unit fluence for 14 MeV neutrons in soft tissue is 6.06×10^{-9} rad cm^2. From Table 10.4, the average quality factor for 14 MeV neutrons

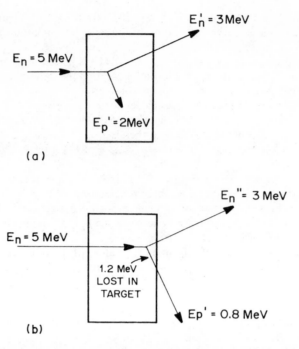

Figure 10.12. Two examples of a single collision of a 5 MeV neutron with a proton in a 1 g target.

is 7.5. Use these two values to estimate the constant fluence rate of 14 MeV neutrons that gives a first-collision dose equivalent of 100 mrem in 40 hr. How do you account for the lower value, 12 neutrons/(cm² sec), given in the last column of Table 10.4?

37. Calculate the first-collision dose per unit fluence that results from the scattering of 10 MeV neutrons by the oxygen in soft tissue. Compare your answer with the value given in Table 10.5.
38. In Fig. 10.11, why does the ratio E_p/E_γ decrease with increasing depth?
39. Figure 10.12 shows two examples of a single collision of a 5 MeV neutron ($E_n = 5$ MeV) with a proton in a 1 g target. In both instances the neutron loses 2 MeV ($E_n' = 3$ MeV) to the proton ($E_p' = 2$ MeV) and escapes from the target. In (a), the recoil proton stops completely in the target. In (b), the collision occurs near the back surface and the proton loses only 1.2 MeV in the target before escaping with an energy $E_p'' = 0.8$ MeV. (a) What is the average kerma in the target in both instances? (b) What is the absorbed dose in both cases?
40. Calculate the specific energy for the examples shown in Figs. 10.12(a) and (b) (see the last problem).

CHAPTER 11
CHEMICAL AND BIOLOGICAL EFFECTS OF RADIATION

11.1 TIME FRAME FOR RADIATION EFFECTS

To be specific, we describe the chemical changes produced by ionizing radiation in liquid water, which are relevant to understanding biological effects. Mammalian cells are typically ~70–85% water, 10–20% proteins, ~10% carbohydrates, and ~2–3% lipids.

Ionizing radiation produces abundant secondary electrons in matter. As discussed in Section 4.2 for heavy particles and in Section 5.7 for electrons, most secondary electrons are produced in water with energies in the range ~30–80 eV. The secondaries slow down very quickly ($\leq 10^{-15}$ sec) to subexcitation energies; i.e., energies below the threshold required to produce electronic transitions (~7.4 eV for liquid water). Various temporal stages of radiation action can be identified, as we now discuss. The time scale for some important radiation effects, summarized in Table 11.1, covers over 20 orders of magnitude.

11.2 PHYSICAL AND PRECHEMICAL CHANGES IN IRRADIATED WATER

The initial changes produced by radiation in water are the creation of ionized and excited molecules, H_2O^+ and H_2O^*, and free, subexcitation electrons. These species are produced in $\leq 10^{-15}$ sec in local regions of a track. Although an energetic charged particle may take longer to stop (Sections 4.8 and 5.6), we shall see that portions of the same

Table 11.1. Time Frame for Effects of Ionizing Radiation

Times	Events
Physical stage $\leq 10^{-15}$ sec	Formation of H_2O^+, H_2O^*, and subexcitation electrons, e^-, in local track regions ($\leq 0.1\ \mu m$)
Prechemical stage ~10^{-15} sec to ~10^{-11} sec	Three initial species replaced by H_3O^+, OH, e_{aq}^-, H, H_2, and O
Chemical stage ~10^{-11} sec to ~10^{-6} sec	The four species H_3O^+, OH, e_{aq}^-, and H diffuse and either react with one another or become widely separated. Intratrack reactions essentially complete by ~10^{-6} sec
Biological stages $\leq 10^{-3}$ sec	Radical reactions with biological molecules complete
≤ 1 sec	Biochemical changes
Hours	Cell division affected
Days	Gastrointestinal and central nervous system changes
Weeks	Lung fibrosis develops
Years	Cataracts and cancer may appear; genetic effects in offspring

track that are separated by more than ~0.1 μm develop independently. Thus we say that the initial physical processes are over in $\leq 10^{-15}$ sec in local track regions.

The water begins to adjust to the sudden physical appearance of the three species even before the molecules can move appreciably in their normal thermal agitation. At room temperature, a water molecule can move an average distance roughly equal to its diameter (2.9 Å) in ~10^{-11} sec. Thus, 10^{-11} sec after passage of a charged particle marks the beginning of the ordinary, diffusion-controlled chemical reactions that take place within and around the particle's path. During this prechemical stage, from ~10^{-15} sec to ~10^{-11} sec, the three initial species produced by the radiation induce changes as follows. First, in about 10^{-14} sec, an ionized water molecule reacts with a neighboring molecule, forming a hydronium ion and a hydroxyl radical:

$$H_2O^+ + H_2O \rightarrow H_3O^+ + OH. \tag{11.1}$$

Second, the excited water molecules get rid of their energy either by losing an electron, thus becoming an ion and proceeding according to the reaction (11.1), or by molecular dissociation:

$$H_2O^* \rightarrow \begin{cases} H_2O^+ + e^- \\ H + OH \\ H_2 + O. \end{cases} \tag{11.2}$$

The vibrational periods of the water molecule are ~10^{-14} sec, which is the time that characterizes the dissociation process. Third, the subexcitation electrons migrate, losing energy by vibrational and rotational excitation of water molecules, and become thermalized by times ~10^{-11} sec. Moreover, the thermalized electrons orient the permanent dipole moments of neighboring water molecules, forming a cluster, called a hydrated electron. We denote the thermalization–hydration process symbolically by writing

$$e^- \rightarrow e_{aq}^-, \tag{11.3}$$

where the subscript aq refers to the fact that the electron is hydrated (aqueous solution). These changes are summarized for the prechemical stage in Table 11.1. Of the six species formed, H_2 does not react further, and O quickly forms hydrogen peroxide, $O + H_2O \rightarrow H_2O_2$, which also does not react further.

11.3 CHEMICAL STAGE

At ~10^{-11} sec after passage of a charged particle in water, the four chemically active species H_3O^+, OH, e_{aq}^-, and H are located near the positions of the original H_2O^+, H_2O^*, and e^- that triggered their formation. Three of the new reactants, OH, e_{aq}^-, and H, are free radicals, i.e., chemical species with unpaired electrons. The reactants begin to migrate randomly about their initial positions in thermal motion. As their diffusion in the water proceeds, individual pairs can come close enough to react chemically. The principal reactions that occur in the track of a charged particle in water during this stage are the following:

$$OH + OH \rightarrow H_2O_2, \tag{11.4}$$

$$OH + e_{aq}^- \rightarrow OH^-, \tag{11.5}$$

$$OH + H \rightarrow H_2O, \tag{11.6}$$

$$H_3O^+ + e_{aq}^- \rightarrow H + H_2O, \tag{11.7}$$

Table 11.2. Diffusion Constants D and Reaction
Radii R for Reactive Species

Species	D $(10^{-5}\ cm^2/sec)$	R (Å)
OH	2	2.4
e_{aq}^-	5	2.1
H_3O^+	8	0.30
H	8	0.42

$$e_{aq}^- + e_{aq}^- + 2H_2O \rightarrow H_2 + 2OH^-, \tag{11.8}$$

$$e_{aq}^- + H + H_2O \rightarrow H_2 + OH^-, \tag{11.9}$$

$$H + H \rightarrow H_2. \tag{11.10}$$

With the exception of (11.7), all of these reactions remove chemically active species, since none of the products on the right-hand sides except H will consume additional reactants. As time passes, the reactions (11.4)–(11.10) proceed until the remaining reactants diffuse so far away from one another that the probability for additional reactions is small. This occurs by $\sim 10^{-6}$ sec, and the chemical development of the track in pure water then is essentially over.

The motion of the reactants during this diffusion-controlled chemical stage can be viewed as a random walk, in which a reactant makes a sequence of small steps in random directions beginning at its initial position. If the measured diffusion constant for a species is D, then, on the average, it will move a small distance λ in a time τ such that

$$\frac{\lambda^2}{6\tau} = D. \tag{11.11}$$

Each type of reactive species can be regarded as having a reaction radius R. Two species that approach each other closer than the sum of their reactive radii have a chance to interact according to Eqs. (11.4)–(11.10). Diffusion constants and reaction radii for the four reactants in irradiated water are shown in Table 11.2.

Example
Estimate how far a hydroxyl radical will diffuse in 10^{-12} sec.

Solution
From Eq. (11.11) with $\tau = 10^{-12}$ sec and from Table 11.2, we find

$$\lambda = (6\tau D)^{1/2} = (6 \times 10^{-12}\ sec \times 2 \times 10^{-5}\ cm^2/sec)^{1/2} = 1.10 \times 10^{-8}\ cm = 1.10\ \text{Å}. \tag{11.12}$$

For comparison, the diameter of the water molecule is 2.9 Å. The answer (11.12) is compatible with our taking the time $\sim 10^{-11}$ sec as marking the beginning of the chemical stage of charged-particle track development.

11.4 EXAMPLES OF CALCULATED CHARGED-PARTICLE TRACKS IN WATER

Before discussing the biological effects of radiation we present some examples of detailed calculations of charged-particle tracks in water. The calculations have been made from the beginning of the physical stage through the end of the chemical stage.

Monte Carlo computer codes have been developed for calculating in complete detail the passage of a charged particle and its secondaries in liquid water. In such computations, an individual particle is allowed to lose energy and generate secondary electrons on a statistical basis, as it does in nature. Where available, experimental values of the energy-loss cross sections are used in the computations. The secondary electrons are similarly transported and are allowed to produce other secondary electrons until the energies of all secondaries reach subexcitation levels (<7.4 eV). Such calculations give in complete detail the position and identity of every reactant H_2O^+, H_2O^*, and subexcitation electron present along the track. These species are allowed to develop according to (11.1), (11.2), and (11.3) to obtain the positions and identities of every one of the reactive species OH, H_3O^+, e_{aq}^-, and H at 10^{-11} sec. The computations then carry out a random-walk simulation of diffusion by letting each reactant take a small jump in a random direction and then checking all pairs to see which are closer than the sum of their reaction radii. Those that can react do so and are removed from further consideration [except when H is produced by (11.7)]. The remainder are jumped again from their new positions and the procedure is repeated to develop the track to later times. The data in Table 11.2 and the reaction schemes (11.4)–(11.10) can thus be used to carry out in complete detail the chemical development of a track.

Three examples of calculated tracks of 5 keV electrons in liquid water at 10^{-11} sec are shown in Fig. 5.6. Each point there represents the location of one of the four reactive species, OH, H_3O^+, e_{eq}^-, or H, shown in Table 11.2. Calculations for the first track in Fig. 5.6 were continued to later times by the Monte Carlo simulation of diffusion and chemical reaction according to (11.4)–(11.10). The results at various later times are shown in Fig. 11.1. Reactions in the track occur rapidly as the reactants begin to diffuse. The initial number $N = 1174$ at 10^{-11} sec decreases progressively through the later frames. Less than half the original number are left at 2.8×10^{-9} sec, and diffusion has appreciably spread them apart. At 2.8×10^{-7} sec, about $\frac{1}{3}$ of the initial number of reactants remain, but all traces of the structure of the original track are gone. Very few subsequent reactions take place, and the chemical stage is essentially complete in the lower right of Fig. 11.1.

A 1 μm segment of the calculated track of a 2 MeV proton, traveling from left to right in liquid water, is shown in Fig. 11.2. In contrast to the 5 keV electron in the last figure, the proton track is virtually straight and its high LET leads to a dense formation of reactants along its path. The relative reduction in the number of reactants and the disappearance of the details of the original track structure by 2.8×10^{-7} sec are, however, comparable. This similarity is due to the fact that intratrack chemical reactions occur only on a local scale of a few hundred angstroms or less, as can be inferred from Figs. 11.1 and 11.2. Separate track segments of this size develop independently of other parts of the track.

These descriptions are borne out by closer examination of the tracks. The middle one-third of the proton track at 10^{-11} sec in Fig. 11.2 is reproduced on a blown-up scale in the upper line of Fig. 11.3. The second line in this figure shows this segment at 2.8×10^{-9} sec, as it develops independently of the rest of the track. On an even more expanded scale, the third and fourth lines in Fig. 11.3 show the last third of the track segment from the top line of the figure at 10^{-11} sec and 2.8×10^{-9} sec. The scale 0.01 μm = 100 Å indicates that most of the chemical development of charged-particle tracks takes place within local regions of a few hundred angstroms or less.

Figure 6.1 shows three examples of 1 μm segments of the tracks of protons and alpha particles, having the same velocities, at 10^{-11} sec. Fast heavy ions of the same velocity have almost the same energy-loss spectrum. Because it has two units of charge, the linear

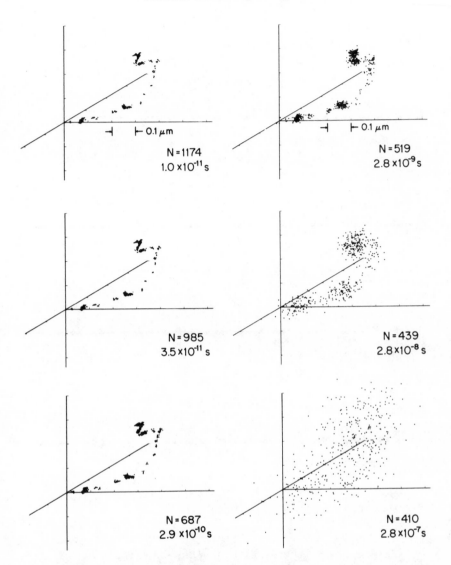

Figure 11.1. Subsequent chemical development of the first track from Fig. 5.6 at the times shown. Start at upper left and read down. Each dot gives the position of a reactive species. The initial number $N = 1174$ of closely spaced reactants at 10^{-11} sec is reduced steadily to 410 widely separated ones at 2.8×10^{-7} sec, after which few additional reactions occur. Details of the initial track structure are gone at 2.8×10^{-7} sec. (Courtesy Oak Ridge National Laboratory, operated by Martin Marietta Energy Systems, Inc., for the Department of Energy)

rate of energy loss (stopping power) for an alpha particle is four times that of a proton at the same speed (cf. Section 4.3). Thus the LET of the alpha particles is about four times that of the protons at each energy.

11.5 CHEMICAL YIELDS IN WATER

When performing such calculations for a track, the numbers of various chemical species present (e.g. OH, e_{aq}^-, H_2O_2, etc.) can be tabulated as functions of time. These chemical yields are conveniently expressed in terms of G values — that is, the number of a given spe-

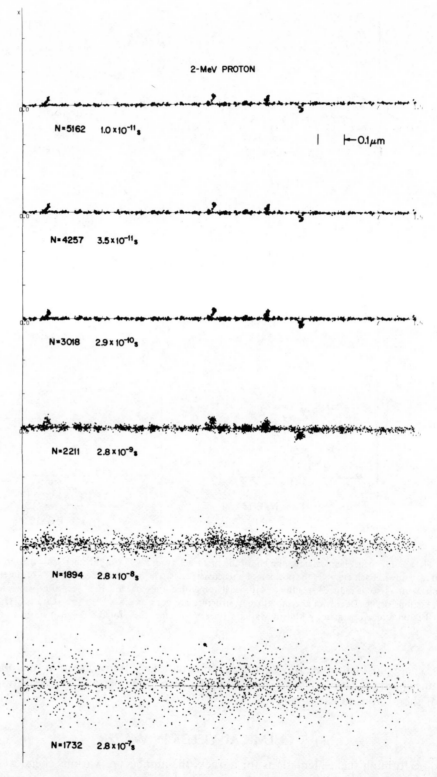

Figure 11.2. Development of a 1 μm segment of the track of a 2 MeV proton, traveling from left to right, in liquid water. (Courtesy Oak Ridge National Laboratory, operated by Martin Marietta Energy Systems, Inc., for the Department of Energy)

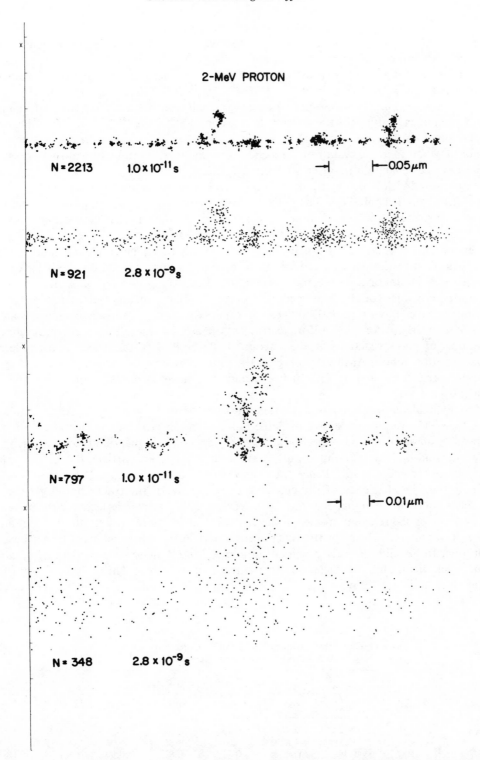

Figure 11.3. Magnified view of the middle one-third of the track segment from Fig. 11.2 at 10^{-11} sec and at 2.8×10^{-9} sec is shown in the upper two lines. The lower two lines show the right-hand third of this segment at these times under still greater magnification. These figures illustrate how most of the chemical development of charged-particle tracks takes place only in local regions of a few hundred angstroms or less in a track. (Courtesy Oak Ridge National Laboratory, operated by Martin Marietta Energy Systems, Inc., for the Department of Energy)

cies produced per 100 eV of energy loss by the original charged particle and its secondaries, on the average, when it stops in the water. Calculated chemical yields can be compared with experimental measurements. To obtain adequate statistics, computations are repeated for a number of different, independent tracks and the average G values are compiled. As seen from reactions (11.4)–(11.10), G values for the reactant species decrease with time. For example, hydroxyl radicals and hydrated electrons are continually used up. G values for the other species, such as H_2O_2 and H_2, increase with time. As mentioned earlier, by about 10^{-6} sec the reactive species remaining in a track have moved so far apart that additional reactions are unlikely. As functions of time, therefore, the G values change little after 10^{-6} sec.

Calculated yields for the principal species produced by electrons of various initial energies are given in Table 11.3. The G values are determined by averaging the product yields over the entire tracks of a number of electrons at each energy. [The last line, for Fe^{3+}, applies to the Fricke dosimeter (Section 9.4). The measured G value for the Fricke dosimeter for tritium beta rays (average energy 5.6 keV), is 12.9.] The table indicates how subsequent changes induced by radiation can be partially understood on the basis of track structure – an important objective in radiation chemistry and radiation biology. One sees that the G values for the four reactive species (the first four lines) are smallest for electrons in the energy range 750–1000 eV. In other words, the intratrack chemical reactions go most nearly to completion for electrons at these initial energies. At lower energies, the number of initial reactants at 10^{-11} sec is smaller and diffusion is more favorable compared with reaction. At higher energies, the LET is less and the reactants at 10^{-11} sec are more spread out than at 750–1000 eV, and thus have a smaller probability of subsequently reacting.

Similar calculations have been carried out for the track segments of protons and alpha particles. The results are shown in Table 11.4. As in Fig. 6.1, pairs of ions have the same speeds, and so the alpha particles have four times the LET of the protons in each case. Several findings can be pointed out. First, for either type of particle, the LET is smaller at the higher energies and hence the initial density of reactants at 10^{-11} sec is smaller. Therefore, the efficiency of the chemical development of the track should get progressively smaller at the higher energies. This decreased efficiency is reflected in the increasing G values for the reactant species in the first four lines (more are left at 10^{-7} sec) and in the decreasing G values for the reaction products in the fifth and sixth lines (fewer are produced). Second, at a given velocity, the reaction efficiency is considerably greater in the track of an alpha particle than in the track of a proton. Third, comparison of Tables 11.3 and 11.4 shows some overlap and some differences in yields between electron

Table 11.3. G Values (Number per 100 eV) for Various Species in Water at 0.28 μsec for Electrons at Several Energies

Species	Electron energy (eV)							
	100	200	500	750	1000	5000	10,000	20,000
OH	1.17	0.72	0.46	0.39	0.39	0.74	1.05	1.10
H_3O^+	4.97	5.01	4.88	4.97	4.86	5.03	5.19	5.13
e_{aq}^-	1.87	1.44	0.82	0.71	0.62	0.89	1.18	1.13
H	2.52	2.12	1.96	1.91	1.96	1.93	1.90	1.99
H_2	0.74	0.86	0.99	0.95	0.93	0.84	0.81	0.80
H_2O_2	1.84	2.04	2.04	2.00	1.97	1.86	1.81	1.80
Fe^{3+}	17.9	15.5	12.7	12.3	12.6	12.9	13.9	14.1

Table 11.4. *G* Values (Number per 100 eV) for Various Species at 10^{-7} sec for Protons of Several Energies and Also for Alpha Particles of the Same Velocities

Species type	Protons (MeV)				Alpha particles (MeV)			
	1	2	5	10	4	8	20	40
OH	1.05	1.44	2.00	2.49	.35	.66	1.15	1.54
H_3O^+	3.53	3.70	3.90	4.11	3.29	3.41	3.55	3.70
e_{aq}^-	.19	.40	.83	1.19	.02	.08	.25	.46
H	1.37	1.53	1.66	1.81	.79	1.03	1.33	1.57
H_2	1.22	1.13	1.02	.93	1.41	1.32	1.19	1.10
H_2O_2	1.48	1.37	1.27	1.18	1.64	1.54	1.41	1.33
Fe^{3+}	8.69	9.97	12.01	13.86	6.07	7.06	8.72	10.31

tracks and heavy-ion track segments. At the highest LET, the reaction efficiency in the heavy-ion track is much greater than that for electrons of any energy.

Electrons, protons, and alpha particles all produce the same species in local track regions at 10^{-15} sec: H_2O^+, H_2O^*, and subexcitation electrons. The chemical differences that result at later times are due entirely to the different spatial patterns of initial energy deposition that the particles have.

11.6 BIOLOGICAL EFFECTS

Depending on the dose, kind of radiation, and observed endpoint, the biological effects of radiation can differ widely. Some occur relatively fast while others may take years to become evident. Table 11.1 includes a summary of the time scale for some important biological effects caused by ionizing radiation. Probably by about 10^{-3} sec, radicals produced by a charged-particle track in a biological system have all reacted. Some biochemical processes are altered almost immediately, in less than about 1 sec. Cell division can be affected in a matter of hours. In higher organisms, damage to the gastrointestinal tract and central nervous system appears within a matter of days or less, and haemopoietic death occurs in about a month. Other kinds of damage, such as lung fibrosis, for example, may take several weeks to develop. Cataracts and cancer occur many years after exposure to radiation. Genetic effects, by definition, are first seen in the next or subsequent generations of an exposed individual.

It is generally assumed that biological effects result from both direct and indirect action of radiation. Direct effects are produced by the initial action of the radiation itself and indirect effects are caused by the later chemical action of free radicals and other radiation products. An example of a direct effect is a strand break in DNA caused by an ionization in the molecule itself. An example of an indirect effect is a strand break that results when an OH radical attacks a DNA sugar at a later time (between $\sim 10^{-11}$ and $\sim 10^{-9}$ sec). The difference between direct and indirect effects is illustrated by Fig. 11.4. The dots in the helical configuration schematically represent the locations of sugars and bases on a straight segment of DNA 200 Å in length in water. The cluster of dots mostly to the right of the helix gives the positions of the reactants at 10^{-11} sec and the subsequent times shown after passage of a 5 keV electron in a straight line perpendicular to the page 50 Å from the center of the axis of the helix. In addition to any transitions produced by the initial passage of the electron or one of its secondaries (direct effects), the reactants produced in the water attack the helix at later times (indirect effects).

The biological effects of radiation can be divided into two general categories, stochas-

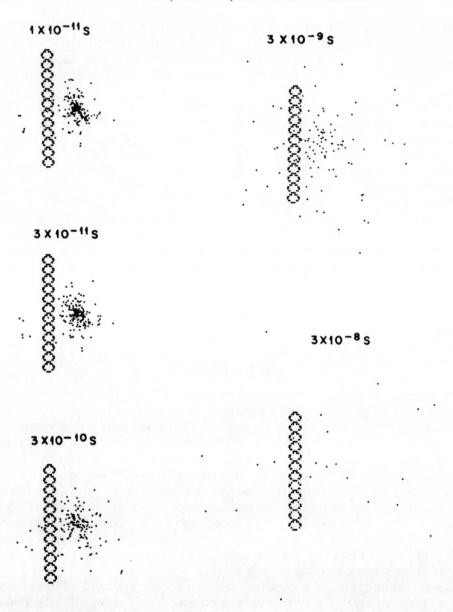

Figure 11.4. Direct and indirect effects of radiation. Double-helical array of dots schematically represents positions of bases and sugars on a 200 Å straight segment of double-stranded DNA. The other dots show the positions of reactants formed in neighboring water from 10^{-11} sec to 3×10^{-8} sec after passage of a 5 keV electron perpendicular to the page 50 Å from the center of the helix. In addition to any direct effects produced initially by passage of the electron, indirect effects also occur later when reactants diffuse and attack the helix. (Courtesy H. A. Wright and R. N. Hamm, Oak Ridge National Laboratory, operated by Martin Marietta Energy Systems, Inc., for the Department of Energy)

tic and nonstochastic. As the name implies, stochastic effects are those that occur in a statistical manner. Cancer is one example. If a large population is exposed to a significant amount of a carcinogen, such as radiation, then an elevated incidence of cancer can be expected. Although one might be able to predict the magnitude of the increased incidence, he cannot say which particular individuals in the population will contract the disease and which will not. Also, since there is a certain natural incidence of cancer without specific exposure to radiation, one will not be completely certain whether a given case was induced or would have occurred without the exposure. In addition, although the expected incidence of cancer increases with dose, the severity of the disease in a stricken individual is not a function of dose. In contrast, nonstochastic effects are those that show a clear causal relationship between dose and effect in a given individual. Usually there is a threshold below which no effect is observed, and the severity generally increases with dose. Skin reddening is a good example of a nonstochastic effect of radiation.

Stochastic effects of radiation have been demonstrated in man and in other organisms only at relatively high doses, where the observed incidence of an effect is not likely due to a statistical fluctuation in the normal level of occurrence. At lower doses, it is not known whether a threshold exists for causing such effects. One cannot say with certainty what the risk is to an individual for an arbitrarily small dose of radiation. As a practical hypothesis, one usually assumes that any amount of radiation, no matter how small, entails some risk. However, there is no agreement among experts on just how risk varies as a function of dose at low doses. We shall return to this subject in Section 11.11 in discussing dose–response curves.

The biological effects of radiation have been investigated with a wide variety of organisms and their components, such as enzymes and nucleic acids. Genetic studies have been carried out with large populations of mice. To establish realistic radiation-protection limits, it is desirable to have information about the effects on man. The extrapolation of data from lower organisms to man is, at best, very uncertain. A considerable body of human data exists, however, as we discuss next.

11.7 SOURCES OF HUMAN DATA

There are three principal sources of information on the effects of radiation on man: (a) persons occupationally exposed to an excessive amount of radiation through negligence or by accident, (b) patients purposefully exposed in medical procedures, and (c) populations exposed to the nuclear weapons at Hiroshima and Nagasaki and to fallout from weapons tests. In many cases, detailed dosimetric work has been coupled with medical observations and epidemiological studies to obtain dose–effect data. Without attempting to be complete, we mention some of the major sources of human data to indicate the various kinds of exposures that have occurred and the effects studied.

Occupational Exposures

Among occupational exposures, important data come from studies made of several hundred radium dial painters, who tipped brushes with their tongues when painting luminous watch dials. This practice occurred until about 1925. Elevated incidence of bone cancer was observed in the workers in later years, and levels of radium in the bone were measured and analyzed after death. A registry of individual workers was compiled. An occupational guide of 0.1 μg for the maximum permissible amount of ^{226}Ra in the body was later established, based on the radium-dial workers' experience. It was then estimated that this level corresponds to an average dose rate of 0.06 rad/wk and a dose-equivalent rate of perhaps between ~0.1 and ~0.6 rem/wk. We shall return to this baseline level in the next chapter on radiation-protection standards (Section 12.3).

Physicians and attendant personnel have received significant exposures from X-ray machines and fluoroscopes in treating patients. Studies, involving thousands of doctors, have been conducted to detect any statistically significant differences in such occurrences as leukemia and life shortening between physicians who regularly use radiation and others, in similar circumstances, who do not. The evidence suggests that long-term exposure to radiation does increase the incidence of leukemia in man. However, such studies entail a large number of variables as well as uncertainties in the dose received, and differences can be ascribed to other factors. Life shortening and leukemia due to radiation have been demonstrated in animal experiments.

Thousands of uranium miners have been studied for lung cancer and other effects. These workers breathe radon and radon daughters in the mine air. The data for miners who have worked underground for a number of years show increased mortality due to respiratory neoplasms and pulmonary fibrosis.

In contrast to the above long-term occupational exposure, often involving thousands of persons, serious acute exposures to small numbers of individuals have resulted from accidents. Several criticality accidents with fatalities have occurred with critical assemblies and research reactors and in the chemical processing of fissionable materials. Other accidents have involved radioactive sources, such as those used for radiography or medical irradiations. These events have provided data on the acute effects of radiation on man.

Medical Exposures

For many years patients have been treated with radiation for diagnostic and therapeutic purposes. A number of researchers have investigated leukemia and cancer in tens and hundreds of thousands of patients exposed in various ways. Separate studies have been made, for example, of patients who received pelvic radium treatment for cancer of the cervix, spinal X-rays for ankylosing spondylitis, and ^{131}I for hyperthyroidism. Large studies of the effects of prenatal and pediatric exposures have also been conducted.

Nuclear Weapons

The atomic bombs of Hiroshima and Nagasaki exposed populations to neutrons, gamma rays, and fission products. The medical histories of tens of thousands of survivors exposed to the radiation — thousands of whom showed acute symptoms — have been documented and followed. Based on information about the air dose of the weapon and an individual's location, shielding and attenuation factors have been estimated, and a tentative dose has been assigned to each survivor. Although many uncertainties exist, the information on the Japanese survivors provides the best data available on the relationship between radiation exposure and leukemia in man. A highly significant increase in leukemia incidence has been found, and the occurrence of various hematological types can be associated with the ages of individuals at the time of irradiation.

Several hundred persons were exposed inadvertently to local fallout from nuclear weapons tests in several instances. The beta and gamma radiation from contamination on the skin and clothing caused severe skin burns to a number of individuals. (In many cases, simple washing or rinsing of the skin after becoming contaminated would have reduced the severity of the injury considerably.)

11.8 THE ACUTE RADIATION SYNDROME

If a person receives a single, large, short-term, whole-body dose of radiation, a number of vital tissues and organs are damaged simultaneously. Radiosensitive cells become depleted because their reproduction is impeded. The effects and their severity will depend

Table 11.5. Acute Radiation Syndrome for Gamma Radiation

Dose (rad)	Symptoms	Remarks
0–25	None	No detectable effects.
25–100	Mostly none. A few persons may exhibit mild prodromal symptoms, such as nausea and anorexia	Bone marrow damaged; decrease in red and white blood-cell counts and platelet count. Lymph nodes and spleen injured; lymphocyte count decreases.
100–300	Mild to severe nausea, malaise, anorexia, infection.	Hematologic damage more severe. Recovery probable, though not assured.
300–600	Severe effects as above, plus hemorrhaging, infection, diarrhea, epilation, temporary sterility.	Fatalities will occur—about 50% in the range 450–500 rad.
More than 600	Above symptoms plus impairment of central nervous system; incapacitation at doses above ~1000 rad.	Death expected.

on the dose and the particular conditions of the exposure. Also, specific responses can be expected to differ from person to person. The complex of clinical symptoms that develop in an individual plus the results of laboratory and bioassay findings are known, collectively, as the acute radiation syndrome.

The acute radiation syndrome can be characterized by four sequential stages. In the initial, or prodromal, period, which lasts until about 48 hr after the exposure, an individual is apt to feel tired and nauseous, with loss of appetite (anorexia) and sweating. The remission of these symptoms marks the beginning of the second, or latent, stage. This period, from about 48 hr to 2 or 3 wk postexposure, is characterized by a general feeling of well being. Then in the third, or manifest illness, stage, which lasts until 6 or 8 wk postexposure, a number of symptoms develop within a short time. Damage to the radiosensitive hematologic system will be evident through hemorrhaging and infection. At high doses, gastrointestinal symptoms will occur. Other symptoms include fever, loss of hair (epilation), lethargy, and disturbances in perception. If the individual survives, then a fourth, or recovery, stage lasts several additional weeks or months.

Depending on the dose received, the acute radiation syndrome can appear in a mild to very severe form. Table 11.5 summarizes typical expectations for different doses of gamma radiation, which because of its penetrating power, gives an approximately uniform whole-body dose.

An acute, whole-body, gamma-ray dose of about 450–500 rad would probably be fatal to about 50% of the persons exposed. This dose is known as the LD50—that is, the dose that is lethal to 50% of a population. More specifically, it is also sometimes called the LD50/30, indicating that the fatalities occur within 30 days.

11.9 DELAYED SOMATIC EFFECTS

We next cite some radiogenic effects that can develop many years after either acute or protracted radiation exposure, even without apparent symptoms at the time of exposure. Such changes are called delayed, or late, somatic effects. Their documentation is complicated by the fact that none of them are uniquely caused by radiation.

Cancer

There is evidence to show that radiation can produce cancer in the blood-forming tissue, skin, bone, lung, thyroid, and connective tissue of man. Several sources of data were mentioned in Section 11.7. Leukemia, or cancer of the blood-forming tissue, is of special importance because it is a radiosensitive endpoint and is relatively rare in natural occurrence. The minimum latent period for most cancers induced by radiation is 10 yr or more. Exceptions to this include cancers resulting from in utero irradiation, leukemia, and bone cancers caused by radium, all of which have been observed within 2–4 yr after exposure. Evidence also suggests that the increased risk for bone cancer and leukemia may become negligible 25–30 yr after irradiation. Risk estimates for radiation-induced cancer are not universally agreed upon by the experts. The majority of the BEIR Committee gave the lifetime risk of cancer mortality, expressed as the number of excess deaths per rad per million persons in the general population, in the range 77–226.†

Degenerative Changes

Specific organs and organ systems can become damaged as a late consequence of radiation. Radiation generally reduces the regenerative capacity of exposed tissue, and so its normal functioning is impaired. As a result, late degeneration effects manifest themselves as diseases of the affected tissues. Most data come from local medical and occupational exposures, often at doses that would be lethal if received by the whole body. Late degenerative changes due to radiation have been demonstrated in the bone, skin, lung, kidney, and gastrointestinal tract.

Life Shortening

As mentioned in Section 11.7, a reduction in life span due to radiation has been observed in animal experiments. This general life shortening is an effect seen in addition to that brought about by the increased incidence of cancer and degenerative changes. Individuals in an exposed population that die of other causes do so in the same pattern as in a control population, but at an earlier age on the average. Thus, only their age-specific death rate is different from the controls. No clear-cut conclusions can be drawn from human studies.

Cataracts

Radiation is known to produce lens opacities, with a latent period of several years. While a few hundred rad of low-LET radiation will produce some changes in the lens, significant cataracts probably require 500–1000 rad. High-LET radiation is considerably more effective.

Teratogenic Effects

Radiation can produce growth and developmental defects in children and fetuses. Fetal tissue, especially, is very radiosensitive (Fig. 11.5). Elevated frequencies of mental retardation and microcephaly were found among the Japanese children who were irradiated in utero at Hiroshima and Nagasaki.

---

†See *The Effects on Populations of Exposure to Low Levels of Ionizing Radiation: 1980,* BEIR III Report, National Academy of Sciences, Washington, D.C. (1980).

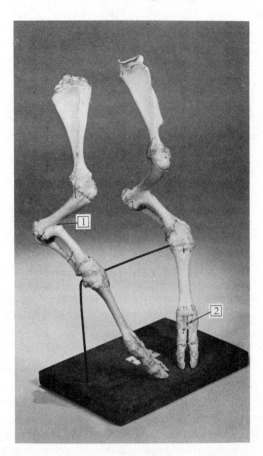

Figure 11.5. Effects of prenatal irradiation (400 rad, whole-body gamma, on 32nd day of gestation) on anatomical development of a calf are seen in severe deformities of the forelimbs at birth: (1) bony ankylosis of the humero-radial joints and (2) deformities of the phalanges. In addition, the posterior surfaces of the limbs are turned inward. Such effects are dose and time specific. Other fetal calves irradiated two days later suffered only minor damage to the phalanges. (Courtesy G. R. Eisele and W. W. Burr, Jr., Medical and Health Sciences Division, Oak Ridge Associated Universities, Oak Ridge, TN)

Fertility

Temporary to permanent sterility in humans can result from a single exposure of the gonads to radiation at levels from about 200 rad to 800 rad or more. Sterility can appear as an acute symptom. Studies with laboratory animals have shown that the late effect of irradiation of the gonads is an acceleration of regressive changes associated with aging.

11.10 GENETIC EFFECTS

Mueller discovered the mutagenic property of ionizing radiation in 1927. Like a number of chemical substances, radiation can alter the genetic information contained in a germ cell or zygote (fertilized ovum). Although mutations can be produced in any cell of the body, only these can transmit the alterations to future generations. Genetic changes may be inconsequential to an individual of a later generation or they may pose a serious handicap.

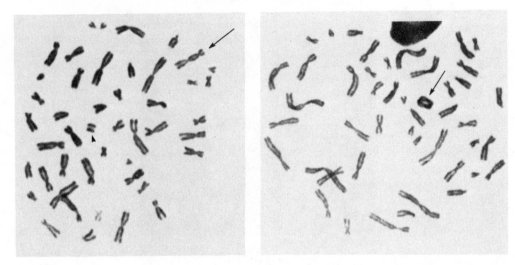

Figure 11.6. Radiation-induced chromosome aberrations in human lymphocytes. Left: chromosome-type dicentric (⟋) and accompanying acentric fragment (⟍). Right: chromosome-type centric ring (⟋). The accompanying acentric fragment is not included in the metaphase spread. (Courtesy H. E. Luippold and R. J. Preston, Oak Ridge National Laboratory, operated by Martin Marietta Energy Systems, Inc., for the Department of Energy)

The mutation is called a point mutation when there is a change at a single gene locus. Genetic damage also results from chromosome breaks produced by radiation or other mutagens. Chromosome fragments can reunite incorrectly or be lost, thus providing altered information upon cell division. Point mutations and chromosome aberrations are generally detrimental to cellular reproduction ability, and offspring show abnormalities. Figures 11.6 and 11.7 show examples of chromosome aberrations and genetic defects produced by radiation in individual organisms.

Both the frequency and type of mutation in offspring depend on the time of mating of the parents after irradiation. If a male mates soon thereafter, it is likely that the germ cell involved was mature when exposed. Under these conditions, there is a relatively high probability that a mutation will be of a dominant-lethal type. If mating occurs late after exposure, then the exposed germ cell was likely immature, and a relatively high frequency of recessive mutations can be expected.

Although the genetic effects of radiation have been extensively studied in plants and animals, there is no clear evidence that similar mutations occur in man. The effects sought include the incidence of stillbirths, neonatal deaths, malformations, and the sex ratio of offspring. Very large populations would be needed to demonstrate the expected increased incidences of these endpoints. Extensive studies of this type were made on the offspring of Japanese survivors in Hiroshima and Nagasaki. Compared with a control group, all factors were comparable except for significantly different sex ratios in the immediate progeny. Radiogenic mutations in the male would be expected to increase the relative number of male births, and such mutations in the female should reduce the number of male births. The observed sex-ratio differences in the Japanese were in this direction. The dose for eventually doubling the mutation rate in man is perhaps of the order of 50–250 rem.†

Experiments with mice have demonstrated that fewer mutations are produced per rad

---

†*The Effects on Populations of Exposure to Low Levels of Ionizing Radiation: 1980*, BEIR III Report, Chapter IV, National Academy of Sciences, Washington, D.C. (1980).

Figure 11.7. Top: normal Drosophila male. Bottom: Drosophila male with four wings resulting from one spontaneous and two X-ray-induced mutations. (Source: E. B. Lewis, California Institute of Technology. Reprinted with permission from J. Marx, Genes That Control Development, *Science, 213,* 1485–1488 (1981). Copyright 1981 by the American Association for the Advancement of Science.)

at low rates than at high dose rates. The existence of this dose-rate dependence suggests that repair mechanisms play a role in reducing mutagenic effects of radiation.

11.11 DOSE–RESPONSE RELATIONSHIPS

Biological effects of radiation can be quantitatively described in terms of dose–response relationships, i.e., the incidence or severity of a given effect, expressed as a function of

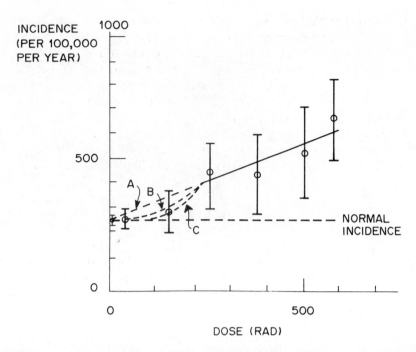

Figure 11.8. Example of a dose–response curve, showing the incidence of an effect (e.g., incidence of certain cancers per 100,000 population per year) as a function of dose. Circles show measured values with associated error bars. Solid line at high doses is drawn to extrapolate linearly (dashed curve A) to the level of normal incidence at zero dose. Dashed curve B shows a quadratic extrapolation to zero dose. Dashed curve C corresponds to a threshold at about 75 rad.

dose. These relationships are conveniently represented by plotting a dose–response curve, such as that shown in Fig. 11.8. The ordinate gives the observed degree of some biological effect under consideration (e.g., the incidence of certain cancers in animals per 100,000 population per year) at the dose level given by the abscissa. The circles show data points with error bars that represent a specified confidence level (e.g., 80%). At zero dose, one typically has a natural, or spontaneous, level of incidence, which is known from a large population of unexposed individuals. Often the numbers of individuals exposed at higher dose levels are relatively small, and so the error bars there are large. As a result, although the trend of increasing incidence with dose may be clearly evident, there is no unique dose–response curve that describes the data. In the figure, a solid straight line, consistent with the observations, has been drawn at high doses. The line is constructed in such a way that it intersects the ordinate at the level of natural incidence when a linear extension (dashed curve A) to zero dose is made. In this case, we say that a linear dose–response curve, extrapolated down to zero dose, is used to represent the effect.

Curves with other shapes can usually be drawn through biological dose–effect data. Also, extrapolations to low doses can be made in a number of ways. Sometimes there are theoretical reasons for assuming a particular dose dependence, particularly at low doses. The dashed curve B in Fig. 11.8 shows a quadratic dependence. Both curves A and B imply that there is always some increased incidence of the effect due to radiation, no matter how small the dose. In contrast, the extrapolation shown by the curve C implies that there is a threshold of about 75 rad for inducing the effect.

Another important kind of dose–response relationship is illustrated by the survival of cells exposed to different doses of radiation. The endpoint studied is cell inactivation, or killing, in the sense of cellular reproductive death, or loss of a cell's ability to proliferate indefinitely. Large cell populations can be irradiated and then diluted and tested for colony formation. Cell survival can be measured over three and sometimes four orders of magnitude. It provides a clear, quantitative example of a cause-and-effect relationship for the biological effects of radiation.

Cell inactivation is conveniently represented by plotting the natural logarithm of the surviving fraction of irradiated cells as a function of the dose they receive. A linear semi-log survival curve, such as that shown in Fig. 11.9, implies exponential survival of the form

$$S = S_0 e^{-D/D_0}. \tag{11.13}$$

Here S is the number of surviving cells at dose D, S_0 is the original number of cells irradiated, and D_0 is the negative reciprocal of the slope of the curve in Fig. 11.9. Analogous to the reciprocals of λ in Eq. (3.67) and μ in Eq. (7.31), it is called the mean lethal dose; D_0 is, therefore, the average dose absorbed by each cell before it is killed. The surviving fraction when $D = D_0$ is, from Eq. (11.13),

$$\frac{S}{S_0} = e^{-1} = 0.37. \tag{11.14}$$

For this reason, D_0 is also called the "D-37" dose.

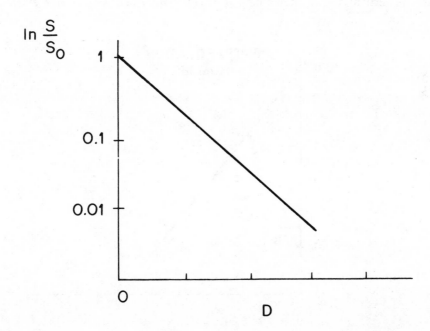

Figure 11.9. Semilogarithmic plot of surviving fraction S/S_0 as a function of dose D showing exponential survival characterized by straight line.

Survival curves are observed to have various shapes, many of which can be accounted for on the basis of formal models for cell inactivation. Exponential survival results, for example, from the assumption that a cell is killed by a single "hit" from radiation. The actual physical event that is called a hit need not be explicitly specified. Mathematically, the relative number of unhit cells at a dose D in Eq. (11.13) is equivalent to the relative number of unscattered photons at depth x in Eq. (7.31). In both cases, the probability of the "survival" of a given individual decreases exponentially in proportion to the amount of the "risk" experienced.

The model that leads to exponential cell survival is described as a single-target, single-hit model; that is, the inactivation of a cell requires that a single event, or hit, occur in a single target within the cell. Another model that gives a survival curve with a different shape, also observed in many experiments, is the multitarget, single-hit model. In this case, n identical targets are ascribed to a cell and all n targets must be hit at least once in order to inactivate the cell. The resultant survival curve, shown in Fig. 11.10, is described by writing

$$S = S_0 [1 - (1 - e^{-D/D_0})^n].\tag{11.15}$$

[This equation reduces to Eq. (11.13) when $n = 1$.] The curve has a shoulder that begins with zero slope at low doses, reflecting the fact that more than one target must be hit in a cell to inactivate it. As the dose increases, more and more cells accumulate targets that are already hit, and so the slope increases. At sufficiently high doses, surviving cells are unlikely to have more than one remaining unhit target. Their response then takes on the characteristics of single-target, single-hit survival, and additional dose produces an exponential decrease. Thus, at high doses, S/S_0 becomes linear with slope $-1/D_0$, as indicated in Fig. 11.10. When this straight portion of the curve is extrapolated backwards, it intersects the vertical axis at the value n [Eq. (11.15)], which is called the extrapolation

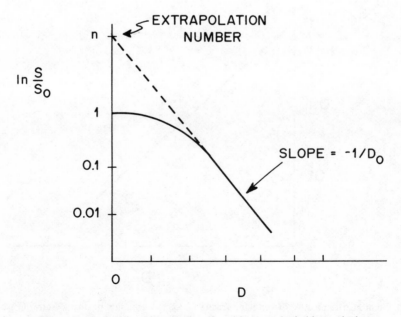

Figure 11.10. Semilogarithmic plot of multitarget, single-hit survival.

number. Many mammalian cell-survival curves have the appearance of multitarget, single-hit response. However, because the underlying mechanisms of cell inactivation are unknown, the extrapolation number is not to be taken literally as the number of targets per cell. For fitted experimental data it is frequently nonintegral. Cell populations can be heterogeneous, having subpopulations in different phases of the reproductive cycle and with different sensitivities.

Additional models include variations of the multitarget, single-hit one in which inactivation results from the hitting of any $m < n$ of the n targets. Also, single-target, multi-hit models have been proposed in which the cell has a single target which must be hit a given number $n > 1$ times to inactivate the cell.

11.12 FACTORS AFFECTING DOSE–RESPONSE RELATIONSHIPS

Relative Biological Effectiveness (RBE)

Generally, dose–response curves depend on the type of radiation used and on the biological endpoint studied. As a rule, radiation of high LET is more effective biologically than radiation of low LET. Different radiations can be contrasted in terms of their relative biological effectiveness (RBE) compared with X-rays. If a dose D of a given type of radiation produces a specific biological endpoint, then RBE is defined as the ratio

$$RBE = \frac{D_X}{D},$$
(11.16)

where D_X is the X-ray dose needed under the same conditions to produce the same endpoint. If, for example, 8000 rad of X-rays or 650 rad of alpha particles of a certain energy reduces the survival of a certain strain of haploid yeast cells to 10%, then the RBE is $8000/650 = 12.3$ for alpha particles of that energy for 10% survival of that strain of haploid yeast cells. The RBE will generally be different for other survival levels, for different strains of cells, other organisms, and for other energies or kinds of radiation. The use of X-rays, which have low LET, as the standard for comparison (sometimes 250 kVp X-rays are specified) leads to RBE values that are ≥ 1 in most instances.

The dependence of RBE on the LET of radiation often follows one of four general patterns shown in Fig. 11.11. Curve A characterizes an endpoint, like enzyme inactivation, that involves a small target, showing "single-hit" response. Except at very high LET, the number of molecules inactivated is proportional to the dose, independently of its microscopic spatial distribution, and RBE = 1. At high LET, some dose is "wasted" by overkill, the radiation becomes less efficient per unit dose than X-rays, and the RBE curve drops below unity. Curve B is typical of chromosome aberrations and the survival of simple bacterial cells. With increasing LET, the radiation becomes somewhat more efficient and then the RBE falls off at high LET, again because of overkill. Curve C exemplifies the behavior of many kinds of mammalian cells and curve D illustrates the inactivation of dried seeds and spores.

Dose Rate

The dependence of dose–response relationships on dose rate has been demonstrated for a large number of biological effects. At the end of Section 11.10 we mentioned the role of repair mechanisms in reducing the mutation frequency per rad in mice when the dose rate

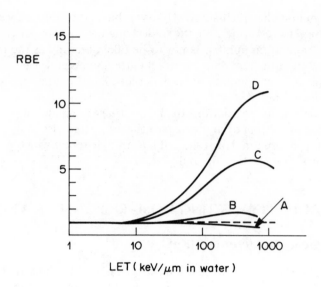

Figure 11.11. General kinds of variation found for RBE as a function of LET. Typical shapes are shown for inactivation of (a) enzymes, (b) simple bacterial cells, (c) mammalian cells, and (d) dried seeds and spores.

is lowered. Another example of dose–rate dependence is shown in Fig. 11.12. Mice were irradiated with ^{60}Co gamma rays at dose rates ranging up to several thousand rad/hr and the LD50 determined. It was found that LD50 = 800 rad when the dose rate was several hundred rad/hr or more. At lower dose rates the LD50 increased steadily, reaching approximately 1600 rad at a rate of 10 rad/hr. Evidently, animal cells and tissues can repair enough of the damage caused by radiation at low dose rates to survive what would be lethal doses if received in a shorter period of time.

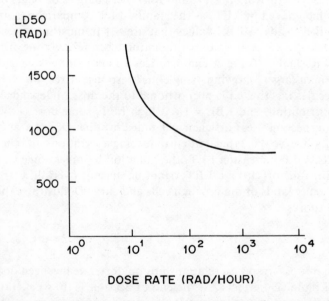

Figure 11.12. Dependence of LD50 on dose rate for mice irradiated with ^{60}Co gamma rays. [Based on J. F. Thomson and W. W. Tourtellotte, *Am. J. Roentg. Rad. Ther. Nucl. Med.* **69**, 826 (1953)]

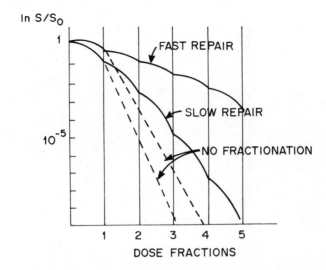

Figure 11.13. Semilogarithmic plot of surviving fraction, S/S_0, of cell populations with relatively fast and slow repair rates vs. the number of dose fractions given with recovery time in between. Dashed curves show responses without fractionation. Dose fractionation is used in radiotherapy when normal tissue has a faster repair rate than the tumor, in order to enhance the differential killing of the two kinds of cells.

Dose fractionation is common practice in treating tumors with radiation. The treatments entail massive doses to reduce the survival probability of tumor cells to a very low level. If administered at one time, such doses would also destroy the healthy tissue traversed by the radiation. The body's normal tissue, however, usually has the ability to repair itself faster than the cells in a tumor. Therefore, a large, but tolerable, dose is administered and then the tissues are allowed a period of time to partially recover, the normal tissue doing so at a faster rate. Then a second dose is given, followed by another recovery period, another irradiation, and so on. The results of dose fractionation are shown schematically in Fig. 11.13 in terms of two cell populations having different repair rates. Each time an increment of dose is added after an elapsed recovery period, the curve with the larger shoulder (fast recovery) continues with a smaller slope than the other. After repeated fractions are administered, the difference in the relative survival fractions of the two populations becomes much greater than if the total dose were administered in a single irradiation.

Oxygen Enhancement Ratio (OER)

Dissolved oxygen in tissue acts as a radiosensitizing agent. This so-called oxygen effect, which is almost invariably observed in radiobiology, is illustrated by Fig. 11.14. The curves show the survival of cells irradiated under identical conditions, except that one culture contains dissolved O_2 from the air and the other is purged with N_2. The effect of oxygen can be expressed quantitatively by the oxygen enhancement ratio (OER), defined as the ratio of the slopes of the straight portions of the O_2 and N_2 curves on the semilogarithmic plot:

$$\text{OER} = \frac{\text{slope of } O_2 \text{ curve (straight portion)}}{\text{slope of } N_2 \text{ curve (straight portion)}}. \tag{11.17}$$

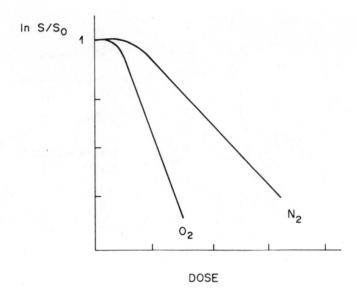

Figure 11.14. Cell survival in the presence of dissolved oxygen (O_2) and after purging with nitrogen (N_2).

Since the slopes give the change in the survival fraction per unit dose, the OER defined in this way expresses the relative sensitivity of the cells in the presence and absence of oxygen. OER values are typically 2–3 for X-rays, gamma rays, and fast electrons; around 1.7 for fast neutrons; and close to unity for alpha particles.

The existence of the oxygen effect provides strong evidence of the importance of indirect effects in producing biological lesions (Section 11.6). It would also be expected that dissolved oxygen is most effective with low- rather than high-LET radiation, because intratrack reactions compete to a lesser extent for the initial radiation products.

The oxygen enhancement ratio is defined in a different way in the ICRU's 1979 Report 30:

Oxygen Enhancement Ratio (OER): The ratio of the dose required under conditions of hypoxia to that under conditions in air to produce the same level of effect.

According to this definition, one would obtain the OER from Fig. 11.14 by taking the ratio of doses at a given survival level. Unless the curves had no shoulders, the OER would depend upon the survival level chosen. Also, without shoulders, the ICRU definition and Eq. (11.17) are equivalent.

Since rapidly dividing tumor cells often have a poor supply of blood and oxygen, the oxygen effect has important implications for radiotherapy. Most of the enhancement brought about by oxygen is realized at concentrations considerably below saturation. Some success has been achieved by elevating the level of dissolved O_2 in persons undergoing radiotherapy. While their normal tissues as well as the tumor attain a higher level of O_2, the relative enhancement of the radiation effectiveness in the tumor is greater because of its initially low level.

11.13 PROBLEMS

1. What initial changes are produced directly by ionizing radiation in water (at $\sim 10^{-15}$ sec)?
2. What reactive species exist in pure water at times $> 10^{-11}$ sec after irradiation?
3. Do all of the reactive species (Problem 2) interact with one another?

4. Estimate how far an H_3O^+ ion will diffuse, on the average, in water in 5×10^{-12} sec.
5. Estimate the average time it takes for an OH radical to diffuse 400 Å in water.
6. If an OH radical in water diffuses an average distance of 3.5 Å in 10^{-11} sec, what is its diffusion constant?
7. Estimate how close an H_3O^+ ion and a hydrated electron must be to interact.
8. How far would a water molecule with thermal energy (0.025 eV) travel in 10^{-12} sec in a vacuum?
9. If a 20 keV electron stops in water and an average of 352 molecules of H_2O_2 are produced, what is the G value for H_2O_2 for electrons of this energy?
10. If the G value for hydrated electrons produced by 20 keV electrons is 1.13, how many of them are produced, on the average, when a 20 keV electron stops in water?
11. What is the G value for ionization in a gas if $W = 30$ eV/ip (Section 9.1.2)?
12. Use Table 11.3 to find the average number of OH radicals produced by a 500 eV electron in water.
13. For what physical reason is the G value for H_2 in Table 11.3 smaller for 20 keV electrons than for 1 keV electrons?
14. Why do the G values for the reactant species H_3O^+, OH, H, and e_{aq}^- decrease between 10^{-11} sec and 10^{-6} sec? Are they constant after 10^{-6} sec? Explain.
15. For 5 keV electrons, the G value for hydrated electrons is 8.4 at 10^{-11} sec and 0.89 at 2.8×10^{-7} sec. What fraction of the hydrated electrons react during this period of time?
16. (a) Why are the yields for the reactive species in Tables 11.4 for protons greater than those for alpha particles of the same speed? (b) Why are the relative yields of H_2 and H_2O_2 smaller?
17. Distinguish between the "direct" and "indirect" effects of radiation. Give a physical example of each.
18. Give examples of two stochastic and two nonstochastic biological effects of radiation.
19. What are the major symptoms of the acute radiation syndrome?
20. What are the principal late somatic effects of radiation? Are they stochastic or nonstochastic?
21. Why do experiments that seek to quantify dose-effect relationships at low doses require large exposed and control populations?
22. If cell survival is described by the function $S/S_0 = e^{-0.031D}$, where S/S_0 is the relative number of cells surviving dose D, what is the mean lethal dose?
23. If 4100 rad reduces the exponential survival of cells to a level of 1%, what is the mean lethal dose?
24. For multitarget, single-hit survival with $D_0 = 750$ rad and an extrapolation number $n = 4$, what fraction of cells survive a dose of 1000 rad?
25. Repeat Problem 24 for $D_0 = 750$ rad and $n = 3$.
26. Repeat Problem 24 for $D_0 = 500$ rad and $n = 4$.
27. What interrelationship do the extrapolation number, the magnitude of D_0, and the size of the shoulder have in a multitarget, single-hit cell survival model?
28. Why does survival in a multitarget, single-hit model become exponential at high doses?
29. (a) Sketch a linear plot of the exponential survival curve from Fig. 11.9. (b) Sketch a linear plot for the multitarget, single-hit curve from Fig. 11.10. What form of curve is it?
30. A multitarget, single-hit survival model requires hitting n targets in a cell at least once each to

Table 11.6. Data for Problem 32

Dose (rad)	Surviving fraction
10	0.993
25	0.933
50	0.729
100	0.329
200	0.0458
300	0.00578
400	0.00072

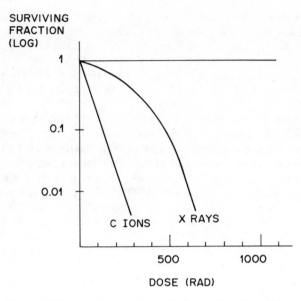

Figure 11.15. Surviving fraction of cells as a function of dose.

cause inactivation. A single-target, multi-hit model requires hitting a single target in a cell n times to produce inactivation. Show that these two models are inherently different in their response. (For example, at high dose consider the probability that hitting a target will contribute to the endpoint.)

31. One can describe the exponential survival fraction, S/S_0, by writing $S/S_0 = e^{-pD}$, where D is the number of "hits" per unit volume (proportional to dose) and p is a constant, having the dimensions of volume. Show how p can be interpreted as the target size (or, more rigorously, as an upper limit to the target size in a single-hit model).

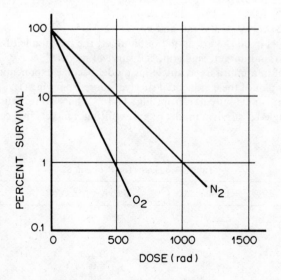

Figure 11.16. Cell-survival curves.

32. The cell-survival data in Table 11.6 fit a multitarget, single-hit survival curve. Find the slope at high doses and the extrapolation number. Write the equation that describes the data.

33. What factors can modify dose–effect relationships?

34. Figure 11.15 shows the surviving fraction of cells as a function of dose when exposed to either X-rays or carbon ions in an experiment. From the curves, estimate the RBE of the carbon ions for 1% survival and for 50% survival. What appears to happen to the RBE as one goes to lower and lower doses?

35. Explain why radiation is used in cancer therapy, even though it kills normal cells.

36. Estimate the oxygen enhancement ratio from the cell-survival curves in Fig. 11.16. The curves labeled O_2 and N_2 refer to the presence of oxygen or nitrogen (i.e., the absence of oxygen).

37. Are the curves in Fig. 11.16 more typical of results expected with high-LET or low-LET radiation? Why?

CHAPTER 12
RADIATION PROTECTION CRITERIA AND STANDARDS

12.1 OBJECTIVE OF RADIATION PROTECTION

Man benefits greatly from the use of X-rays, radioisotopes, and fissionable materials in medicine, industry, research, and power generation. However, the realization of these gains entails the exposure of persons to radiation in the procurement and normal use of sources as well as from accidents that occur. Since any radiation exposure presumably involves some risk to the individuals involved, the levels of exposures allowed should be worth the result that is achieved. In principle, therefore, the objective of radiation protection is to balance the risks and benefits from activities that involve radiation. If the standards are too lax, the risks may be unacceptably large; if the standards are too stringent, the activities may be prohibitively expensive or impractical, to the overall detriment to society.

As an example of the need to balance risks and benefit, we consider the use of radioactive iodine for the diagnosis of hyperthyroidism. A patient drinks a solution tagged with radioiodine, and the rate of thyroid uptake of the element is measured by monitoring the radiation that the thyroid emits. The uptake is abnormally fast for patients with the disease. When such a procedure is employed, there is a conscious or implied medical judgment that the expected benefit to the patient outweighs any harm that the radiation does to him. Moreover, it is also implied that the benefits of the procedure to society are worth the exposures of other persons, including members of the public, that result from the fabrication, transportation, storage, administration, and ultimate disposal of the material. The exposure guidelines for these activities should keep the risks of harm from radiation within the levels of other risks that society allows in its activities. Moreover, the guidelines should also call for keeping doses as low as feasible without unduly restricting the benefits that this use of radioiodine has to offer.

The balancing of risks and benefits in radiation protection cannot be carried out in an exact manner. The risks from radiation are not precisely known, particularly at the low levels of allowed exposures. Moreover, the "benefits" are usually not easily measurable and often involve matters that are personal value judgments. Because of the existence of legal radiation-protection standards, in use everywhere, their acceptance rests with society as a whole rather than with particular individuals or groups.

12.2 ELEMENTS OF RADIATION-PROTECTION PROGRAMS

Different uses of ionizing radiation warrant the consideration of different exposure guidelines. Medical X-rays, for example, are generally under the control of the physician, who makes a medical judgment as to their being warranted. Specific radiation-protection standards, such as those recommended by the International Commission on Radiological Protection (ICRP) and the National Council on Radiation Protection and Measurements (NCRP) and discussed in Chapter 1, have been traditionally applied to the "peaceful uses of atomic energy," the theory being that these activities justify the exposure limits being specified. In contrast, different exposure criteria might be appropriate for military or national-defense purposes, where the risks involved and the objectives are of an entirely different nature than those for medical and other peaceful uses of radiation.

We shall concentrate principally on the radiation-protection recommendations of the ICRP and NCRP, which are very similar. They apply to practices employed in connection with the peaceful uses of atomic energy and with a number of medical procedures. The maximum levels of exposure permitted are deemed acceptable in view of the benefits to mankind, as judged by various authorities and agencies who, in the end, have the legal responsibility for radiation safety. Since, in principle, the benefits justify the exposures, the limits apply to an individual worker or member of the public independently of any medical or dental radiation exposure he or she might receive.

Different permissible exposure criteria are usually applied to different groups of persons. Certain levels are permitted for persons who work with radiation. These guidelines are referred to as "occupational" or "on-site" radiation-protection standards. Other levels, often one-tenth of the allowable occupational values, apply to members of the general public. These are referred to as "nonoccupational" or "off-site" guides. Several philosophical distinctions can be drawn in setting occupational and nonoccupational standards. In routine operations, radiation workers are exposed in ways that they and their employers have some control over. The workers are also compensated for their jobs and are free to seek other employment. Members of the public, in contrast, are exposed involuntarily to the gaseous and liquid effluents that are permitted to escape from a site where radioactive materials are handled. In addition, off-site exposures usually involve a larger number of persons as well as individuals in special categories of concern, such as children and pregnant women. (Special provisions are also made for occupational radiation exposure of women of child-bearing age.)

On a worldwide scale, the potential genetic effects of radiation have been addressed in setting radiation standards. Exposure of a large fraction of the world's population to even a small amount of radiation represents a genetic risk to mankind that can be passed on indefinitely to succeeding generations. In contrast, the somatic risks are confined to the persons actually exposed.

An essential facet of the application of maximum permissible exposure levels to radiation-protection practices is the ALARA (As Low As Reasonably Achievable) philosophy. The ALARA concept gives primary importance to the principle that exposures should always be kept as low as practicable. The maximum permissible levels are not to be considered as "acceptable," but, instead, they represent the levels that should not be exceeded.

Another consideration in setting radiation-protection standards is the degree of control or specificity that the criteria may require. The ICRP and NCRP have generally made recommendations for the maximum dose equivalent (Section 10.2) for individual workers or members of other groups in a certain length of time, e.g., a year or three months. Without requiring the specific means to achieve this end, the recommendations allow maximum flexibility in their application. Many federal and international agencies, however, have very specific regulations that must be met in complying with the ICRP and NCRP dose-equivalent limits.

12.3 RADIATION PROTECTION BASED ON DOSE EQUIVALENT

Radiation-protection standards have evolved through the years and are still undergoing changes, additions, and revisions today. Both the ICRP and NCRP recognize the need to continually review the basic principles and scientific data on which their recommendations are based as well as the requirements of society for guidance in various areas of radiation safety. For many years the ICRP and NCRP recommended annual and quarterly limits for the dose equivalent to various organs of the body. We shall refer to such a regime as a dose-equivalent system of radiation protection. It is the kind of system recommended in the NCRP's 1971 Report No. 39 and is the subject of the present section. In 1977 the ICRP issued its current Publication 26, which emphasizes the concepts of risk and detriment. We shall refer to the ICRP system, which is described in the next section, as a combined dose-equivalent/risk system. The NCRP is currently revising its recommendations for a system of radiation protection based solely on risk. As of this writing, a new NCRP report with the Council's recommended risk system is nearing completion.

To begin the detailed discussion of the dose-equivalent system of radiation protection, we consider the recommendations for whole-body occupational exposure as given in NCRP Report No. 39.† Dose-equivalent H is used in the Report as we defined it in Section 10.2: $H = QD$, where no modifying factors other than the quality factor Q are used to multiply the absorbed dose D. The basic approach of using H for radiation protection is to limit the annual and quarterly dose equivalents to various critical organs of the body, the numerical values of the limits depending on the particular organs. Report 39 states (p. 89), "The maximum permissible prospective dose equivalent for whole body irradiation from all occupational sources shall be 5 rems in any one year. For the purposes of this recommendation, the critical organs are considered to be the gonads, the lens of the eye, and red bone marrow." These organs are thus the ones that limit the exposure of the whole body. The Council emphasizes that all numerical limits are guides, and that retrospective dose equivalent of increments of up to 15 rem in any 1 yr may not be improper if they are well distributed over time. ("Retrospective" refers to the actual dose equivalent, after the fact, in contrast to "prospective," which is that used for planning purposes.) To keep the long-term dose equivalent at an average of 5 rem/yr, they make the following recommendation (p. 89): "Long-term accumulation of combined whole body dose equivalent (assessed in the gonads, lens of the eye or red bone marrow) shall not exceed 5 rems multiplied by the number of years of age beyond 18, i.e., maximum accumulated dose equivalent = $(N - 18) \times 5$ rems, where N is the age in years and is greater than 18."

The allowable 5 rem/yr for occupational whole-body exposure can be compared with other levels of radiation in order to put it into some perspective. As mentioned in Section 11.7, human data led to the establishment of an occupational limit for the amount of radium in the body (0.1 μg) that implied an annual dose equivalent estimated to be about 5–30 rem. No serious injury was known to occur with 10 times that amount. In another context, as discussed in Section 11.8 (cf. Table 11.5), no noticeable effects are expected as a result of an acute whole-body dose equivalent of 25 rem. Other human data are available, as described in the last chapter. The 5 rem/yr occupational whole-body limit is well below a level at which one would expect to see any radiation effects in the worker popula-

---

†In defining "occupational" exposures, Report 39 specifically excludes contributions from (a) natural background radiation, (b) man-made devices outside the work environment (e.g., television sets, fallout), and (c) exposures in the healing arts (e.g., medical and dental X-rays).

tion. On the other hand, it is considerably greater than natural background and population-averaged medical exposures, which each contribute an average of about 0.1 rem/yr to a person's exposure in the United States (cf. Table 1.1).

The principal dose-equivalent limiting recommendations given in NCRP Report 39 are summarized in Table 12.1. The annual limit for most organs of the body is 15 rem, which was also an earlier limit for the whole body. Higher limits apply to the extremities. The recommended limit for female workers who might be pregnant is 0.5 rem/yr during gestation. As already mentioned, the off-site levels for individuals in the general population are one-tenth the occupational whole-body limit. Also, for genetic reasons, an average of 0.17 rem/yr is the recommended limit for the whole population, amounting to 5 rem in 30 yr. Special limits are also suggested for application to accident situations.

12.4 RADIATION PROTECTION BASED ON COMBINED DOSE-EQUIVALENT/RISK

As already mentioned, the dose-equivalent/critical-organ system of radiation protection has been superseded by the combined dose-equivalent/risk system proposed in the ICRP's Publication 26 and subsequent publications. This system is in current general use, and we describe its principal features now.

After an introductory section, ICRP Publication 26 discusses the objectives of radiation protection.

Radiation protection is concerned with the protection of individuals, their progeny and mankind as a whole, while still allowing necessary activities from which radiation exposure might result.

Table 12.1. Dose-Equivalent Limiting Recommendations
from NCRP Report No. 39 (1971)

Exposure category	Recommended limits
Occupational	
whole body (prospective)	5 rem/yr
whole body (retrospective)	10–15 rem in any year
whole body (to age N yr)	$5(N{-}18)$ rem
skin	15 rem/yr
hands	75 rem/yr (25/quarter)
forearms	30 rem/yr (10/quarter)
other organs, tissues, and organ systems	15 rem/yr (5/quarter)
fertile women	0.5 rem in gestation period
Nonoccupational	
members of the public	0.5 rem/yr
students	0.1 rem/yr
Population	
genetic	0.17 rem/yr average
somatic	0.17 rem/yr average
Emergency — life saving	
individual	100 rem
hands and forearms	200 rem, additional
Emergency — less urgent	
individual	25 rem
hands and forearms	100 rem, total

The report goes on to discuss stochastic and nonstochastic effects of radiation (Section 11.6). It continues,

> The aim of radiation protection should be to prevent detrimental non-stochastic effects and to limit the probability of stochastic effects to levels deemed to be acceptable. An additional aim is to ensure that practices involving radiation exposure are justified.
>
> The prevention of non-stochastic effects would be achieved by setting dose-equivalent limits at sufficiently low values so that no threshold dose would be reached, even following exposure for the whole of a lifetime or for the total period of working life. The limitation of stochastic effects is achieved by keeping all justifiable exposures as low as is reasonably achievable, economic and social factors being taken into account, subject always to the boundary condition that the appropriate dose-equivalent limits shall not be exceeded.

The Commission goes on to state that it recommends a system of dose limitation, with the following main features:

(a) no practice shall be adopted unless its introduction produces a positive net benefit;
(b) all exposures shall be kept as low as reasonably achievable, economic and social factors being taken into account; and
(c) the dose equivalent to individuals shall not exceed the limits recommended for the appropriate circumstances by the Commission.

Publication 26 then details five basic concepts that underlie the ICRP recommendations.

The *detriment* in a population is the "mathematical 'expectation' of the harm incurred, taking into account not only the probability of each type of deleterious effect, but also the severity of the effect." It is noted that "there may be deleterious effects not associated with health, such as the need to restrict the use of some areas or products." The Commission does not specify how this concept or numbers related to it are to be obtained.

The *dose equivalent* definition is the one given in Section 10.2:

$$H = DQN, \tag{12.1}$$

where H is the dose equivalent, D is the absorbed dose, Q is the quality factor, and N is the product of all other modifying factors specified by the Commission (e.g., for dose-rate effects). In ICRP 26, the assignment $N = 1$ is made. The value of the quality factor is based on LET (Table 10.1). When the distribution of absorbed dose is not known as a function of LET, then Publication 26 recommends use of the average quality factors given in Table 12.2. The units of dose and dose equivalent used in ICRP 26 are the SI units, gray (Gy) and sievert (Sv), defined in Section 10.2.

The concept of *collective dose equivalent* S is introduced for application to exposed populations. It is defined by the expression

$$S = \sum_i H_i P_i, \tag{12.2}$$

where H_i is the *per capita* dose equivalent to the whole body or a specified organ or tissue in the number of persons P_i in the ith subgroup of the exposed population. The col-

Table 12.2. Average Quality Factors, $\bar{Q}$, Recommended in ICRP Publication 26 when Dose Distribution in LET Is Not Known

Type of radiation	$\bar{Q}$
X-rays, gamma rays, electrons	1
Neutrons, protons, and heavier singly-charged particles	10
Alpha particles and particles with multiple or unknown charge	20

lective dose equivalent in a population subgroup would be the same, for example, if 10^5 persons each received a whole-body dose equivalent of 0.0001 Sv or if 10^4 persons each received 0.001 Sv. The product of the number of people exposed and their average dose equivalent in both instances is the same. In earlier language, the concept of "man-rem" was used to express the integral dose equivalent taken over a population. In ICRP 26, the Commission suggests that an additional small detriment to health that results from a practice which increases the collective dose equivalent by a small increment ΔS is approximately proportional to ΔS.

The Commission defines the *dose-equivalent commitment* H_c as the infinite time integral of the average *per-capita* dose-equivalent rate $\bar{\dot{H}}(t)$ in a given organ for a specified population:

$$H_c = \int_0^\infty \bar{\dot{H}}(t) \, dt. \tag{12.3}$$

The dose-equivalent commitment is used to give a measure of the total dose equivalent that will result in time to an average individual in a population as a result of a given decision or practice.

The *committed dose equivalent*, H_{50}, to a given organ or tissue from the single intake of radioactive material into the body is defined as the integrated dose equivalent accumulated over the next 50 yr from that intake. This can be written

$$H_{50} = \int_{t_0}^{t_0+50} \dot{H}(t) \, dt, \tag{12.4}$$

where t_0 is the time of intake and $\dot{H}(t)$ is the resultant dose-equivalent rate in the organ or tissue at time t, in years. The period of 50 yr is used to represent a working lifetime. The committed dose equivalent may be considered as a special case of dose-equivalent commitment.

For radiation protection purposes, the Commission assumes for stochastic effects that there is a linear relationship without threshold between dose and the probability of an effect.

> The simple summation of doses received by a tissue or organ as a measure of the total risk, and the calculation of the collective dose equivalent, as an index of the total detriment to a population, are valid only on the basis of this assumption and that the severity of each type of effect is independent of dose. (ICRP 26, p. 6)

Under the previous dose-equivalent protection system, discussed in the last section, a dose-equivalent limit was specified for each organ, independently of the dose equivalent that another organ might have received. This procedure did not take into account the increased risk presumably associated with the irradiation of several tissues or organs. The new procedure recommended in ICRP 26 now "takes account of the total risk attributable to the exposure of all tissues irradiated." A number of tissues and organs are discussed explicitly in Publication 26, on the basis of their radiosensitivity and the seriousness of damage and its treatability. These include the gonads, red bone marrow, bone, lung, pulmonary lymphoid tissue, thyroid, breast, lens of the eye, and skin. Risk factors for the different tissues are discussed; many are age or sex dependent. For protection purposes, the Commission regards the use of a single, average risk factor for each tissue as warranted, both for workers and for members of the general public. Recommended tissue weighting factors, w_T, which are explained below, are given in Table 12.3. The Commission assumes a mortality risk factor of about 10^{-2} Sv^{-1} for radiation-induced cancers for the purpose of radiation protection involving individuals. (The aver-

Table 12.3. Tissue Weighting Factors w_T Recommended in ICRP Publication 26

Tissue	w_T
Gonads	0.25
Breast	0.15
Liver	0.12
Lung	0.12
Thyroid	0.03
Bone surfaces	0.03
Remainder	0.30†
Sum	1.00

†The Commission recommends that the value $w_T = 0.06$ be applied to each of the five remaining organs or tissues that receive the highest dose equivalents and that the exposures of all other remaining tissues then be neglected.

age risk factor for hereditary effects in the next few generations is substantially less.) The acceptability of the average level of risk in radiation work can be judged by comparing it with the average risk in other occupations having high standards of safety, in which the average annual mortality rate from occupational hazards does not exceed 10^{-4}.

In Publication 26, the ICRP retained its previously recommended annual dose-equivalent limits for uniform irradiation of the whole body: 50 mSv (5 rem) for radiation workers and 5 mSv (0.5 rem) for individuals in the general public. (A population dose limit is no longer recommended in ICRP Publication 26, since each man-made exposure is to be justified by its benefits.) This baseline value is denoted by the symbol $H_{wb,L} = 50$ mSv. For stochastic effects, the recommended dose-equivalent limit is based on the principle that the risk should be the same, whether the whole body is irradiated uniformly or nonuniformly. If several tissues T receive dose equivalents H_T, then the total risk to the individual will not exceed that resulting from 50 mSv of uniform, whole-body irradiation provided

$$\sum_T w_T H_T \le H_{wb,L}, \qquad (12.5)$$

where the w_T are the factors given in Table 12.3. (The w_T represent the proportion of the total stochastic risk that results from the dose in tissue T when the whole body is irradiated; their sum in Table 12.3 is unity.) In applying (12.5), the sum can be understood to apply also to whole-body irradiation, for which $w_T = 1$.

Example

A worker has received 28 mSv whole-body dose equivalent from external sources and 40 mSv to the liver from an internally deposited radioisotope. These are his only occupational exposures during a working year. How much additional whole-body dose equivalent would have been permissible during that year? What maximum dose equivalent would have been permissible to the lung, if that were the only additional occupational exposure entailed during that year?

Solution

According to the ICRP recommendation, the total risk shall not exceed that equivalent to an annual, uniform, whole-body dose equivalent of 50 mSv. Since the worker received 28 mSv whole body ($w_T = 1$) plus 40 mSv to the liver ($w_T = 0.06$, Table 12.3, footnote), the relation (12.5) implies that his total risk is equivalent to that resulting from a whole-body dose equivalent of

$$\sum_T w_T H_T = 1 \times 28 + 0.06 \times 40 = 30.4 \text{ mSv}. \tag{12.6}$$

Therefore, he could have received an additional $50 - 30.4 = 19.6$ mSv whole-body dose equivalent during that year. If, on the other hand, only the lung were exposed additionally, then, since $w_T = 0.12$ for this organ, the lung dose equivalent could have been as high as $19.6/0.12 = 163$ mSv. For protection purposes, this lung dose equivalent is regarded as entailing the same average risk to an individual as 19.6 mSv of whole-body exposure.

Example
What would the answers to the last problem be, based on the previously used dose-equivalent/critical organ system described in Section 12.3?

Solution
We refer to Table 12.1. The allowable whole-body annual limit of 50 mSv implies that the worker could have received another $50 - 28 = 32$ mSv, provided none of the individual organ dose-equivalent limits is exceeded. The liver is included under "other organs, tissues, and organ systems," and therefore is allowed up to 150 mSv. The additional 32 mSv of whole-body exposure would bring the total liver dose equivalent up to $28 + 40 + 32 = 100$ mSv, which would be acceptable. Note that this protection system does not further limit the whole-body dose equivalent by virtue of the additional liver dose, as reflected by the factor 0.12×40 in Eq. (12.6) of the current system. Similarly, the lung could receive 150 mSv annually. Since the worker received 28 mSv to the lung from whole-body irradiation, an additional dose-equivalent of $150 - 28 = 122$ mSv would be permissible. This limit is also independent of the irradiation of the liver or any other individual organ systems.

For nonstochastic effects, the ICRP recommends an annual limit of 0.5 Sv to all tissues except the lens of the eye, for which the recommended limit is 0.15 Sv. These limits are intended to prevent nonstochastic effects from occurring and to constrain any extreme exposure that still satisfies the limitation on stochastic effects. For example, if only the thyroid were irradiated, the relation (12.5) would imply an annual limit of $50/0.03 = 1670$ mSv. This would, however, not be permitted in view of the limiting dose equivalent for the prevention of nonstochastic effects.

Publication 26 also addresses other categories of exposure, such as special planned exposures in excess of the recommended limits, occupational exposure of women of reproductive capacity, pregnant women, and exposures from accidents and emergencies. It also discusses ICRP principles and philosophy for operational radiation protection in the application of its recommendations.

12.5 DERIVED AND AUTHORIZED LIMITS

As we described, the NCRP and ICRP have issued recommendations that primarily limit organ dose equivalents or risk to individuals. In so doing, these organizations have allowed flexibility for implementing specific radiation-protection practices. However, detailed criteria are a practical necessity to assure compliance with the primary NCRP and ICRP limits. Often very specific requirements are laid down by regulatory agencies that have responsibility for certain operations. For example, when radioactive materials are shipped, various authorities have detailed requirements that must be met. Depending on the radionuclide and its activity, regulations specify its containment, packaging, labeling, maximum dose rate at the surface, and other factors. Such specific criteria are based on conservative models and assumptions so that, if complied with, the dose equivalent to any individual will almost certainly be well within the recommended limits.

The following descriptions are from ICRP Publication 26.

In many practical situations it will be convenient to make use of a derived limit, calculated with the aid of a model, which provides a quantitative link between a particular measurement and the recommended dose-equivalent limit. In deriving such a limit the intention should be to establish a figure such that adherence to it will provide virtual certainty of compliance with the Commission's recommended dose-equivalent limits. However, failure to adhere to the derived limit will not necessarily imply a failure to achieve compliance with the Commission's recommendations and may require only a more careful study of the circumstances....

In practical radiation protection, it is often necessary to provide limits associated with quantities other than dose equivalent, committed dose equivalent, or intake, and relating, for example, to environmental conditions. When these limits are related to the basic limits by a defined model of the situation and are intended to reflect the basic limits, they are called derived limits. Derived limits may be set for quantities such as dose-equivalent rate in a workplace, contamination of air, contamination of surfaces and contamination of environmental materials....

Limits laid down by a competent authority or by the management of an institution are called authorized limits. These should, in general, be below derived limits though, exceptionally, they may be equal to them. The process of optimization may be used in the establishment of authorized limits and they apply only in limited circumstances.... Where an authorized limit exists it will always take precedence over a derived limit.

Use can be made of derived limits or authorized limits in the application of individual monitoring programs. Such limits are essential in the monitoring of operations where, in general, only part of the dose-equivalent limit or secondary limit can be committed to the operation.

Only in a few circumstances can the results of programs of monitoring of the workplace be used to estimate the dose equivalents or intakes of individual workers. The use of derived or authorized limits is essential in the interpretation of environmental monitoring programs.

The ICRP and NCRP have established a reference, or "standard," man with metabolic models for various chemical elements taken into the body via inhalation or ingestion. Models for the respiratory and gastrointestinal systems are used to calculate concentrations of specific radionuclides in air, water, and food as derived limits for controlling the dose equivalent to various organs of the body. Internal radiation dosimetry is the subject of Chapter 14.

12.6 ADDITIONAL INFORMATION

The reader is referred to the entire comprehensive reports published by the International Commission on Radiological Protection (ICRP) and the National Council on Radiation Protection and Measurements (NCRP). ICRP Publication 26 and NCRP Report 39, discussed in this chapter, put forth the general criteria for limiting radiation exposure to persons in various categories of exposure. Other reports of these two organizations address a wide variety of radiation problems, such as the storage, use, and disposal of unsealed sources in hospitals and medical research establishments (ICRP Publication 25, 1977); the release of radionuclides into the environment (ICRP 29, 1979); the inhalation of radon daughters by workers (ICRP 32, 1981); dental X-ray protection (NCRP 35, 1970); medical exposure of pregnant women (NCRP 54, 1977); radiation exposures from consumer products (NCRP 56, 1977); instrumentation and monitoring for radiation protection (NCRP 57, 1978); tritium in the environment (NCRP 62, 1979); mammography (NCRP 66, 1980); ^{129}I releases from nuclear-power generation (NCRP 75, 1983); and occupational and environmental exposures from radon and radon daughters in the United States (NCRP 78, 1984).

Other organizations that provide publications concerned with radiation-protection practices, standards and criteria include the following: American National Standards Institute (ANSI); Food and Drug Administration (FDA); International Atomic Energy Agency (IAEA); International Commission on Radiation Units and Measurements (ICRU); International Labor Organization (ILO); Society of Nuclear Medicine–Medical Internal Radiation Dose (MIRD) Committee; U.S. Environmental Protection Agency

(EPA); U.S. Nuclear Regulatory Commission (NRC); and the U.S. Department of Energy. In the United States, specific legal radiation-exposure guides are published in the Federal Register under Title 10, Part 20 of the Code of Federal Regulations.

12.7 PROBLEMS

1. In earlier years, fluoroscopes were available in stores for inspecting how well shoes fit. What are the benefits and risks from this use of X-rays?
2. Welds in metal structures can be inspected with gamma rays (e.g., from ^{137}Cs) to detect flaws not visible externally. The production, transport, and use of such sources entails exposures of workers and the public to some radiation. Give an example to show how banning the use of gamma rays for this purpose could be of greater detriment to society than the radiation exposures it entails.
3. Discuss risks and benefits associated with the development of nuclear power. What risk and benefit factors are there, apart from the potential health effects of radiation?
4. Discuss the risks and benefits associated with having nuclear submarines.
5. A proposal is made to test a new dental-hygiene procedure for children that is said to have the potential of greatly reducing tooth decay. Dental X-rays of several thousand children would have to be made periodically during a 5 yr study in order to perfect and evaluate the procedure. Discuss the rationale on which a decision could be made either to implement or to reject the proposal.
6. A gamma source is used for about an hour every day in a laboratory room. If the source, when not in use, is kept in its container in the room, the resulting dose equivalent to persons working there is well within the allowable limit. Alternatively, the source could be stored in the unoccupied basement, two floors below, with virtually no exposure to personnel. How would the ALARA principle apply in this example?
7. According to the NCRP Report 39, what is the annual occupational dose-equivalent limit for the liver? If a worker receives 2 rem whole-body exposure to penetrating gamma rays during a year, what is the maximum allowed dose equivalent for the liver in that same year?
8. The maximum permissible annual occupational whole-body dose equivalent recommended in NCRP Report 39 is 5 rem. In making this recommendation, what did the Council consider as the "critical organs"?
9. According to NCRP Report 39, what is the recommended maximum dose equivalent that a worker theoretically could have accumulated at age 45?
10. Estimate the dose equivalent to a worker who receives a dose of 2 mGy from fast neutrons.
11. What is the total dose equivalent to the lung of a worker whose lung receives a 6 mGy dose from alpha radiation from an internally deposited radionuclide plus a 10 mGy dose from external gamma radiation?
12. Why can the committed dose equivalent be regarded as a special case of the dose-equivalent commitment?
13. According to ICRP Publication 26, how much dose equivalent to the red bone marrow alone entails a risk equivalent to that from 50 mSv of whole-body irradiation?
14. Based on ICRP Publication 26, a 15 mSv dose equivalent to the gonads entails a certain risk to an individual. What level of dose equivalent to the lung entails the same risk?
15. Based on ICRP Publication 26, what whole-body dose equivalent would result in the same risk to a worker as that from the combined internal and external irradiation in Problem 11?
16. How much whole-body dose equivalent would be allowed to a worker who has received 106 mSv to the gonads, if the total risk to him is not to exceed the annual limit recommended in ICRP Publication 26 for stochastic effects?
17. What is the maximum annual dose equivalent to the thyroid recommended in ICRP Publication 26? Is this limit dictated by that for stochastic or nonstochastic effects?
18. Are larger dose-equivalent limits permitted to certain organs by ICRP Publication 26 or by NCRP Report 39? Give one or more examples to illustrate your answer.
19. Show that the 5 rem/yr whole-body dose-equivalent limit is equivalent to an average rate of 2.5 mrem/hr for a 40 hr work week, 50 wk/yr.

CHAPTER 13
EXTERNAL RADIATION PROTECTION

We now describe procedures for limiting the dose received from radiation sources outside the human body. In the next chapter we discuss protection from radionuclides that can enter the body.

13.1 DISTANCE, TIME, AND SHIELDING

In principle, one's dose in the vicinity of an external radiation source can be reduced by increasing the distance from the source, by minimizing the time of exposure, and by the use of shielding. Distance is often employed simply and effectively. For example, tongs are used to handle radioactive sources in order to minimize the dose to the hands as well as the rest of the body. Limiting the duration of an exposure significantly is not always feasible, because a certain amount of time is usually required to perform a given task. Sometimes, though, practice runs beforehand without the source can reduce exposure times when an actual job is carried out.

While distance and time factors can be employed advantageously in external radiation protection, shielding provides a more reliable way of limiting personnel exposure by limiting the dose rate. In principle, shielding alone can be used to reduce dose rates to desired levels. In practice, however, the amount of shielding employed will depend on a balancing of practical necessities such as cost and the benefit expected.

In this chapter we describe methods for determining appropriate shielding for the most common kinds of external radiation: gamma rays, X-rays from diagnostic and therapeutic machines, beta rays with accompanying bremsstrahlung, and neutrons.

13.2 GAMMA-RAY SHIELDING

In Section 7.7 we discussed attenuation coefficients and described the transmission of photons through matter under conditions of "good" and "poor" geometry. The relative intensity I/I_0 of monoenergetic photons transmitted without interaction through a shield of thickness x is given by Eq. (7.32),

$$I = I_0 e^{-\mu x}, \tag{13.1}$$

where μ is the linear attenuation coefficient. If the incident beam is broad, as in Fig. 7.8, or the shield very thick, then the measured intensity will be greater than that described by Eq. (13.1) because scattered photons will also be detected. Such conditions of poor geometry usually apply to the thick shields required for protection from gamma-ray sources. The increased transmission of photon intensity over that measured in good geometry can be taken into account by writing

$$I = B I_0 e^{-\mu x}, \tag{13.2}$$

where B is called the buildup factor ($B \geq 1$). For a given shielding material, thickness, photon energy, and source geometry, B can be obtained from measurements or calculations.

Table 13.1 gives buildup factors for water and several elements for monoenergetic photons from a point isotropic source. The thickness of a shield for which the photon intensity in a narrow beam is reduced to $1/e$ of its original value is called the relaxation length. One relaxation length, therefore, is numerically equal to $1/\mu$. The dependence of B in Table 13.1 on shield thickness is expressed by its variation with the number of relaxation

Table 13.1. Buildup Factors B for a Point Isotropic Source†

Material	MeV	Relaxation length, μx						
		1	2	4	7	10	15	20
Water	0.255	3.09	7.14	23.0	72.9	166	456	982
	0.5	2.52	5.14	14.3	38.8	77.6	178	334
	1.0	2.13	3.71	7.68	16.2	27.1	50.4	82.2
	2.0	1.83	2.77	4.88	8.46	12.4	19.5	27.7
	3.0	1.69	2.42	3.91	6.23	8.63	12.8	17.0
	4.0	1.58	2.17	3.34	5.13	6.94	9.97	12.9
	6.0	1.46	1.91	2.76	3.99	5.18	7.09	8.85
	8.0	1.38	1.74	2.40	3.34	4.25	5.66	6.95
	10.0	1.33	1.63	2.19	2.97	3.72	4.90	5.98
Aluminum	0.5	2.37	4.24	9.47	21.5	38.9	80.8	141
	1.0	2.02	3.31	6.57	13.1	21.2	37.9	58.5
	2.0	1.75	2.61	4.62	8.05	11.9	18.7	26.3
	3.0	1.64	2.32	3.78	6.14	8.65	13.0	17.7
	4.0	1.53	2.08	3.22	5.01	6.88	10.1	13.4
	6.0	1.42	1.85	2.70	4.06	5.49	7.97	10.4
	8.0	1.34	1.68	2.37	3.45	4.58	6.56	8.52
	10.0	1.28	1.55	2.12	3.01	3.96	5.63	7.32
Iron	0.5	1.98	3.09	5.98	11.7	19.2	35.4	55.6
	1.0	1.87	2.89	5.39	10.2	16.2	28.3	42.7
	2.0	1.76	2.43	4.13	7.25	10.9	17.6	25.1
	3.0	1.55	2.15	3.51	5.85	8.51	13.5	19.1
	4.0	1.45	1.94	3.03	4.91	7.11	11.2	16.0
	6.0	1.34	1.72	2.58	4.14	6.02	9.89	14.7
	8.0	1.27	1.56	2.23	3.49	5.07	8.50	13.0
	10.0	1.20	1.42	1.95	2.99	4.35	7.54	12.4
Lead	0.5	1.24	1.42	1.69	2.00	2.27	2.65	(2.73)
	1.0	1.37	1.69	2.26	3.02	3.74	4.81	5.86
	2.0	1.39	1.76	2.51	3.66	4.84	6.87	9.00
	3.0	1.34	1.68	2.43	2.75	5.30	8.44	12.3
	4.0	1.27	1.56	2.25	3.61	5.44	9.80	16.3
	5.0	1.21	1.46	2.08	3.44	5.55	11.7	23.6
	6.0	1.18	1.40	1.97	3.34	5.69	13.8	32.7
	8.0	1.14	1.30	1.74	2.89	5.07	14.1	44.6
	10.0	1.11	1.23	1.58	2.52	4.34	12.5	39.2
Uranium	0.5	1.17	1.30	1.48	1.67	1.85	2.08	–
	1.0	1.31	1.56	1.98	2.50	2.97	3.67	–
	2.0	1.33	1.64	2.23	3.09	3.95	5.36	(6.48)
	3.0	1.29	1.58	2.21	3.27	4.51	6.97	9.88
	4.0	1.24	1.50	2.09	3.21	4.66	8.01	12.7
	6.0	1.16	1.36	1.85	2.96	4.80	10.8	23.0
	8.0	1.12	1.27	1.66	2.61	4.36	11.2	28.0
	10.0	1.09	1.20	1.51	2.26	3.78	10.5	28.5

†Tables 13.1 and 13.2 are from U.S. Public Health Service, *Radiological Health Handbook*, Publ. No. 2016, Bureau of Radiological Health, Rockville, MD (1970). For concrete, use average of values for Al and Fe.

lengths μx. Buildup factors for concrete can be obtained from Table 13.1 as the arithmetic average of the values for Al and Fe:

$$B_{\text{concrete}} = \frac{1}{2} (B_{\text{Al}} + B_{\text{Fe}}).$$ (13.3)

Table 13.2 gives buildup factors for a broad, parallel beam of monoenergetic photons.

Tables 13.1 and 13.2 can be used with Eq. (13.2) to calculate the shielding thickness x necessary to reduce gamma-ray intensity from a value I_0 to I. Since the exponential attenuation factor $e^{-\mu x}$ and the buildup factor B both depend on x, which is originally unknown, the appropriate thickness for a given problem usually has to be found by making successive approximations until Eq. (13.2) is satisfied. An initial (low) estimate of the amount of shielding needed can be obtained by solving Eq. (13.2) for x with assumed narrow-beam geometry, i.e., with $B = 1$. One can then add some additional shielding and see whether the values of B and the exponential for the new thickness satisfy Eq. (13.2).

Table 13.2. Buildup Factors B for a Parallel Broad Beam

Material	MeV	Relaxation length, μx					
		1	2	4	7	10	15
Water	0.5	2.63	4.29	9.05	20.0	35.9	74.9
	1.0	2.26	3.39	6.27	11.5	18.0	30.8
	2.0	1.84	2.63	4.28	6.96	9.87	14.4
	3.0	1.69	2.31	3.57	5.51	7.48	10.8
	4.0	1.58	2.10	3.12	4.63	6.19	8.54
	6.0	1.45	1.86	2.63	3.76	4.86	6.78
	8.0	1.36	1.69	2.30	3.16	4.00	5.47
Iron	0.5	2.07	2.94	4.87	8.31	12.4	20.6
	1.0	1.92	2.74	4.57	7.81	11.6	18.9
	2.0	1.69	2.35	3.76	6.11	8.78	13.7
	3.0	1.58	2.13	3.32	5.26	7.41	11.4
	4.0	1.48	1.90	2.95	4.61	6.46	9.92
	6.0	1.35	1.71	2.48	3.81	5.35	8.39
	8.0	1.27	1.55	2.17	3.27	4.58	7.33
	10.0	1.22	1.44	1.95	2.89	4.07	6.70
Tin	1.0	1.65	2.24	3.40	5.18	7.19	10.5
	2.0	1.58	2.13	3.27	5.12	7.13	11.0
	4.0	1.39	1.80	2.69	4.31	6.30	—
	6.0	1.27	1.57	2.27	3.72	5.77	11.0
	10.0	1.16	1.33	1.77	2.81	4.53	9.68
Lead	0.5	1.24	1.39	1.63	1.87	2.08	—
	1.0	1.38	1.68	2.18	2.80	3.40	4.20
	2.0	1.40	1.76	2.41	3.36	4.35	5.94
	3.0	1.36	1.71	2.42	3.55	4.82	7.18
	4.0	1.28	1.56	2.18	3.29	4.69	7.70
	6.0	1.19	1.40	1.87	2.97	4.69	9.53
	8.0	1.14	1.30	1.69	2.61	4.18	9.08
	10.0	1.11	1.24	1.54	2.27	3.54	7.70
Uranium	0.5	1.17	1.28	1.45	1.60	1.73	—
	1.0	1.30	1.53	1.90	2.32	2.70	3.60
	2.0	1.33	1.62	2.15	2.87	3.56	4.89
	3.0	1.29	1.57	2.13	3.02	3.99	5.94
	4.0	1.25	1.49	2.02	2.94	4.06	6.47
	6.0	1.18	1.37	1.82	2.74	4.12	7.79
	8.0	1.13	1.27	1.61	2.39	3.65	7.36
	10.0	1.10	1.21	1.48	2.12	3.21	6.58

Typically, one might add to the initial estimate a half-value layer (HVL) of shielding or a fraction thereof. A half-value layer h is the thickness that reduces the intensity by one-half. For monoenergetic photons, h can be estimated from narrow-beam conditions by writing

$$\frac{1}{2} = e^{-\mu h}, \tag{13.4}$$

giving

$$h = \frac{\ln 2}{\mu} = \frac{0.693}{\mu}. \tag{13.5}$$

It follows that h is equivalent to 0.693 relaxation lengths.

Some examples will illustrate gamma-ray shielding calculations.

Example
Calculate the thickness of a lead shield needed to reduce the exposure rate 1 m from a 10 Ci point source of ^{42}K to 2.5 mR/hr. The decay scheme of the β^- emitter is shown in Fig. 13.1. The daughter ^{42}Ca is stable.

Solution
With no shielding, the exposure rate at 1 m is given by Eq. (10.29):

$$\dot{\Gamma} = 0.5CE = 0.5 \times 10 \times (0.18 \times 1.52) = 1.37 \text{ R/hr.} \tag{13.6}$$

We make an initial estimate of the shielding required to reduce this to 2.5 mR/hr on the basis of narrow-beam geometry. From Fig. 7.9 we obtain for the mass attenuation coefficient of lead for 1.52 MeV photons $\mu/\rho = 0.047$ cm^2/g. Since $\rho = 11.4$ g/cm^3, we find for the linear attenuation coefficient $\mu = 0.047 \times 11.4 = 0.536$ cm^{-1}. To reduce the unshielded exposure rate of 1370 mR/hr to 2.5 mR/hr under conditions of good geometry would require lead of thickness x given by

$$2.5 = 1370e^{-0.536x}, \tag{13.7}$$

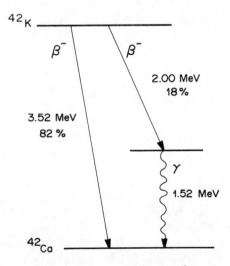

Figure 13.1. Decay scheme of ^{42}K.

or $x = 11.8$ cm. To obtain a better estimate, we add one half-value layer to this and see what exposure rate results when the buildup factor is considered. From Eq. (13.5), one HVL is $h = 0.693/0.536 = 1.29$ cm; and so we try the thickness $x = 11.8 + 1.3 = 13.1$ cm, or $\mu x = 0.536 \times 13.1 = 7.02$ relaxation lengths. To find B, we interpolate linearly in Table 13.1 between the entries 3.02 and 3.66 for 1.0 MeV and 2.0 MeV photons in Pb at 7 relaxation lengths. We obtain $B = 3.35$. The exposure rate at 1 m, including buildup, is then given by Eq. (13.2),

$$I = 3.35 \times 1370e^{-7.02} = 4.10 \text{ mR/hr.} \tag{13.8}$$

This rate is too high, and so we must add more shielding. As seen from Eq. (13.2) and Table 13.1, increasing μx gives greater exponential attenuation, but increases the buildup factor. However, B increases much more slowly than the exponential factor decreases. As a next estimate, we add an additional thickness y which will reduce the exposure rate by exponential attenuation alone to some value less than the desired 2.50 mR/hr — say, to 2.40 mR/hr. Thus, writing

$$e^{-y} = \frac{2.40}{4.10} \tag{13.9}$$

gives $y = 0.536$ relaxation lengths. The new estimated shield thickness becomes $7.02 + 0.54 = 7.56$ relaxation lengths. The corresponding buildup factor is obtained by two-dimensional linear interpolation in Table 13.1:

	7	7.56	10
1.0	3.02		3.74
1.52	3.35	3.53	4.31
2.0	3.66		4.84.

With $B = 3.53$, the exposure rate at 1 m is now

$$I = 3.53 \times 1370e^{-7.56} = 2.52 \text{ mR/hr,} \tag{13.10}$$

which is close to the design value. We round off upward and use 7.60 relaxation lengths as the solution to the problem. The thickness of lead shielding needed is $x = 7.60/\mu = 7.60/0.536 = 14.2$ cm. A shield of this thickness can be interposed anywhere between the source and the point of exposure. Usually, shielding is placed close to a source to realize the greatest solid-angle protection.

Until now we have discussed monoenergetic photons. When photons of different energies are present, separate calculations at each energy are usually needed, since their attenuation coefficients and buildup factors are different.

Example
A 144 Ci point source of ^{24}Na is to be stored at the bottom of a pool of water. The radionuclide emits two photons per disintegration with energies 2.75 MeV and 1.37 MeV in decaying by β^- emission to stable ^{24}Mg. How deep must the water be if the exposure rate at a point 6 m directly above the source is not to exceed 20 mR/hr? What is the exposure rate at the surface of the water right above the source?

Solution
The mass attenuation coefficients for the two photon energies can be obtained from Fig. 7.11. Since we are dealing with water, they are numerically equal to the linear attenuation coefficients. Thus, $\mu_1 = 0.043$ cm^{-1} and $\mu_2 = 0.061$ cm^{-1}, respectively, for the 2.75 MeV and 1.37 MeV photons. The approach we use is to consider the harder photons first and find a depth of water that will reduce their exposure rate to a level somewhat below 20 mR/hr, and then see what additional exposure rate results from the softer photons. The final depth can be adjusted to make the total 20 mR/hr. The exposure rate from the 2.75 MeV photons at a distance $d = 6$ m with no shielding is

$$\dot{\Gamma}_{2.75} = \frac{0.5CE}{d^2} = \frac{0.5 \times 144 \times 2.75}{6^2} = 5.50 \text{ R/hr.} \qquad (13.11)$$

To reduce this to 20 mR/hr under conditions of good geometry requires a water depth $\mu_1 x$ given by

$$20 = 5500e^{-\mu_1 x}, \qquad (13.12)$$

or $\mu_1 x = 5.62$ relaxation lengths. From Table 13.1 we can see that the buildup factor for 2.75 MeV photons in a shield of this thickness is in the neighborhood of 5 or 6. This amount of buildup can be roughly compensated by 2.5 HVLs, which we add for our next estimate. From Eq. (13.5), this addition represents $2.5 \times 0.693 = 1.73$ relaxation lengths, resulting in a total estimated water thickness of $5.62 + 1.73 = 7.35$ relaxation lengths. Linear interpolation in Table 13.1 gives the buildup factor $B = 7.11$. Applying Eq. (13.2), we find for the exposure rate from the 2.75 MeV photons

$$\dot{\Gamma}_{2.75} = 7.11 \times 5500e^{-7.35} = 25.1 \text{ mR/hr.} \qquad (13.13)$$

Adding another $h/2$ gives a thickness $7.35 + 0.35 = 7.70$ relaxation lengths and $B = 7.44$. The exposure rate then becomes

$$\dot{\Gamma}_{2.75} = 7.44 \times 5500e^{-7.70} = 18.5 \text{ mR/hr.} \qquad (13.14)$$

For the 1.37 MeV photons, the thickness of this shield in relaxation lengths is larger by the ratio of the attenuation coefficients: $7.70 \times (0.061/0.043) = 10.9$. The buildup factor is, from Table 13.1, 24.8. The exposure rate at 6 m for these photons without shielding is

$$\dot{\Gamma}_{1.37} = \frac{0.5 \times 144 \times 1.37}{6^2} = 2.74 \text{ R/hr.} \qquad (13.15)$$

With the shield it is

$$\dot{\Gamma}_{1.37} = 24.8 \times 2740e^{-10.9} = 1.25 \text{ mR/hr.} \qquad (13.16)$$

The total exposure rate is

$$\dot{\Gamma} = \dot{\Gamma}_{2.75} + \dot{\Gamma}_{1.37} = 18.5 + 1.25 = 19.8 \text{ mR/hr,} \qquad (13.17)$$

which is acceptably close to the design figure. The needed depth of water is, therefore, $7.70/0.043 = 179$ cm. The exposure level at the surface of the water is $19.8(600/179)^2 = 222$ mR/hr.

13.3 SHIELDING IN X-RAY INSTALLATIONS

X-ray machines have three principal uses—as diagnostic, therapeutic, and nonmedical radiographic devices. An X-ray tube is usually housed in a heavy lead casing with an aperature through which the primary, or useful, beam emerges. Typically, the beam passes through metal filters (e.g., Al, Cu) to remove unwanted, less penetrating radiation and is then collimated to reduce its width. The housing, supplied by the manufacturer, should conform to certain specifications in order to limit the leakage radiation that emerges from it during operation. For diagnostic X-ray tubes, the housing is so constructed that the leakage exposure rate at a distance of 1 m from the target of the tube does not exceed 0.1 R/hr when the tube is operated continuously at its maximum rated current and potential. For the housing of a therapeutic machine with a peak tube potential not exceeding 500 keV, the corresponding exposure rate at 1 m should not be greater than 1 R/hr. For operating potentials above 500 keV, the leakage at 1 m should not exceed either 1 R/hr or 0.1% of the useful-beam exposure rate at 1 m from the target, whichever is greater, when the tube is operated at its maximum current and potential. For nonmedical radiographic X-ray machines, the housing should conform at least to the requirements for therapeutic devices.

The shielding provided by the X-ray housing is referred to as source shielding. Additional protection is obtained by the use of structural shielding in an X-ray facility. The basic components of the radiation field considered in the design of structural shielding are shown in Fig. 13.2. A primary protective barrier, such as a lead-lined wall, is fixed in place in any direction in which the useful beam can be pointed. This shield reduces the exposure rate outside the X-ray area in the direction of the primary beam. Locations not in the direct path of the beam are also exposed to photons in two ways. As illustrated in Fig. 13.2, leakage radiation escapes from the housing in all directions. In addition, photons are scattered from exposed objects in the primary beam and from walls, ceilings, and other structures. Secondary protective barriers are needed to reduce exposure rates outside the X-ray area from both leakage and scattered radiation. Sometimes existing structures, such as concrete walls, provide sufficient secondary barriers; otherwise, additional shielding, such as lead sheets, must be added to them.

Generally, structural shielding is designed to limit the average dose equivalent to indi-

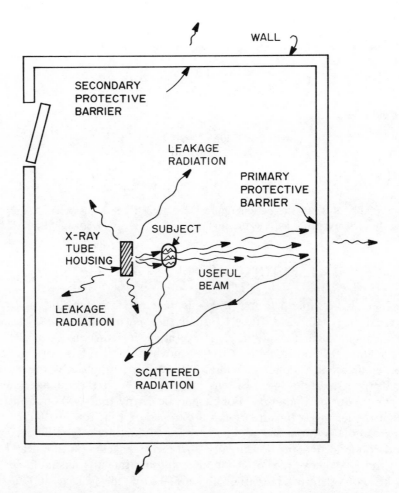

Figure 13.2. Schematic top view of an X-ray room showing the different radiation components considered in the design of structural shielding to provide primary and secondary protective barriers.

viduals outside an X-ray room to 0.1 rem/wk in controlled areas and to 0.01 rem/wk in uncontrolled areas. A controlled area is one in which access and occupancy are regulated in conjunction with operation of the X-ray machine; an uncontrolled area is one over which the operator of the X-ray facility has no jurisdiction. These design rates adhere to the usual annual limits of 5 rem and 0.5 rem to persons for occupational and nonoccupational dose equivalent. For computational purposes, the permissible average exposure rates to individuals may be taken to be 0.1 R/wk and 0.01 R/wk.

Design of Primary Protective Barrier

The attenuation of primary X-ray beams through different thicknesses of various shielding materials has been measured experimentally. The data have been plotted to give empirical attenuation curves, which are used to design primary protective barriers. It is found experimentally that the primary beam intensity transmitted through a shield depends strongly on the peak operating voltage but very little on the filtration of the beam. (The effect of filters on exposure rate is small compared with that of the thicker shields.) In addition, at fixed kVp, the exposure from transmitted photons at a given distance from the X-ray machine is proportional to the time integral of the beam current, usually expressed in milliampere-minutes (mA min). In other words, the total exposure per mA min is virtually independent of the tube operating current itself. These circumstances permit the presentation of X-ray attenuation data for a given shielding material as a family of curves at different kVp values. Measurements are conveniently referred to a distance of 1 m from the target of the tube with different thicknesses of shield interposed.

Attenuation curves measured for lead and concrete at a number of peak voltages (kVp) are shown in Figs. 13.3 through 13.7. The ordinate, K, gives the exposure of the attenuated radiation in R/(mA min) at the reference distance of 1 m. The abscissa gives the shield thickness. Figure 13.3 shows, for example, that behind 2 mm of lead, the exposure 1 m from the target of an X-ray machine operating at 150 kVp is 10^{-3} R/(mA min). If the machine is operated with a beam current of 200 mA for 90 sec, i.e., for $200 \times 1.5 = 300$ mA min, then the exposure at 1 m will be $300 \times 10^{-3} = 0.3$ R behind the 2 mm lead shield. The same exposure results if the tube is operated at 300 mA for 60 sec. The exposure at other distances can be obtained by the inverse-square law; for example, the exposure per mA min at 2 m is $10^{-3}/2^2 = 2.5 \times 10^{-4}$ R/(mA min). The 2 mm of lead shielding can be located anywhere between the X-ray tube and the point of interest.

The amount of shielding needed to provide the primary barrier for an area adjoining an X-ray room can be found from the attenuation curves, once the appropriate value of K has been determined. In addition to the peak voltage, the value of K in a specific application will depend on several other circumstances:

1. The maximum permissible exposure rate to an individual, P, which is usually $P = 0.1$ R/wk or $P = 0.01$ R/wk for controlled and uncontrolled areas.
2. The workload, W, or weekly amount of use of the X-ray machine, expressed in mA min/wk.
3. The use factor, U, or fraction of the workload during which the useful beam is pointed in a direction under consideration.
4. The occupancy factor, T, which takes into account the fraction of the time that an area outside the barrier is likely to be occupied by a given individual. (Average weekly exposure rates may be greater than P in areas not occupied full time by anyone.) In the absence of more specific information, the occupancy factors given in Table 13.3 can be used as guides for shielding design. The allowed average exposure rate in the area is P/T R/wk.

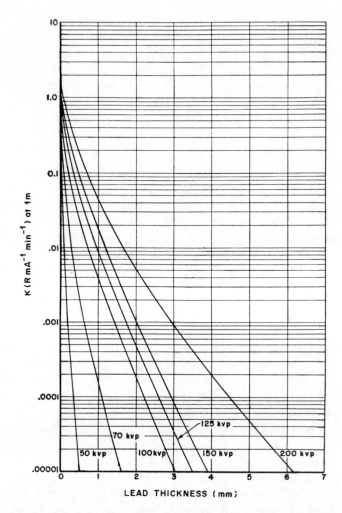

Figure 13.3. Attenuation in lead of X-rays produced by potentials of 50 to 200 kVp (*National Bureau of Standards Handbook 76, 1961, Washington, D.C.*)

5. The distance, d, in meters from the target of the tube to the location under consideration. The curves in Figs. 13.3 through 13.7 give the value of K for a distance of 1 m. At other distances, a factor of d^2 enters in the evaluation of K.

With these considerations included, the value of K can be computed from the formula

$$K = \frac{Pd^2}{WUT} \ . \tag{13.18}$$

With P in R/wk, d in m, and W in mA min/wk, K gives the exposure of the transmitted radiation in R/(mA min) at 1 m.

The role of the various factors in Eq. (13.18) is straightforward. When $d = 1$ m, $W = 1$ mA min/wk, and the useful beam is always pointed ($U = 1$) in the direction of an area of full occupancy ($T = 1$), then it follows that K is numerically equal to P R/wk. If U and T are not unity, then the weekly exposure in the area can be increased to P/UT, which is reflected in a larger value of K and hence a smaller shield thickness. The factor d^2, with

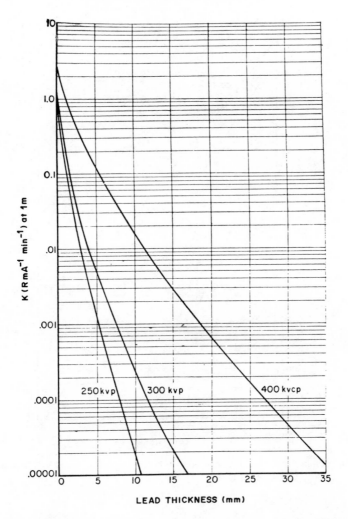

Figure 13.4. Attenuation in lead of X-rays produced by potentials of 250 to 400 kVp (*National Bureau of Standards Handbook 76*, 1961, Washington, D.C.)

d expressed in meters, adjusts *K* for locations other than 1 m. Finally, since exposure is proportional to the workload, one divides by *W* in Eq. (13.18).

Example

A diagnostic X-ray machine is operated at 125 kVp and 220 mA for an average of 90 sec/wk. Calculate the primary protective barrier thickness if lead or concrete alone were to be used to protect an uncontrolled hallway 15 ft from the tube target (Fig. 13.8). The useful beam is directed horizontally toward the barrier $\frac{1}{3}$ of the time and vertically into the ground the rest of the time.

Solution

For the uncontrolled hall, $P = 0.01$ R/wk. The distance to the hall in meters is $d = 15/3.28$, and the workload is $W = 220$ mA × 1.5 min/wk = 330 mA min/wk. The use factor is $U = \frac{1}{3}$ and the occupancy factor (Table 13.3) is $T = \frac{1}{4}$. Equation (13.18) gives

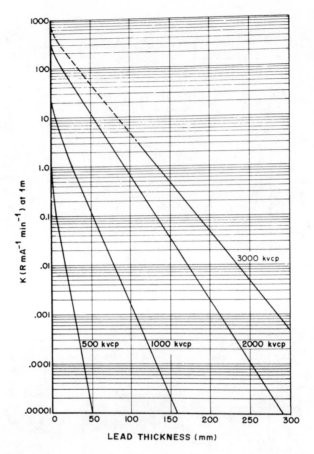

Figure 13.5. Attenuation in lead of X-rays produced by potentials of 500 to 3000 kVp (*National Bureau of Standards Handbook 76*, 1961, Washington, D.C.)

$$K = \frac{0.01 \times (15/3.28)^2}{330 \times 1/3 \times 1/4} = 7.61 \times 10^{-3} \ \frac{R}{mA \ min} \ \text{at 1 m.} \tag{13.19}$$

From Fig. 13.3 we find, for 125 kVp, that the required thickness of lead is about 1.05 mm; from Fig. 13.6, that of concrete is about 3.6 in.

Table 13.3. Occupancy Factors†

Full occupancy $T = 1$	Work areas such as offices, laboratories, shops, wards, nurses' stations; living quarters; children's play areas; and occupied space in nearby buildings.
Partial occupancy $T = \frac{1}{4}$	Corridors, rest rooms, elevators using operators, and unattended parking lots.
Occasional occupancy $T = \frac{1}{16}$	Waiting rooms, toilets, stairways, unattended elevators, janitors' closets, and outside areas used only for pedestrians or vehicular traffic.

†Tables 13.3, 13.4 and 13.5 are from *Structural Shielding Design and Evaluation for Medical Use of X-rays and Gamma Rays of Energies up to 10 MeV*, NCRP Report No. 49, National Council on Radiation Protection and Measurements, Washington, D.C. (1976). Copyright 1976 NCRP; reprinted with permission.

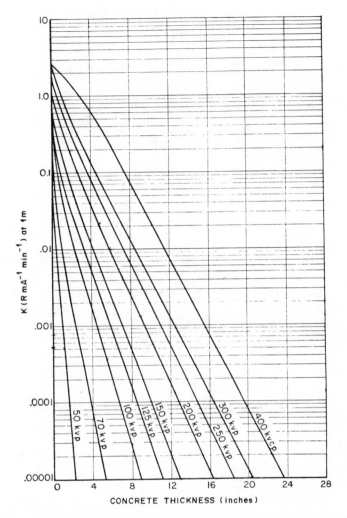

Figure 13.6. Attenuation in concrete of X-rays produced by potentials of 50 to 400 kVp (*National Bureau of Standards Handbook 76*, 1961, Washington, D.C.)

A primary protective barrier can either be erected when a structure is built or it can be provided by adding shielding to an existing structure. The attenuation of most common building materials per g/cm^2 is approximately the same as that of concrete, which has an average density of 2.35 g/cm^3. Table 13.4 gives the range of densities and the average densities of some common commercial materials. X-ray attenuation in these materials can be obtained from the curves in Figs. 13.6 and 13.7 for concrete of equivalent thickness. For example, the attenuation provided by 2 in. of tile (average density 1.9 g/cm^3) is equivalent to that of 2(1.9/2.35) = 1.62 in. of concrete. (If the building materials are of significantly higher atomic number than concrete, this procedure tends to overestimate the amount of shielding needed.) Layers of lead are commonly used with building materials to provide protective barriers. For computing the attenuation it is convenient to have half-value layers for both lead and concrete at different tube operating potentials. These are given in Table 13.5.

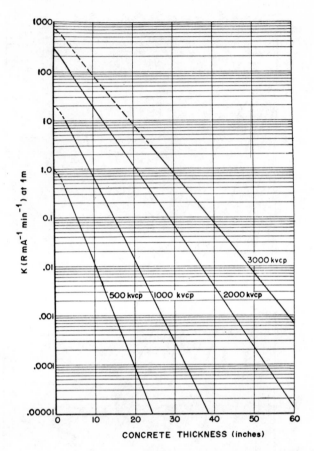

Figure 13.7. Attenuation in concrete of X-rays produced by potentials of 500 to 3000 kVp (*National Bureau of Standards Handbook 76*, 1961, Washington, D.C.)

Example

If, in the previous example, an existing 3 in. sand plaster wall separates the X-ray room and the hallway in Fig. 13.8, find the thickness of lead that must be added to the wall to provide the primary protective barrier.

Table 13.4. Densities of Commercial Building Materials

Material	Range of density (g/cm^3)	Average density (g/cm^3)
Barytes concrete	3.6–4.1	3.6
Brick (soft)	1.4–1.9	1.65
Brick (hard)	1.8–2.3	2.05
Earth (packed)	–	1.5
Granite	2.6–2.7	2.65
Lead	–	11.4
Lead glass	–	6.22
Sand plaster	–	1.54
Concrete	2.25–2.4	2.35
Steel	–	7.8
Tile	1.6–2.5	1.9

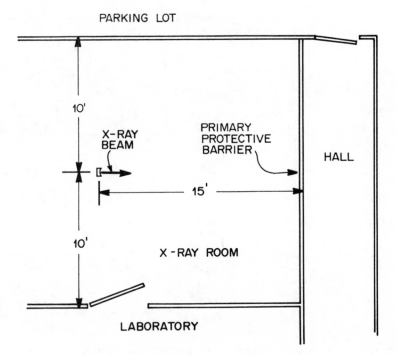

Figure 13.8. Schematic top view of an X-ray facility.

Solution

We found in the last example that 3.6 in. of concrete would provide an adequate primary protective barrier. The 3 in. plaster wall (density = 1.54 g/cm^3) is equivalent to concrete of thickness 3(1.54/2.35) = 2.0 in. Therefore, additional shielding equivalent to 3.6 − 2.0 = 1.6 in. of concrete is needed. From Table 13.5, 1 HVL of concrete for 125 kVp X-rays is 2.0 cm = 0.80 in., and so the additional shielding required is 1.6/0.8 = 2.0 HVLs. Table 13.5 shows that the required thickness of lead to be added to the plaster wall is 2.0 × 0.28 = 0.56 mm.

Design of Secondary Protective Barrier

As illustrated in Fig. 13.2, the secondary barrier is designed to protect areas not in the line of the useful beam from the leakage and scattered radiation. Physically, these two components of the radiation field can be of quite different quality. Therefore, the shielding requirements are computed separately for each and the final barrier thickness is chosen to be adequate for their sum. Because conditions vary greatly, no single method of calculation is always satisfactory; however, the one presented here can be used as a guide. We assume that the leakage and scattered radiations are isotropic. The use factor for them is then unity ($U = 1$).

Leakage Radiation

Limits placed on the manufacturer for the maximum allowed leakage radiation from the housing of diagnostic and therapeutic machines at the reference distance of 1 m were given in the first paragraph of this section. Let Y designate any one of these three limits in R/hr at 1 m. Given Y, the secondary barrier thickness for the leakage radiation is then

Table 13.5. Half-Value Layers for X-Rays (Broad Beams)
in Lead and Concrete

Peak voltage (kVp)	HVL Lead (mm)	HVL Concrete (cm)
50	0.06	0.43
70	0.17	0.84
100	0.27	1.6
125	0.28	2.0
150	0.30	2.24
200	0.52	2.5
250	0.88	2.8
300	1.47	3.1
400	2.5	3.3
500	3.6	3.6
1000	7.9	4.4
2000	12.5	6.4
3000	14.5	7.4
4000	16.0	8.8
6000	16.9	10.4
8000	16.9	11.4
10,000	16.6	11.9

computed as the number of half-value layers needed to restrict the exposures of individuals in other areas to allowed levels. When the tube is operated t min/wk, the weekly exposure in R in an area at a distance of d meters from the tube target is, with no structural shielding present, $Yt/60d^2$. If P denotes the individual weekly maximum permissible exposure in the area, having an occupation factor T, then the required attenuation factor B for the leakage X-ray intensity is given by

$$P = B\frac{YtT}{60d^2}. \tag{13.20}$$

Solving for B and writing $t = W/I$, where W is the work load in mA min/wk and I is the average beam current in mA, we find that

$$B = \frac{60IPd^2}{YWT}. \tag{13.21}$$

The number N of half-value layers that attenuates the radiation by a factor B is given by $B = 2^{-N}$, or

$$N = -\frac{\ln B}{\ln 2} = -\frac{\ln B}{0.693}. \tag{13.22}$$

We should point out that using a certain number of HVLs to obtain the attenuation factor applies strictly only to monoenergetic photons. In contrast, the changing slopes of the attenuation curves in Figs. 13.3 through 13.7 show that the quality of transmitted primary X-rays changes as they penetrate different shield thicknesses. In fact, an X-ray beam hardens and becomes more nearly monochromatic as it penetrates deeper into matter. The leakage radiation, which is filtered and hardened by the lead tube housing, can, for the purpose of computing shielding, be regarded as approximately monochromatic, with its half-value layer depending primarily on the kVp of the tube.

Example

An X-ray therapy machine is installed in the layout shown in Fig. 13.8 in place of the diagnostic device in the previous example. The new unit has a continuous tube current rating of 26 mA at 290 kVp. The average workload in the facility is 24,000 mA min/wk. How many HVLs of shielding would be needed to protect the laboratory from the leakage radiation alone?

Solution

We use Eqs. (13.21) and (13.22). For a therapeutic machine operating below 500 kVp, $Y = 1.0$ R/hr at 1 m. The current is $I = 26$ mA; $P = 0.1$ R/wk, since the laboratory is a controlled area; $d = 10/3.28$ m; $W = 24,000$ mA min/wk; and $T = 1$. The needed attenuation factor is

$$B = \frac{60 \times 26 \times 0.1 \times (10/3.28)^2}{1 \times 24,000 \times 1} = 0.0604. \tag{13.23}$$

From Eq. (13.22), the number of HVLs is

$$N = -\frac{\ln 0.0604}{0.693} = 4.05 \tag{13.24}$$

for protection from the leakage radiation alone. Table 13.5 can be used to find the corresponding thickness of lead or concrete. However, we next consider the scattered radiation before specifying the secondary protective barrier thickness.

Scattered Radiation

For the purpose of estimating shielding thickness, if the tube operating potential is not more than 500 kVp, then the barrier penetrating capability of the scattered X-rays is assumed to be the same as that of the useful beam. (Low-energy photons lose a relatively small fraction of their energy in Compton scattering.) If the potential is greater than 500 kVp, then the scattered photons are treated like the primary photons in a *useful* beam of 500 kVp X-rays. Thus, for X-ray beams of 500 kVp or less, the attenuation curves for the respective peak kilovoltages in Figs. 13.3 through 13.7 are used; for beams of higher kVp, the attenuation curves at 500 kVp are used.

The value of K for the scattered radiation can be determined from the formula (13.18) with two modifications. First, measurements show that the exposure rate of scattered X-rays at a location 1 m from a scatterer and 90° from the primary-beam direction is about 10^{-3} times as large as the incident exposure rate at the scatterer. Therefore, to account for the smaller intensity of the scattered radiation compared with the useful beam, K in Eq. (13.18) is increased by a factor of 1000. Second, the output of an X-ray machine increases greatly at high kVp, and so K must be reduced by an empirical factor f, which is given in Table 13.6. For the scattered radiation, then, we have with $U = 1$ the modified version of Eq. (13.18):

Table 13.6. Values of f for Scattered Radiation in Eq. (13.25)

kVp	f
≤ 500	1
1000	20
2000	300
3000	700

$$K = \frac{1000Pd^2}{fWT}.$$ (13.25)

As before, P is in R/wk, d in meters, W in mA min/wk, and K is the exposure of the transmitted radiation in R/(mA min) at 1 m.

Example
Calculate the number of half-value layers needed to protect the laboratory area in the last example from the scattered radiation alone (Fig. 13.8 with the therapy unit installed).

Solution
We apply Eq. (13.25): $P = 0.1$ R/wk, $d = 10/3.28$ m, $f = 1$ for the 290 kVp unit (Table 13.6), $W = 24{,}000$ mA min/wk, and $T = 1$. Thus,

$$K = \frac{1000 \times 0.1 \times (10/3.28)^2}{1 \times 24{,}000 \times 1} = 0.0387.$$ (13.26)

Using Fig. 13.4, we estimate the thickness of lead to be about 2.5 mm. From Table 13.5, the HVL is about 1.35 mm; and so, for the scattered radiation alone, $2.5/1.35 = 1.85$ HVLs of shielding would be required. Alternatively, Fig. 13.6 with $K = 0.0387$ indicates that about 6 in. of concrete would be needed. The HVL from Table 13.5 is 3.0 in., implying that 2 HVLs are needed. Ideally, one should obtain the same number of HVLs for either lead or concrete; however, the above two values agree to within the precision with which the graphs can be read and the HVL determined by linear interpolation in Table 13.5. It should also be kept in mind that the many underlying simplifications already made do not warrant greater precision. We choose the more conservative answer of 2 HVLs for shielding the scattered radiation alone.

Having computed the number of HVLs for the leakage and scattered radiations separately, one must design the secondary protective barrier to be adequate for both together. To do this, the following rule is applied. If the barrier thicknesses for leakage and scattered radiations are found to be within 3 HVLs of one another, 1 HVL can be added to the larger to obtain a sufficient secondary barrier thickness. If the two differ by more than 3 HVLs, then the thicker one alone will suffice.

Example
What thickness of lead must be added to an existing 2.5 in. plaster wall between the X-ray room and the laboratory in Fig. 13.8 to provide an adequate secondary protection barrier for the therapy unit considered in the last two examples?

Solution
We found that 4 HVLs would be needed for the leakage alone and 2 HVLs for the scattered radiation alone. By the above rule, the total secondary protective barrier thickness that is needed is $4 + 1 = 5$ HVLs. The existing wall provides some of the barrier. Its concrete-equivalent thickness is $2.5(1.54/2.35) = 1.64$ in., as found from the densities given in Table 13.4. From Table 13.5, as we saw in the last example, the half-value layer of concrete for the 290 kVp X-rays is 3.0 cm. Therefore, the plaster wall provides $1.64 \times 2.54/3.0 = 1.39$ HVLs; and so the amount of additional shielding needed is $5 - 1.39 = 3.61$ HVLs. The required thickness of lead (1 HVL = 1.35 mm) to be added to the plaster wall, therefore, is $3.61 \times 1.35 = 4.87$ mm.

The exposure rates from leakage and scattered radiations are usually comparable for X-ray machines that operate under 500 kVp. However, the leakage radiation is more penetrating, and, therefore may require a greater barrier thickness, as was the case in the last example (cf. Problem 10).

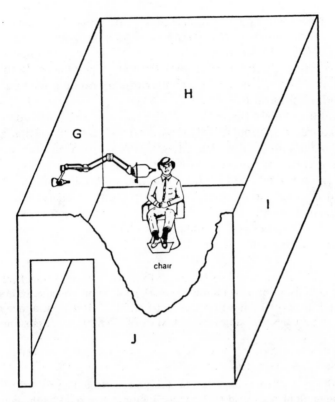

Figure 13.9. Dental X-ray installation. Useful beam is almost always directed towards walls G, H, I, or floor. Wall J, which the dental chair faces, and the ceiling receive little exposure to the direct beam. [Reprinted with permission from *Dental X-Ray Protection*, NCRP Report No. 35, National Council on Radiation Protection and Measurements, Washington, D.C. (1970)]

Some special considerations can be applied in the design of the primary protective barriers for dental X-ray installations.† Figure 13.9 illustrates such a facility. The patient is seated facing wall J. When an X-ray picture is made, the useful beam will likely be directed toward wall G, H, I or the floor. Wall J and the ceiling will rarely receive the direct beam. Furthermore, the radiation fields for different types of examinations and for different patients usually do not overlap on the primary barriers. In addition, the useful beam will be partially attenuated by the patient's head. For these reasons, a use factor of $\frac{1}{16}$ is commonly assigned to walls G, H, I and the floor. A use factor of zero is given to wall J and the ceiling for primary barrier design. All of these surfaces must be considered, however, in secondary shielding for the leakage and scattered radiations.

13.4 PROTECTION FROM BETA EMITTERS

Beta (including positron) emitters present two potential external radiation hazards, namely, the beta rays themselves and the bremsstrahlung they produce in the source and in adjacent materials. In addition, annihilation photons are always present with positron

---

†*Dental X-Ray Protection*, NCRP Report No. 35, National Council on Radiation Protection and Measurements, Washington, D.C. (1970).

sources. Beta particles can be stopped in a shield surrounding the source if it is thicker than their range. To minimize bremsstrahlung production, this shield should have low atomic number [cf. Eq. (5.14)]. It, in turn, can be enclosed in another material (preferably of high atomic number) that is thick enough to attenuate the bremsstrahlung intensity to the desired level. For a shielded beta emitter bremsstrahlung may be the only significant external radiation hazard.

The bremsstrahlung shield thickness can be calculated in approximate fashion by the following procedure. Equation (5.14) is used to estimate the radiation yield, letting $T = T_{max}$ be the maximum beta-particle energy. This assumption overestimates the actual bremsstrahlung intensity, because most of the photons have energies much lower than the upper limit T_{max}. To roughly compensate, one ignores buildup in the shielding material and uses the linear attenuation coefficient for photons of energy T_{max} to estimate the bremsstrahlung shield thickness. Since the bremsstrahlung spectrum is hardened by passing through the shield, the exposure rate around the source is calculated by using the air absorption coefficient for photons of energy T_{max}.

Example
Design a suitable container for a 10 Ci source of ^{32}P in a 50 ml aqueous solution, such that the exposure rate at a distance of 1.5 m will not exceed 1 mR/hr. ^{32}P decays to the ground state of ^{32}S by emission of beta particles with an average energy of 0.70 MeV and a maximum energy of 1.71 MeV.

Solution
We choose a bottle made of some material, such as polyethylene (density = 0.93 g/cm^3), with elements of low atomic number to hold the aqueous solution. It should be thick enough to stop the beta particles of maximum energy. From Fig. 5.3, the range for T_{max} = 1.71 MeV is about 0.80 g/cm^2. The thickness of the polyethylene bottle should, therefore, be at least 0.80/0.93 = 0.86 cm. To estimate the bremsstrahlung yield by Eq. (5.14), we need the effective atomic number of the medium in which the beta particles lose their energy. Most of the energy will be lost in the water, a small part being absorbed in the container walls. The effective atomic number for water is

$$Z_{eff} = \frac{2}{18} \times 1 + \frac{16}{18} \times 8 = 7.22. \tag{13.27}$$

The estimated fraction of the beta-particle energy that is converted into bremsstrahlung is, by Eq. (5.14) with $Z = Z_{eff}$ and $T = T_{max}$,

$$Y \cong \frac{6 \times 10^{-4} \times 7.22 \times 1.71}{1 + 6 \times 10^{-4} \times 7.22 \times 1.71} = 7.4 \times 10^{-3}. \tag{13.28}$$

The rate of energy emission by the 10 Ci source of beta particles with an average energy of 0.70 MeV is

$$E_\beta = 10 \times 3.7 \times 10^{10} \times 0.70 = 2.59 \times 10^{11} \text{ MeV/sec}. \tag{13.29}$$

The rate of energy emission in the form of bremsstrahlung photons is therefore

$$YE_\beta = 7.4 \times 10^{-3} \times 2.59 \times 10^{11} = 1.92 \times 10^9 \text{ MeV/sec}. \tag{13.30}$$

We next compute the exposure rate from the unshielded bremsstrahlung, treated as coming from a point source at a distance of 1.5 m. Following the above procedure, we use the mass absorption coefficient of air for 1.71 MeV photons, which is (from Fig. 7.12) $\mu_A/\rho = 0.026$ cm^2/g. Since $\rho = 0.00129$ g/cm^3, it follows that the linear absorption coefficient is $\mu_A = 3.4 \times 10^{-5}$ cm^{-1}. The rate of energy absorption in a thin spherical shell of radius r and thickness t centered around the source is $YE_\beta\mu_A t$. Since the mass of air in the shell is $4\pi r^2 t\rho$, the rate of energy absorption per unit mass in the shell of air is given by $YE_\beta\mu_A/4\pi r^2\rho$. The exposure rate of the unshielded bremsstrahlung at a distance $r = 1.5$ m is therefore

$$\dot{\Gamma} = \frac{1.92 \times 10^9 \text{MeV/sec} \times 1.6 \times 10^{-6} \text{ erg/MeV} \times 3600 \text{ sec/hr} \times 3.4 \times 10^{-5} \text{ cm}^{-1}}{4\pi(150 \text{ cm})^2 \times 0.00129 \text{ g/cm}^3 \times 87.8 \text{ erg/(g/R)}} \quad (13.31)$$

$$= 1.17 \times 10^{-2} \text{ R/hr} = 11.7 \text{ mR/hr}. \quad (13.32)$$

Lead is a convenient material for the bremsstrahlung shield. As specified above, one ignores buildup and uses the linear attenuation coefficient for photons of energy T_{max} to compute the bremsstrahlung shield thickness. Figure 7.9 gives for 1.71 MeV photons in lead $\mu/\rho = 0.048$ cm^2/g, and so $\mu = 0.048 \times 11.4 = 0.55$ cm^{-1}. The thickness x needed to reduce the exposure rate to 1 mR/hr is given by

$$1 = 11.7e^{-0.55x}, \quad (13.33)$$

or $x = 4.5$ cm. A lead container of this thickness could be used to hold the polyethylene bottle.

13.5 NEUTRON SHIELDING

Photon shielding design is simplified by a number of factors that do not apply to computations for neutrons. Whereas photon cross sections vary smoothly with atomic number and energy, neutron cross sections can change irregularly from element to element and have complicated resonance structures as functions of energy. In addition, photon cross sections are generally better known than those for neutrons. Elaborate computer codes, using Monte Carlo and other techniques, are available for calculating neutron interactions and transport in a variety of materials. Sometimes circumstances permit useful estimations to be made by simpler means. In this section we present only a general discussion of neutron shielding.

Basically, a neutron shield acts to moderate fast neutrons to thermal energies, principally by elastic scattering, and then absorb them. Most effective in slowing down neutrons are the light elements, particularly hydrogen [cf. Eq. (8.3)]. Many hydrogenous materials, such as water and paraffin, make efficient neutron shields. However, water shields have the disadvantage of needing maintenance; also, evaporation can lead to a potentially dangerous loss of shielding. Paraffin is flammable. Concrete (ordinary or heavy aggregate) or earth is the neutron shielding material of choice in many applications. Often temporary neutron shielding must be provided in experimental areas around a reactor or an accelerator. Movable concrete blocks are convenient for this purpose. One must exercise care to assure that cracks, access ports, and ducts in such shielding do not permit the escape of neutrons. Vertical cracks should be staggered. (Natural sagging of concrete blocks under the force of gravity usually precludes the existence of horizontal cracks.) Generally, surveys are desirable to check temporary neutron shielding before and during extensive use.

Hydrogen captures thermal neutrons through the reaction ^{1}H(n, γ)^{2}H with a cross section of 0.33 barns. Other materials, like cadmium, have a very high (n, γ) neutron capture cross section (2450 barns) and are therefore frequently used as neutron absorbers. Hydrogen and cadmium have the disadvantage of emitting energetic (2.22 MeV and 9.05 MeV) capture gamma rays, which might, themselves, require shielding. Other nuclides, such as ^{10}B and ^{6}Li, capture thermal neutrons through an (n, α) reaction without emission of appreciable gamma radiation. In addition to possible health-physics problems that may arise from capture gamma rays, these shields can acquire induced radioactivity through neutron capture or other reactions.

Examples of neutron attenuation in a hydrogenous material are provided by the depth–dose curves in Chapter 10 for monoenergetic neutrons normally incident on tissue slabs. Figure 10.11 for 5 MeV neutrons, for instance, shows that the absorbed dose decreases by an order of magnitude over 30 cm. The energy spectrum of the neutrons

Table 13.7. Macroscopic Neutron Removal Cross Sections and Attenuation Lengths in Several Materials[†]

Material	Macroscopic removal cross section Σ_r (cm^{-1})	Attenuation length $1/\Sigma_r$ (cm)
Water	0.103	9.7
Paraffin	0.106	9.4
Iron	0.1576	6.34
Concrete (6% H_2O by weight)	0.089	11.3
Graphite (density 1.54 g/cm^3)	0.0785	12.7

[†]Data in part from *Protection Against Neutron Radiation*, NCRP Report No. 38, National Council on Radiation Protection and Measurements, Washington, D.C. (1971).

changes with the penetration depth as the original 5 MeV neutrons are moderated. The relative number of thermal neutrons at different depths can be seen from the dose curve labeled E_γ for the ^{1}H(n, γ)^{2}H thermal-neutron capture reaction. The thermal neutron density builds up to a maximum at about 10 cm and thereafter falls off as the total density of neutrons decreases by absorption. In paraffin, the half-value layer for 1 MeV neutrons is about 3.2 cm and that for 5 MeV neutrons is about 6.9 cm.

Neutron shielding can sometimes be estimated by a simple "one-velocity" model that employs neutron removal cross sections. Such shielding must be sufficiently thick and the neutron source energies so distributed that only the most penetrating neutrons in a narrow energy band contribute appreciably to the dose beyond the shield. The neutron dose can then be represented by an exponential function of shield thickness. Conditions must also be such that the slowing-down distance from the most penetrating energies down to 1 MeV is short. In addition, the shield must contain enough hydrogen to assure a short average transport distance from 1 MeV down to thermal energy and the point of absorption. The removal cross sections for various elements are roughly three-quarters of the total cross sections (except ~0.9 for hydrogen). Most measurements of removal cross sections have been made with fission-neutron sources and shields of such a thickness that the principal component of dose arises from source neutrons in the energy range 6–8 MeV. Table 13.7 gives macroscopic removal cross sections, Σ_r, and attenuation lengths, $1/\Sigma_r$, in some shielding and reactor materials.

A number of approximate formulas for neutron shielding, based on removal cross sections, have been developed for reactor cores having various shapes and other characteristics. We will not attempt to cover them here. A simple, useful formula is available, however, for radioactive neutron sources.[†] Because of the small intensities compared with fission sources, relatively thin shields are needed. Therefore, the scattered neutrons contribute significantly to the dose outside the shield, and their effect can be represented by a buildup factor B. The dose-equivalent rate $\dot{D}$ outside a shield of thickness T at a distance R cm from a point source of strength S neutrons/sec is given by

$$\dot{D} = \frac{BSqe^{-\Sigma_r T}}{4\pi R^2},$$ (13.34)

where Σ_r is the removal cross section and q is the dose-equivalent rate per unit neutron fluence rate (e.g., rem/hr per neutron/cm^2 sec) for neutrons of the source energy. The

---

[†]"Protection Against Neutron Radiation up to 30 Million Electron Volts," *Handbook 63*, National Bureau of Standards, Washington, D.C. (1957).

factor q can be obtained from Table 10.4; $\dot{D}$ and q will have the same units for dose equivalent. For Po–Be and Po–B sources with a water or paraffin shield at least 20 cm thick, $B \cong 5$.

Example
Calculate the dose-equivalent rate 1.6 m from an unshielded 0.82 Ci ^{210}Po–Be source, which emits 2.05×10^6 neutrons/sec. By what factor is the rate reduced by a 25 cm water shield? What is the dose-equivalent rate behind 50 cm of water?

Solution
In Eq. (13.34) the presence of the shield introduces the factors $B \exp(-\Sigma_r T)$. For the unshielded source, $\dot{D}_0 = Sq/4\pi R^2$ and $S = 2.05 \times 10^6$ neutrons/sec. Table 8.2 shows the average energy to be 4.2 MeV, for which Table 10.4 indicates that about 16 neutrons/(cm² sec) give a dose-equivalent rate of 2.5 mrem/hr. Therefore, we have $q = 2.5/16 = 0.156$ mrem/hr; and so

$$\dot{D}_0 = \frac{2.05 \times 10^6 \times 0.156}{4\pi(160)^2} = 0.994 \text{ mrem/hr} \tag{13.35}$$

for the unshielded source. With $\Sigma_r = 0.103$ cm^{-1} from Table 13.7, $B = 5$, and $T = 25$ cm, the dose-equivalent rate is reduced by the factor

$$Be^{-\Sigma_r T} = 5e^{-0.103 \times 25} = 0.381. \tag{13.36}$$

The rate with 50 cm of water interposed is

$$\dot{D} = \frac{5 \times 2.05 \times 10^6 \times 0.156}{4\pi(160)^2} e^{-0.103 \times 50} = 0.0288 \text{ mrem/hr.} \tag{13.37}$$

Note that Eq. (13.34) with $B \cong 5$ applies to shields thicker than 20 cm, as was the case here. Note also that one should generally be concerned with gamma-ray shielding where neutrons are present. In this example, however, ^{210}Po is a weak gamma emitter (0.001%) and decays to stable ^{206}Pb.

13.6 PROBLEMS

1. Calculate the thickness of lead shielding needed to reduce the exposure rate 2.5 m from a 16 Ci point source of ^{137}Cs to 1.0 mR/hr.
2. Repeat Problem 1 for a concrete shield.
3. A small 5 Ci ^{42}K source is placed inside an iron pipe (on the axis) having an inside diameter of 1 in. and an outside diameter of 2.5 in. What is the exposure rate opposite the source at a point 6 ft away from the center of the pipe?
4. How thick must a spherical lead container be in order to reduce the exposure rate 1 m from a small 100 mCi ^{24}Na source to 2.5 mR/hr?
5. An unshielded 1600 Ci ^{60}Co source is to be used in a room at the spot x shown in Fig. 13.10. Calculate the thickness t of concrete that is needed to limit the exposure rate to 10 mR/hr outside the wall.
6. A broad, parallel beam of 500 keV photons is normally incident on a uranium sheet that is 1.5 cm thick. If the exposure rate in front of the sheet is 1.08 mR/min, what is it behind the sheet?
7. What thickness of lead shielding is needed around a 2000 Ci point source of ^{60}Co to reduce the exposure rate to 10 mR/hr at a distance of 2 m?
8. The front of a 6 cm aluminum slab is located 2 m from an 8 Ci ^{137}Cs point source, as shown in Fig. 13.11. The back of a parallel thick shield, which completely absorbs direct radiation, is 1.7 m from the source. The shield has a cylindrical aperture of area 1 cm² with the source on its axis.
   (a) Calculate the exposure rate at a point P 2.8 m from the source on the cylindrical axis.
   (b) What is the exposure rate at P with the thick shield removed?

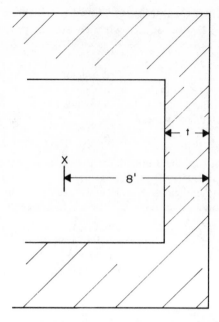

Figure 13.10. Diagram of room to be used with an unshielded 1600 Ci ^{60}Co source.

9. (a) Why are the semilogarithmic curves in Fig. 13.3 nonlinear?
   (b) Why do the slopes decrease with increasing thickness?
   (c) Why, on the other hand, are the curves in Fig. 13.5 essentially linear after the initial depths?
10. (a) Why is the leakage radiation from an X-ray machine generally more penetrating than the scattered radiation?
    (b) Using the density equivalent of concrete for a material of high atomic number tends to overestimate the amount of shielding needed for X-rays. Why?
11. A 200 kVp diagnostic X-ray machine is installed in the position shown in Fig. 13.8. Its average weekly workload of 250 mA min is divided between 150 mA min when the useful beam is pointed horizontally in the direction of the hall and 100 mA min when pointed horizontally in the direction of the unattended parking lot.
    (a) Calculate the thickness of concrete needed for the primary protective barrier for the hall.
    (b) Calculate the thickness of concrete needed for the primary protective barrier for the parking lot.
12. If a $1\frac{1}{2}$ in. plaster wall exists between the X-ray room and the hall in the last problem (Fig. 13.8), what additional thickness of lead will provide an adequate primary barrier for the hall?
13. (a) If a $2\frac{1}{4}$ in. concrete wall separates the X-ray room and the unattended parking lot in Problem 11 (Fig. 13.8), how much additional lead shielding is needed to make the primary protective barrier?
    (b) What changes, if any, should be made in the design of this primary barrier if the parking lot has an attendant 24 hr per day?
14. (a) Calculate the number of half-value layers needed to shield the laboratory from the leakage radiation in Problem 11 (Fig. 13.8).
    (b) Repeat for the scattered radiation.
    (c) If a $\frac{3}{4}$ in. plaster wall separates the X-ray room and the laboratory, how much additional lead shielding is needed to make a secondary protective barrier?
15. Figure 13.12 shows a schematic top view of an X-ray facility. A diagnostic machine, operated

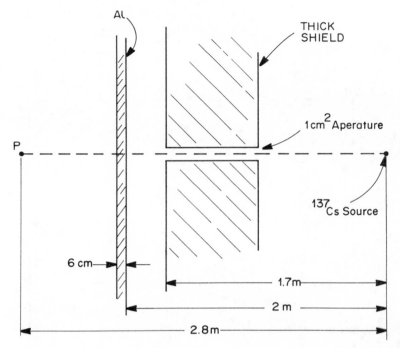

Figure 13.11. Diagram of a room with aluminum slab and parallel thick shield used with an 8 Ci ^{137}Cs point source.

at 150 kVp with a maximum current of 120 mA, is used an average of 22.1 min/day, 5 day/wk. The horizontal beam is always pointed in the direction of the sidewalk. Calculate the thickness of lead shielding needed for the primary protective barrier.

16. (a) Calculate the thickness of additional lead shielding needed in Problem 15 (Fig. 13.12) to make a secondary protective barrier for the book store.
    (b) Calculate the thickness of additional lead shielding needed for the hallway.

17. What thickness t must the plaster wall between the X-ray room and the laboratory have in Problem 15 (Fig. 13.12) to provide an adequate secondary protective barrier?

18. A law firm is located in an office directly below the 2 in. concrete floor of a dentist's office. No other shielding separates the two businesses. The dentist's staff has routinely operated a 100 kVp X-ray machine with an average weekly workload of 100 mA min for the past 4 yr. One of the law partners has gone bald in the two years since the firm moved into their present office. Should the dentist fear a law suit? Explain.

19. A dental X-ray machine that will operate at 85 kVp with a weekly workload of 50 mA min is to be placed 4 ft from the outside of a (soft) brick wall where a public sidewalk is located. How thick must the brick wall be to provide an adequate primary protective barrier?

20. A 30 mL solution containing 2 Ci of ^{90}Sr in equilibrium with ^{90}Y is to be put into a small glass bottle, which will then be placed in a lead container having walls 1.5 cm thick.
    (a) How thick must the walls of the glass bottle be in order to prevent any beta rays from reaching the lead?
    (b) Estimate the bremsstrahlung dose rate at a distance of 1.75 m from the center of the lead container.

21. A small vial containing 200 Ci of ^{32}P in aqueous solution is enclosed in an 8 mm aluminum can as shown in Fig. 13.13. Calculate the thickness of lead shielding needed to reduce the exposure rate to 2.5 mR/hr at a distance 2 m from the can. Treat the vial as a point source and assume that all of the bremsstrahlung is produced by beta particles slowing down in the water.

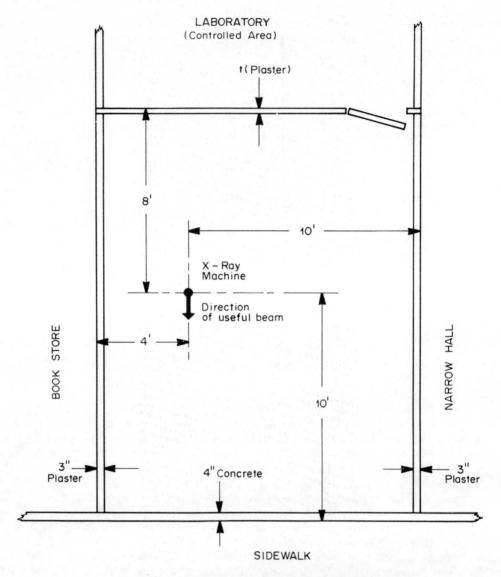

Figure 13.12. Schematic top view of an X-ray facility.

Figure 13.13. Small vial containing 200 Ci of ^{32}P in aqueous solution enclosed in an 8 mm aluminum can.

22. (a) Estimate the dose-equivalent rate at a distance of 80 cm from a ^{210}Po–B point source that emits 2.2×10^7 neutrons/sec and is shielded by 30 cm of water.
    (b) How thick would the water shield have to be to reduce the dose-equivalent rate to 2.5 mrem/hr?

23. (a) What is the maximum number of neutrons/sec that a ^{210}Po–Be point source can emit if it is to be stored behind 65 cm of paraffin and the dose-equivalent rate is not to exceed 10 mrem/hr at a distance of 1 m?
    (b) By what factor would a 31 cm shield reduce the dose-equivalent rate?

CHAPTER 14
INTERNAL DOSIMETRY AND
RADIATION PROTECTION

14.1 OBJECTIVES AND METHODOLOGY

In this chapter we deal with radionuclides that can enter the body via inhalation, ingestion, wounds, or other means. As with external radiation, procedures have been developed to keep the dose equivalent from internal emitters within allowed levels. The primary limits for all exposures are those recommended for the various body organs and the whole body by the ICRP in Publication 26 (Section 12.4). In principle, one can develop secondary limits for the allowable intakes of various radionuclides that would, by themselves, provide the limiting risk to an individual, based on an assumed model for calculating the dose equivalents in various organs. The allowable intakes can, in turn, be used to derive air, food, and water concentrations for use as guides for internal radiation protection.

This methodology, which is applied to control the dose equivalent from internal emitters, is the subject of the present chapter. It is based on calculations made with various metabolic models for a "reference," or "standard," man under specific given conditions. The data for reference man are given in the 1975 *Report of the Task Group on Reference Man*, ICRP Publication 23. We shall concentrate on the procedures used for occupational exposures, based on reference man, as described in ICRP Publication 30 and its Supplements.

14.2 REFERENCE MAN

In 1975 the ICRP issued Publication 23, a 480 page report by the Task Group on Reference Man of Committee 2. The report's three major parts are entitled "Anatomical Values for Reference Man," "Gross and Elemental Content of Reference Man," and "Physiological Data for Reference Man." The report gives extensive data in these categories on characteristics that are likely to be significant for assessing radiation dose from sources either inside or outside the body. "Typical" values are selected for reference man, based in part on information collected from many sources in the literature and on experimental data measured by some Task Group members. Publication 23 states, "Reference Man is defined as being between 20–30 years of age, weighing 70 kg, is 170 cm in height, and lives in a climate with an average temperature of from 10° to 20°C. He is a Caucasian and is a Western European or North American in habitat and custom." While relatively few individuals in any population will closely resemble reference man (however he is defined) the concept provides an important basis for internal dosimetry. When applied to a practical situation, adjustments can be made for individual differences on the basis of the specified procedures and assumptions used for obtaining internal dose estimates with reference man.

The work on reference man represents a continuing, long-term effort. Many refinements and improvements have been made since the 1975 issuance of ICRP Publication 23. In addition to the adult male, reference data have been compiled for an adult female and for children and infants of various ages.

Some of the organ and tissue masses used in ICRP Publication 30 for occupational exposures are listed in Table 14.1. Except for the ovaries and uterus, they are those of the 70 kg reference man given in Publication 23. As described in Section 14.4, internal-dose calculations include the irradiation of one organ by a source located in another organ. Different source- and target-organ masses are sometimes used, for example, for parts of the digestive tract, depending on whether they contain a radioactive source or whether they are irradiated from sources elsewhere.

Detailed dimensions and substructures for organs are available in mathematical form for reference man. Figure 14.1 shows one version of a model, called the Snyder–Fisher phantom, used to represent an adult human for internal-dosimetry calculations. As shown in Fig. 14.2, such detailed phantoms have been fabricated and built in order that calculated results can be compared with actual measurements.

Specific models are also used for the transport and residence of inhaled or ingested radioactive substances in different parts of the body. We shall describe some of the principal dosimetric models below, concentrating on those for the respiratory system for the inhalation of radionuclides and the gastrointestinal tract for ingestion. The ICRP has compiled extensive metabolic data for the elements on their uptake, retention, and distribution in the organs and tissues of the body.

14.3 SECONDARY INTAKE LIMITS FOR WORKERS: ALI AND DAC

The basic occupational limits given in ICRP Publication 26 for stochastic and non-stochastic effects were discussed in Section 12.4. For stochastic effects, the annual dose equivalent values H_T in various tissues T of the body, given weighting factors w_T (Table 12.3), must satisfy condition (12.5), namely

Table 14.1. Organ and Tissue Masses Used in ICRP Publication 30 for Workers

Source organs	Mass (g)	Target organs	Mass (g)
Ovaries	11	Ovaries	11
Testes	35	Testes	35
Muscle	28,000	Muscle	28,000
Red marrow	1500	Red marrow	1500
Lungs	1000	Lungs	1000
Thyroid	20	Thyroid	20
Stomach content	250	Bone surface	120
Small intestine content	400	Stomach wall	150
Upper large intestine content	220	Small intestine wall	640
Lower large intestine content	135	Upper large intestine wall	210
Kidneys	310	Lower large intestine wall	160
Liver	1800	Kidneys	310
Pancreas	100	Liver	1800
Cortical bone	4000	Pancreas	100
Trabecular bone	1000	Skin	2600
Skin	2600	Spleen	180
Spleen	180	Thymus	20
Adrenals	14	Uterus	80
Bladder content	200	Adrenals	14
Total body	70,000	Bladder wall	45

$$\sum_T w_T H_T \le H_{\text{wb,L}} = 0.05 \text{ Sv}, \tag{14.1}$$

where $H_{\text{wb,L}} = 0.05$ Sv is the recommended annual limit for uniform irradiation of the whole body. For nonstochastic effects from the intake of radioactive materials ICRP Publication 30 applies an annual dose-equivalent limit of 0.5 Sv to *all* tissues, since no case is known in which lens opacity is the limiting factor for the intake of radionuclides.

A secondary quantity, called the annual limit on intake, or ALI, of a radionuclide has been defined for use in meeting the basic limits. If I (in Bq) is the annual intake of a specified radionuclide by ingestion or inhalation, and $\hat{H}_{50,T}$ is the committed dose equivalent (Section 12.4) per unit activity of intake (Sv/Bq) in tissue T from the radionuclide, then the ALI is defined as the greatest value of I that satisfies the two conditions

$$I\sum_T w_T \hat{H}_{50,T} \le 0.05 \text{ Sv} \tag{14.2}$$

and

$$I\hat{H}_{50,T} \le 0.5 \text{ Sv}. \tag{14.3}$$

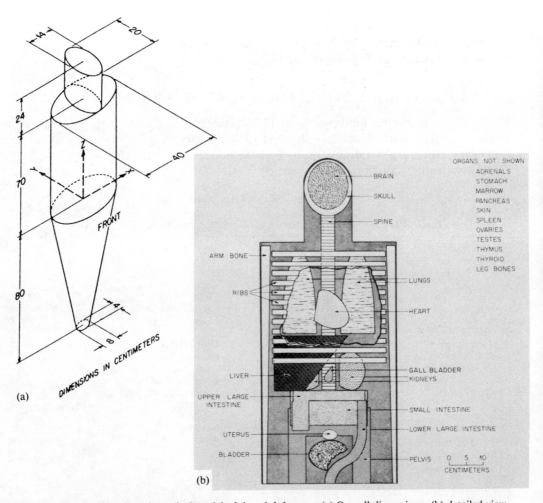

Figure 14.1. Snyder–Fisher mathematical model of the adult human. (a) Overall dimensions; (b) detailed view, showing a number of organs. (Courtesy Oak Ridge National Laboratory, operated by Martin Marietta Energy Systems, Inc., for the Department of Energy)

Figure 14.2. Life-size model built to represent Snyder–Fisher phantom. Various chemical mixtures are used to simulate tissue composition in each region. Source organs can be filled with solution of desired radionuclide. (Courtesy Oak Ridge National Laboratory, operated by Martin Marietta Energy Systems, Inc., for the Department of Energy)

The committed dose equivalent, defined by Eq. (12.4), is the total dose equivalent that results over a subsequent 50 yr working period from the single intake of a radionuclide. The conditions (14.2) and (14.3) satisfy the basic criteria on annual risk by specifying that one year's intake will not give an *integrated* dose equivalent over a worker's life in excess of the annual limit.

Values of $\hat{H}_{50,T}$ and ALI are derived based on the definition of reference man, and tabulated for specific radionuclides. The presence of daughter radionuclides is taken into account in their derivation. The tabulations are used for the control of exposures from internal emitters. When more than one radionuclide is inhaled or ingested, or when external radiation is also present, then the intakes based on the ALI for the individual nuclides are applied at reduced levels in order to meet the basic exposure limits from all sources of occupational radiation. In the analysis of the exposure of a particular individual, allowance can also be made for any differences between known biological parameters for the individual and those used for reference man in the derivation of the $\hat{H}_{50,T}$ and ALI.

The derived air concentration, DAC, is defined by the ICRP as that concentration of a radionuclide (in Bq/m^3) which, if breathed by reference man for one working year, would result in the ALI by inhalation. To this end, the breathing rate for reference man, carrying out "light activity," is taken to be 0.02 m^3/min and the working year, 50 wk of 40 hr each, or 2000 hr. These assumptions imply that

$$DAC = \frac{ALI \; (Bq)}{(0.02 \; m^3/min) \times 2000 \; hr \times (60 \; min/hr)} = \frac{ALI}{2400} \frac{Bq}{m^3}. \tag{14.4}$$

The values of the DAC for radionuclides are tabulated for use as guides. In particular applications, suitable allowance is made for mixtures of radionuclides and for the presence of external radiation. Allowance should also be made for other circumstances, such as for work that requires heavy breathing. In this regard, ICRP Publication 30 emphasizes "that the ALI is the overriding limit and the derived limit DAC should always be used circumspectly."

14.4 COMMITTED DOSE EQUIVALENT FOR INTERNAL EXPOSURES

As set forth by conditions (14.2) and (14.3), committed dose equivalent plays a key role in internal radiation protection. In this section we outline its calculation, which is a major task in internal dosimetry.

Consider the single intake of a radionuclide which results in an absorbed dose $\bar{D}_{50,i}$, averaged over an organ or tissue, from radiation of type i during the next 50 yr. The committed dose equivalent from this intake can be expressed as

$$H_{50} = \sum_i Q_i \bar{D}_{50,i}, \tag{14.5}$$

where the sum goes over all radiation types i and the Q_i are appropriate quality factors, such as the averages given in Table 12.2.

Generally, a given organ or tissue will be irradiated both by radionuclides that reside in it and by radionuclides contained in other tissues. For given target and source organs T and S, which may be the same, one writes for each type of radiation i,

$$H_{50}(T \leftarrow S)_i = Q_i \bar{D}_{50}(T \leftarrow S)_i. \tag{14.6}$$

Here $\bar{D}_{50}(T \leftarrow S)_i$ is the 50 yr absorbed dose, averaged over the target organ T, due to radiation of type i from the source organ S, and $H_{50}(T \leftarrow S)_i$ is the resulting committed dose equivalent. For each complete decay, or transformation, of a radionuclide of type j in S, we denote by $SEE(T \leftarrow S)_i$ the energy in MeV/g that is absorbed per unit mass in T from radiation of type i, weighted by the appropriate average quality factor. The dose equivalent, in Sv (J/kg), is then

$$SEE(T \leftarrow S)_i \frac{MeV}{g} \times \frac{1.6 \times 10^{-13} \; J/MeV}{10^{-3} \; kg/g} = 1.6 \times 10^{-10} \; SEE(T \leftarrow S)_i \; Sv. \tag{14.7}$$

The quantity $\mathrm{SEE}(T \leftarrow S)_i$ is called the specific effective energy for radiation of type i in S. We shall say more about its computation in the next section. If U_S atoms of radionuclide j in S decay over the 50 yr period following intake, then we have

$$H_{50}(T \leftarrow S)_i = 1.6 \times 10^{-10}[U_S\, \mathrm{SEE}(T \leftarrow S)_i]_j\, \mathrm{Sv}. \tag{14.8}$$

Summing over all types i of radiation emitted by radionuclide j gives, for the committed dose equivalent in T from j in S,

$$H_{50}(T \leftarrow S)_j = 1.6 \times 10^{-10}\left[U_S \sum_i \mathrm{SEE}(T \leftarrow S)_i\right]_j\, \mathrm{Sv}. \tag{14.9}$$

When S contains daughters or other radionuclides, we must sum over all types of radionuclides j in S to obtain the committed dose equivalent in T from all sources in S:

$$\sum_j H_{50}(T \leftarrow S)_j = 1.6 \times 10^{-10} \sum_j \left[U_S \sum_i \mathrm{SEE}(T \leftarrow S)_i\right]_j\, \mathrm{Sv}. \tag{14.10}$$

Finally, the total committed dose equivalent $H_{50,T}$ in the target organ or tissue T is the sum of the contributions from all radionuclides in all source organs or tissues S:

$$H_{50,T} = 1.6 \times 10^{-10} \sum_s \sum_j \left[U_S \sum_i \mathrm{SEE}(T \leftarrow S)_i\right]_j\, \mathrm{Sv}. \tag{14.11}$$

The committed dose equivalent, given by Eq. (14.11), depends upon several considerations. First, the region for averaging dose in a tissue or organ needs exact specification, which the ICRP provides. For the skin, the target is the basal layer of the epidermis, taken to be at a depth of 70 μm. For bone, the target is assumed to be the cells lying within 10 μm of the bone surfaces. For the gastrointestinal tract, the mucosal layer is assumed to be the target. The position of the sensitive cells in other cases is not specified in further detail. Nonuniform distributions of dose can be expected to be important for radionuclides that concentrate in sensitive microvolumes and emit radiations of very short range. The ICRP keeps these questions under continuing review and development. Second, the committed dose equivalent (14.11) depends on the $\mathrm{SEE}(T \leftarrow S)_i$, and, third, on the number of transformations U_S in the source organs over 50 yr. The evaluation of these quantities will be described in the next two sections.

14.5 SPECIFIC EFFECTIVE ENERGY AND SPECIFIC ABSORBED FRACTION

The specific effective energy $\mathrm{SEE}(T \leftarrow S)_i$ that enters the expression (14.11) for the committed dose equivalent was introduced after Eq. (14.6). Expressed in MeV/g, it is the quality-factor-weighted energy absorbed per gram in T, giving the dose equivalent there from radiation of type i per transformation of radionuclide of type j in S. For the radionuclide j, the total $\mathrm{SEE}(T \leftarrow S)_j$ per transformation is found by summing over all radiations emitted. It can be expressed by writing

$$\mathrm{SEE}(T \leftarrow S)_j = \frac{1}{M_T} \sum_i Y_i E_i\, \mathrm{AF}(T \leftarrow S)_i Q_i\, \mathrm{MeV/g}. \tag{14.12}$$

Here M_T is the mass of the target organ in grams, Y_i is the yield of radiation of type i per decay of radionuclide j, E_i is the average energy in MeV of radiation of type i per decay, and Q_i is the appropriate quality factor. The quantity $\mathrm{AF}(T \leftarrow S)_i$, called the absorbed fraction, is the fraction of the emitted energy of radiation of type i that is absorbed in T per disintegration of radionuclide j in S.

For most organs it is assumed that the energies of all alpha and beta particles are absorbed in the source organ. (The exceptions, which we shall not go into, are mineral bone and the contents of the gastrointestinal tract.) Thus $AF(T \leftarrow S)_i = 0$ in Eq. (14.12) unless T and S are the same when i denotes alpha or beta radiation. Then also $AF(T \leftarrow S)_i = 1$, in which case the contribution to the SEE can be easily evaluated from the decay-scheme data, as the following example shows.

Example

What is the contribution of the beta radiation from ^{131}I in the thyroid to the specific effective energy values for various target organs? The mass of the thyroid is 20 g. What will be the effect on the value of the SEE when the photons from ^{131}I are included?

Solution

In this example the source organ is the thyroid and the only type of radiation i that we are to consider initially is beta rays. Because their range is small compared with the size of the 20 g thyroid, we assume that the beta particles are completely absorbed in the source organ. Therefore, AF(other organs $\leftarrow$ thyroid)$_{(\beta^-)} = 0$ and the corresponding contributions to the SEE(other organs $\leftarrow$ thyroid)$^{131}I_{(\beta^-)} = 0$. (The absorbed fractions for the other target organs are not zero for the gamma rays emitted by ^{131}I in the thyroid. The SEE for the gamma photons are discussed below.) It remains to compute the SEE for the beta rays in the thyroid itself as the target organ. The various factors that enter Eq. (14.12) are determined as follows. The thyroid mass $M_T = 20$ g is given. The yields Y_i and energies E_i per transformation can be obtained from Appendix D. We assume that the mean beta-particle energy is one-third the maximum. For the two main beta decays shown, we write $Y_1 = 0.006$, $E_1 = 0.806/3 = 0.269$ MeV, $Y_2 = 0.994$, and $E_2 = 0.606/3 = 0.202$ MeV. The absorbed fraction AF(thyroid $\leftarrow$ thyroid) $= 1$, and $Q_i = 1$ (Table 12.2). Thus we obtain from Eq. (14.12).

$$SEE(\text{thyroid} \leftarrow \text{thyroid})^{131}I_{(\beta^-)} = \frac{1}{20} (0.006 \times 0.269 \times 1 \times 1 + 0.994 \times 0.202 \times 1 \times 1)$$

$$= 0.010 \text{ MeV/g} \tag{14.13}$$

per transformation. (As a rule, no more than one or two significant figures are retained in results of computations for internal emitters.) Including the photons from ^{131}I will make all of the SEE(other organs $\leftarrow$ thyroid) $\neq 0$. Since most of the photon energy emitted inside the small thyroid will escape from the organ, the specific effective energy SEE(thyroid $\leftarrow$ thyroid)$^{131}I_{(\gamma)}$ is very much smaller than that for the β^-. Therefore

$$SEE(\text{thyroid} \leftarrow \text{thyroid})^{131}I = SEE(\text{thyroid} \leftarrow \text{thyroid})^{131}I_{(\beta^-)} + SEE(\text{thyroid} \leftarrow \text{thyroid})^{131}I_{(\gamma)}$$

$$\cong 0.010 \text{ MeV/g} \tag{14.14}$$

per transformation. As described in the remainder of this section, this is the correct value as obtained from the detailed calculations.

The absorbed fractions for gamma rays cannot be evaluated in a simple way. Their values depend in a complicated fashion on the photon energy; the size, density, and relative positions of the source and target organs; and on the specific intervening tissues. The Medical Internal Radiation Dose (MIRD) Committee of the Society of Nuclear Medicine has made extensive calculations of the specific absorbed fractions (absorbed fraction per gram of target) for a number of source and target organs in reference man. Monte Carlo techniques are employed in which the transport of many individual photons through the body is carried out by computer codes and the resulting data compiled to obtain the specific absorbed fractions. Calculations have been performed both for monoenergetic photons and for the spectra of photons emitted by a number of radionuclides.

Table 14.2 shows an example of specific absorbed fractions in a number of target

Table 14.2. Specific Absorbed Fraction (per gram per transformation) of Photon Energy
in Several Target Organs and Tissues for Monoenergetic Photon Source in Thyroid
(from ICRP Publication 23)

Target	Photon energy (MeV)		
	0.010	0.100	1.00
Stomach wall	2.07 E-25	1.90 E-07	4.62 E-07
Small intestines plus contents	4.58 E-35	1.97 E-08	1.38 E-07
Lungs	1.52 E-13	3.67 E-06	3.83 E-06
Ovaries	2.33 E-23	1.09 E-08	9.62 E-08
Red marrow	2.68 E-09	4.87 E-06	2.57 E-06
Testes	2.48 E-28	7.87 E-10	2.46 E-08
Thyroid	4.29 E-02	1.44 E-03	1.54 E-03
Total body	1.43 E-05	4.71 E-06	4.26 E-06

organs for photons of several energies emitted from the thyroid as source organ. The table illustrates the effect of the decrease in the linear absorption coefficient (Section 7.7) with increasing photon energy over the range considered. The specific absorbed fractions for the thyroid as target decrease with the greater probability of escape of the higher-energy photons from the source organ. The same holds true for the total body. In contrast, the greater penetrability of the higher-energy photons leads to an increase in the specific absorbed fractions in tissues outside the thyroid. Since 0.010 MeV photons are so strongly absorbed, the increase is dramatic at 0.100 MeV.

The decay schemes used for the calculation of the specific effective energies and specific absorbed fractions are usually very detailed. The ICRP's 1983 Publication 38, *Radionuclide Transformations, Energy and Intensity of Emissions*, contains the physical data used for Publication 30 and its Supplements. In place of the simple decay scheme from Appendix D that we used for ^{131}I in the example given earlier in this section, the ICRP calculations use the complex decay data shown in Fig. 14.3. The yields and average energies are given for five modes of β^- decay, a sixth mode (β^-5) contributing less than 0.1% $\sum Y_i E_i$ for the beta particles. A total of nine gamma photons are included, along with five internal-conversion electrons and the $K_{\alpha1}$ and $K_{\alpha2}$ daughter xenon X-rays. For the particular example chosen above, however, the estimate [Eq. (14.14)] turns out to agree with the value obtained from the detailed scheme shown in Fig. 14.3. The close agreement results because (1) we were not far off in estimating $\sum Y_i E_i$ from Appendix D for the beta particles plus conversion electrons and (2) most of the photon energy escapes from the small thyroid. As is more often the case, however, self-absorption of photons in an organ is important and simple estimates are not reliable. When the source and target organs are different, then the detailed Monte Carlo calculations offer the only feasible way of obtaining all of the needed specific absorbed fractions reliably. As mentioned in connection with Fig. 14.2, some absorbed-fraction computations have been checked experimentally.

14.6 NUMBER OF TRANSFORMATIONS IN SOURCE ORGANS OVER 50 yr

As stated at the end of Section 14.4, the remaining factor to determine in Eq. (14.11) is U_S, the number of atoms of a radionuclide that decay in each source organ S during the 50-yr period over which the committed dose equivalent is defined following single intake. This number is equal to the time integral of the activity of the radionuclide in the organ

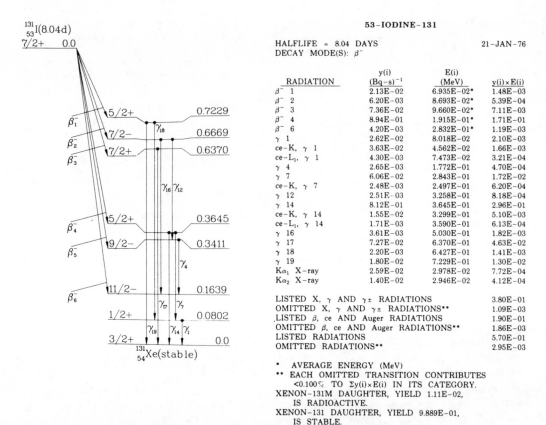

$^{131}_{53}$I(8.04d)

7/2+   0.0

53-IODINE-131

HALFLIFE = 8.04 DAYS   21-JAN-76
DECAY MODE(S): β^-

RADIATION	y(i) (Bq–s)$^{-1}$	E(i) (MeV)	y(i)×E(i)
β^- 1	2.13E–02	6.935E–02*	1.48E–03
β^- 2	6.20E–03	8.693E–02*	5.39E–04
β^- 3	7.36E–02	9.660E–02*	7.11E–03
β^- 4	8.94E–01	1.915E–01*	1.71E–01
β^- 6	4.20E–02	2.832E–01*	1.19E–03
γ 1	2.62E–02	8.018E–02	2.10E–03
ce–K, γ 1	3.63E–02	4.562E–02	1.66E–03
ce–L$_1$, γ 1	4.30E–03	7.473E–02	3.21E–04
γ 4	2.65E–03	1.772E–01	4.70E–04
γ 7	6.06E–02	2.843E–01	1.72E–02
ce–K, γ 7	2.48E–03	2.497E–01	6.20E–04
γ 12	2.51E–03	3.258E–01	8.18E–04
γ 14	8.12E–01	3.645E–01	2.96E–01
ce–K, γ 14	1.55E–02	3.299E–01	5.10E–03
ce–L$_1$, γ 14	1.71E–03	3.590E–01	6.13E–04
γ 16	3.61E–03	5.030E–01	1.82E–03
γ 17	7.27E–02	6.370E–01	4.63E–02
γ 18	2.20E–03	6.427E–01	1.41E–03
γ 19	1.80E–02	7.229E–01	1.30E–02
Kα_1 X–ray	2.59E–02	2.978E–02	7.72E–04
Kα_2 X–ray	1.40E–02	2.946E–02	4.12E–04

LISTED X, γ AND $\gamma\pm$ RADIATIONS	3.80E–01
OMITTED X, γ AND $\gamma\pm$ RADIATIONS**	1.09E–03
LISTED β, ce AND Auger RADIATIONS	1.90E–01
OMITTED β, ce AND Auger RADIATIONS**	1.86E–03
LISTED RADIATIONS	5.70E–01
OMITTED RADIATIONS**	2.95E–03

*   AVERAGE ENERGY (MeV)
** EACH OMITTED TRANSITION CONTRIBUTES
    <0.100% TO Σy(i)×E(i) IN ITS CATEGORY.
XENON–131M DAUGHTER, YIELD 1.11E–02,
    IS RADIOACTIVE.
XENON–131 DAUGHTER, YIELD 9.889E–01,
    IS STABLE.

Figure 14.3. Decay-scheme data for ^{131}I as given in ICRP Publication 38. (Reprinted with permission from *Radionuclide Transformations*, ICRP Publ. 38, p. 453. Copyright 1983 by International Commission on Radiological Protection, Sutton, England)

over this period. Radioactive materials are transported to most tissues and organs via the body fluids, which they enter following inhalation or ingestion. In this section we consider the mathematical model for the transfer of a radionuclide from the body fluids to an organ, tissue, or part thereof. The respiratory-system and gastrointestinal-tract (GI) models used by the ICRP to calculate the earlier entry of the radionuclide into the body and then into the body fluids are described in the next two sections. The specific metabolic data needed to carry out all of the computations are given in detail by the ICRP.

Following inhalation or ingestion, a nuclide enters the body fluids, represented by a transfer compartment (a) in Fig. 14.4. Compartment a is linked to other compartments, b, c,..., i,..., representing various tissues and organs in the body, from which the nuclide can later be excreted. A radionuclide undergoes metabolic clearance from these compartments, as well as radioactive decay. Unless otherwise specified, the metabolic half-life used for the transfer compartment a is 0.25 day. Radioactive transformations that occur in this compartment are assumed to be distributed uniformly in the 70 kg reference man. For simplicity, excretion in the model does not involve transport through compartment a, although, realistically, the body fluids are involved. For this reason, the calculated amount of a radionuclide in compartment a at some time after inhalation or ingestion cannot be used as an estimate of the amount of the radionuclide in the body fluids at that time.

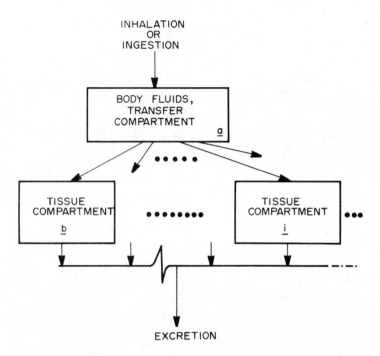

INHALATION
OR
INGESTION

BODY FLUIDS,
TRANSFER
COMPARTMENT
$\underline{a}$

TISSUE
COMPARTMENT
$\underline{b}$

TISSUE
COMPARTMENT
$\underline{i}$

EXCRETION

Figure 14.4. Mathematical model usually used for transfer of a radionuclide from the body fluids (compartment a) to various organs and tissues and its subsequent excretion. [From data in *Annals of the ICRP*, Vol. 2, No. 3/4, ICRP Publ. 30, Part I, International Commission on Radiological Protection, Sutton, England (1979)]

We let $\dot{I}(t)$ represent the rate at which a radionuclide enters the body fluids (compartment a) at time t after its inhalation or ingestion into the body. (This rate is computed from the models described in the next two sections.) If $q_a(t)$ is the activity of the radionuclide in compartment a, λ_R is its radioactive decay constant, and λ_a is the metabolic clearance rate of that element from a, we then have

$$\frac{dq_a(t)}{dt} = \dot{I}(t) - \lambda_R q_a(t) - \lambda_a q_a(t). \tag{14.15}$$

If b represents the fraction of the element that goes to compartment b when it leaves a, then the activity $q_b(t)$ of the radionuclide in b at time t satisfies the equation

$$\frac{dq_b(t)}{dt} = b\lambda_a q_a(t) - \lambda_R q_b(t) - \lambda_b q_b(t), \tag{14.16}$$

where λ_b is the metabolic clearance rate from b. Similar equations describe the activities $q_i(t)$ of the radionuclide in other tissue compartments. With given initial conditions and specific metabolic data, the equations can be solved for the activities $q_i(t)$ in the various organs and tissues. The 50 yr integrals of the $q_i(t)$ then give the numbers of transformations U_S in the source organs or tissues i. We now carry out the calculation for a simple case.

We treat a single amount of activity, $I = q_a(0)$, of a radionuclide introduced instantaneously into the transfer compartment a at time $t = 0$. In addition, we consider only a single tissue compartment, b. Equations (14.15) and (14.16) can be written

$$\frac{dq_a}{dt} = -(\lambda_R + \lambda_a)q_a \tag{14.17}$$

and

$$\frac{dq_b}{dt} = b\lambda_a q_a - (\lambda_R + \lambda_b)q_b. \tag{14.18}$$

Equation (14.17) is mathematically identical to Eq. (3.47) for radioactive decay. Applying the solution (3.52), we have

$$q_a(t) = Ie^{-(\lambda_R + \lambda_a)t}. \tag{14.19}$$

Equation (14.18) is similar to Eq. (3.84). The solution, like Eq. (3.85) with the same initial condition, is

$$q_b(t) = \frac{b\lambda_a I}{\lambda_b - \lambda_a}(e^{-(\lambda_R + \lambda_a)t} - e^{-(\lambda_R + \lambda_b)t}), \tag{14.20}$$

as can be verified by direct substitution. The activity in b thus builds up from its initial value $q_b(0) = 0$ to a maximum, from which it thereafter declines, like the activity A_2 in Figs. 3.17 or 3.18. The integrals of the activities from $t = 0$ to any subsequent time T are

$$U_a(T) = \int_0^T q_a(t)\,dt = -\left.\frac{I}{\lambda_R + \lambda_a}e^{-(\lambda_R + \lambda_a)t}\right|_0^T \tag{14.21}$$

$$= \frac{I}{\lambda_R + \lambda_a}(1 - e^{-(\lambda_R + \lambda_a)T}) \tag{14.22}$$

and

$$U_b(T) = \int_0^T q_b(t)\,dt \tag{14.23}$$

$$= \frac{b\lambda_a I}{\lambda_b - \lambda_a}\left(\frac{1 - e^{-(\lambda_R + \lambda_a)T}}{\lambda_R + \lambda_a} - \frac{1 - e^{-(\lambda_R + \lambda_b)T}}{\lambda_R + \lambda_b}\right). \tag{14.24}$$

The numbers of transformations U_a and U_b in compartments a and b for a radionuclide with decay constant λ_R can thus be evaluated explicitly for the committed dose equivalent ($T = 50$ yr), given the metabolic parameters λ_a, λ_b, and b.

Example
Use the two-component model just described for a radionuclide having a half-life of 0.430 day. The metabolic half-life in the body fluids (compartment a) is 0.25 day, the fraction $b = 0.22$ of the nuclide goes to organ b when it leaves a, the metabolic half-life in b is 9.8 years, and the mass of b is 144 g. Calculate the number of transformations U_a and U_b of the radionuclide in the two compartments during the 50 yr following the single entrance of 1 Bq of the radionuclide into compartment a at time $t = 0$. If the radionuclide emits beta particles with an average energy $E = 0.255$ MeV, what are the resulting committed dose equivalents for organ b and for the total body?

Solution
We use Eqs. (14.22) and (14.24) to calculate U_a and U_b. From the given radioactive and metabolic half-lives, we have the following decay rates:

$$\lambda_R = \frac{0.693}{0.43} = 1.6 \text{ day}^{-1}, \tag{14.25}$$

$$\lambda_a = \frac{0.693}{0.25} = 2.8 \text{ day}^{-1} \tag{14.26}$$

$$\lambda_b = \frac{0.693}{9.8 \times 365} = 1.9 \times 10^{-4} \text{ day}^{-1}. \tag{14.27}$$

(We shall express time in days and retain only two significant figures in the computations.) The time period is $T = 50 \times 365 = 1.8 \times 10^4$ day, and so the exponential terms in Eqs. (14.22) and (14.24) are negligible compared with unity. The initial activity in compartment a is

$$I = 1 \text{ Bq} = 1 \text{ sec}^{-1} \times 86,400 \text{ sec/day} = 8.6 \times 10^4 \text{ day}^{-1}. \tag{14.28}$$

Using (14.22), we find

$$U_a = \frac{8.6 \times 10^4 \text{ day}^{-1}}{(1.6 + 2.8) \text{ day}^{-1}} = 2.0 \times 10^4. \tag{14.29}$$

Using Eq. (14.24) (with λ_b neglected compared with λ_R), we obtain

$$U_b = \frac{b\lambda_a I}{\lambda_b - \lambda_a} \left(\frac{\lambda_b - \lambda_a}{(\lambda_R + \lambda_a)(\lambda_R + \lambda_b)} \right) \tag{14.30}$$

$$= \frac{0.22 \times 2.8 \times 8.6 \times 10^4}{(1.6 + 2.8)(1.6)} = 7.5 \times 10^3. \tag{14.31}$$

The committed dose equivalent for each tissue or organ is given by Eq. (14.11) with the specific effective energy calculated from Eq. (14.12). As with the earlier example (Section 14.5) involving ^{131}I beta particles, $AF = 1$ when the source and target are the same and $AF = 0$ otherwise. Also, $Q = 1$ and $YE = 0.26$. Transformations in compartment a are considered to be uniformly distributed over the whole body, having the mass $M = 70,000$ g. From Eq. (14.12) we have for the whole body (WB)

$$\text{SEE(WB} \leftarrow \text{WB)} = \frac{1}{70,000} (0.26 \times 1 \times 1) = 3.7 \times 10^{-6} \text{ MeV/g} \tag{14.32}$$

per transformation. For compartment b, having the given mass 144 g,

$$\text{SEE(b} \leftarrow \text{b)} = \frac{1}{140} (0.26 \times 1 \times 1) = 1.9 \times 10^{-3} \text{ MeV/g} \tag{14.33}$$

per transformation. The committed dose equivalents are, by Eq. (14.11),

$$H_{50,\text{WB}} = 1.6 \times 10^{-10} U_a \text{ SEE(WB} \leftarrow \text{WB)} \tag{14.34}$$

$$= 1.6 \times 10^{-10} \times 2.0 \times 10^4 \times 3.7 \times 10^{-6} = 1.2 \times 10^{-11} \text{ Sv} \tag{14.35}$$

and

$$H_{50,\text{b}} = 1.6 \times 10^{-10} U_b \text{ SEE(b} \leftarrow \text{b)} \tag{14.36}$$

$$= 1.6 \times 10^{-10} \times 7.5 \times 10^3 \times 1.9 \times 10^{-3} = 2.3 \times 10^{-9} \text{ Sv}. \tag{14.37}$$

The last number represents the dose equivalent in organ b that is delivered by radionuclides in b after they have left a. Since organ b is also irradiated as a part of the whole body by radionuclides in a, the total committed dose equivalent in b is the sum of (14.37) and (14.35). The latter contribution, however, is negligible.

The half-life of a radionuclide in a compartment is governed by its metabolic-clearance and radioactive-decay rates, λ_M and λ_R. Both processes are characterized independently by their respective half-lives:

$$T_R = \frac{0.693}{\lambda_R} \tag{14.38}$$

and

$$T_M = \frac{0.693}{\lambda_M}. \tag{14.39}$$

As seen in Eqs. (14.17) and (14.18), the effective rate of disappearance λ_{EFF} from a compartment is given by the sum $\lambda_R + \lambda_M$. Using the last two equations, we may write

$$\lambda_{EFF} = \lambda_R + \lambda_M = 0.693\left(\frac{1}{T_R} + \frac{1}{T_M}\right) = \frac{0.693}{T_{EFF}}, \tag{14.40}$$

where the effective half-life is given by

$$T_{EFF} = \frac{T_R T_M}{T_R + T_M}. \tag{14.41}$$

Note that T_{EFF} cannot be larger than T_R or T_M.

Example
What were the effective half-lives of the radionuclide in compartments a and b in the last example?

Solution
From the given data, $T_R = 0.43$ day, $T_a = 0.25$ day, and $T_b = 9.8 \times 365 = 3600$ days. Therefore, the effective half-lives are

$$T_{EFF,a} = \frac{0.43 \times 0.25}{0.43 + 0.25} = 0.16 \text{ day} \tag{14.42}$$

and

$$T_{EFF,b} = \frac{0.43 \times 3600}{0.43 + 3600} = 0.43 \text{ day}. \tag{14.43}$$

In ICRP Publication 30, the committed dose equivalent per unit intake, the annual limit on intake (ALI), and the derived air concentration (DAC) all refer to the intake of a specified radionuclide alone. If the radionuclide decays into radioactive daughters, then these are also included in the calculations of committed dose equivalent. The computations include either specific metabolic data for the daughters or the assumption that they follow the transport of the parent. Values of the number of transformations U_S in source organs S for a radionuclide are computed together with values U_S', U_S'', etc. for the daughter radionuclides that build up in the body during the 50 yr following intake of the parent.

14.7 DOSIMETRIC MODEL FOR THE RESPIRATORY SYSTEM

The ICRP model for the respiratory system, shown in Fig. 14.5, is divided into three major parts—the nasal passage (NP), the trachea and bronchial tree (TB), and the pulmonary parenchyma (P). In addition, a pulmonary lymphatic system (L) is included for the removal of dust from the lungs. The direct deposition of inhaled material, which occurs in the first three regions, varies with the particle-size distribution of the inhaled material. Basic calculations are made for reference man with an assumed log-normal distribution

Region	Compart-ment	Class					
		D		W		Y	
		T day	F	T day	F	T day	F
N–P ($D_{N-P}=0.30$)	a	0.01	0.5	0.01	0.1	0.01	0.01
	b	0.01	0.5	0.40	0.9	0.40	0.99
T–B ($D_{T-B}=0.08$)	c	0.01	0.95	0.01	0.5	0.01	0.01
	d	0.2	0.05	0.2	0.5	0.2	0.99
P ($D_P=0.25$)	e	0.5	0.8	50	0.15	500	0.05
	f	n.a.	n.a.	1.0	0.4	1.0	0.4
	g	n.a.	n.a.	50	0.4	500	0.4
	h	0.5	0.2	50	0.05	500	0.15
L	i	0.5	1.0	50	1.0	1 000	0.9
	j	n.a.	n.a.	n.a.	n.a.	∞	0.1

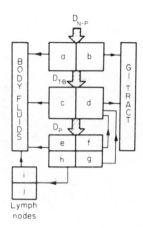

Figure 14.5. Compartmental model for the respiratory system. Table gives initial deposition fractions, half-times, and removal fractions for each compartment and class of material for aerosol with an AMAD of 1 μm (see text). [Reprinted with permission from *Annals of the ICRP*, Vol. 2, No. 3/4, ICRP Publ. 30, Part I, p. ii (Errata), International Commission on Radiological Protection, Sutton, England (1979). Copyright 1979 by ICRP]

of particle diameters having an assumed activity median aerodynamic diameter (AMAD) of 1 μm. The fractions D of inhaled material that are initially deposited in the three regions are then assumed to be $D_{NP}=0.30$, $D_{TB}=0.08$, and $D_P=0.25$. A procedure is given for making particle-size corrections for other values of the AMAD. The deposition fractions for other sizes are given in Fig. 14.6. The model is thus applicable to the inhalation of radioactive aerosols, or particulates. Inhalation of a radioactive gas is treated separately (Section 14.10).

The model of the respiratory system describes the initial deposition and subsequent transport of inhaled radioactive aerosols through various compartments of the system and into the body fluids and the GI tract. [A substance in the GI tract can also enter the body fluids (next section).] It has been found that the dose in the NP region can be neglected for most particle sizes. The target tissue assumed for the lung, therefore, is that of the combined TB, P, and L regions, having a total mass of 1000 g. The committed dose equivalent to the lung has two components, one from the radioactive materials residing there and another from photons emitted by materials that are cleared from the lung and transported to other sites in the body.

ICRP Publication 30 classifies inhaled radioactive materials as D, W, or Y (days, weeks, or years), depending on their retention time in the pulmonary region. Class D materials have a half-time of less than 10 days; W materials, a half-time from 10 days to 100 days; and Y, greater than 100 days. As seen in Fig. 14.5, the four major regions of the model are each subdivided into two or four compartments, each associated with a particular pathway of clearance. For the three classes D, W, and Y, the table in Fig. 14.5 gives the half-time T used in each compartment and the fraction F of material that leaves it at that implied rate. Compartments a, c, and e are associated with the uptake of material from the respiratory system into the body fluids. Compartments b, d, f, and g are associated with the physical transport of particles (e.g., by mucociliary action and swallowing) into the GI tract. Compartment h in the P region provides the pathway to the lymph system L, where some material can be further translocated via i to the body fluids or else retained indefinitely in j. (Compartment j is used only for class Y materials.)

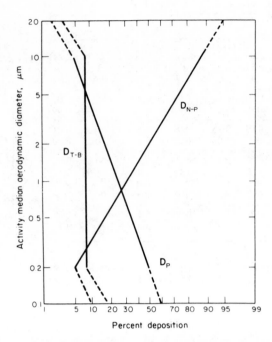

Figure 14.6. Deposition fractions in regions of the respiratory model for aerosols of different AMAD (see text) between 0.2 μm and 10 μm. Dashed lines show provisional extensions of curves outside this range. [Reprinted with permission from *Annals of the ICRP*, Vol. 2, No. 3/4, ICRP Publ. 30, Part I, p. i (Errata), International Commission on Radiological Protection, Sutton, England (1979). Copyright 1979 by ICRP]

Given the rate of inhalation $\dot{I}(t)$ of a radionuclide, ten differential equations, similar to Eqs. (14.15) and (14.16), are used to describe the activities $q_a(t), \ldots, q_j(t)$ in each of the ten compartments shown in Fig. 14.5. The equation describing a, for example, can be written

$$\frac{dq_a}{dt} = \dot{I}D_{NP}F_a - \lambda_a q_a - \lambda_R q_a, \tag{14.44}$$

where $D_{NP} = 0.30$; F_a and T_a are obtained directly from the values of F and T in the table in Fig. 14.5 for class D, W, or Y; $\lambda_a = 0.693/T_a$; and λ_R is the radioactive decay constant. The classifications D, W, or Y for different chemical forms of a radionuclide are provided by the ICRP along with the metabolic data. Radioactive daughters are included in the calculations and are assumed to have the same metabolic behavior as the original parent. While Eq. (14.44) involves only a single activity, q_a, others are more complicated. For compartment d, for example, we have

$$\frac{dq_d}{dt} = \dot{I}D_{TB}F_d + \lambda_f q_f + \lambda_g q_g - \lambda_d q_d - \lambda_R q_d, \tag{14.45}$$

which couples the activity q_d in d to those in f and g. Given a set of initial conditions, the system of ten linear, coupled, differential equations is solved to obtain the activities in each of the compartments a–j as functions of time. The rate of transfer of the inhaled radionuclide into the body fluids as a function of time is then given by

$$BF(t) = \lambda_a q_a(t) + \lambda_c q_c(t) + \lambda_e q_e(t) + \lambda_i q_i(t). \tag{14.46}$$

Similarly, the rate of transfer into the GI tract is

$$G(t) = \lambda_b q_b(t) + \lambda_d q_d(t). \tag{14.47}$$

The respiratory-system model thus completely specifies the deposition, retention, and removal of inhaled materials in various components of the pulmonary–lymph system. It is used to calculate the number of transformations U for the committed dose equivalent to the lung and to calculate source terms for the body fluids and the GI tract.

14.8 DOSIMETRIC MODEL FOR THE GASTROINTESTINAL TRACT

The ICRP dosimetric model for the GI tract is shown in Fig. 14.7. Each of the four sections consists of a single compartment: the stomach (ST), small intestine (SI), upper large intestine (ULI), and lower large intestine (LLI). There are two pathways out of the SI. One leads to the ULI and the other to the body fluids, the only route by which ingested materials are assumed to reach the body fluids. The metabolic rate constants λ are specified by the ICRP for the various chemical elements. The fraction f_1 of a stable element that reaches the body fluids after ingestion is given by

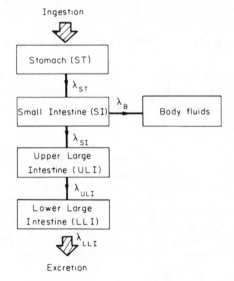

Section of GI tract	Mass of walls* (g)	Mass of contents* (g)	Mean residence time (day)	λ day^{-1}
Stomach (ST)	150	250	1/24	24
Small Intestine (SI)	640	400	4/24	6
Upper Large Intestine (ULI)	210	220	13/24	1.8
Lower Large Intestine (LLI)	160	135	24/24	1

*From ICRP Publication 23 (1975).

Figure 14.7. Dosimetric model for the gastrointestinal system. Table gives masses of the sections and their contents and clearance-rate data. [Reprinted with permission from *Annals of the ICRP*, Vol. 2, No. 3/4, ICRP Publ. 30, Part I, p. 33, International Commission on Radiological Protection, Sutton, England (1979). Copyright 1979 by ICRP]

$$f_1 = \frac{\lambda_B}{\lambda_{SI} + \lambda_B}. \tag{14.48}$$

Values of f_1 are given in the metabolic data of the ICRP (Section 14.11).

As with the lung, each of the four compartments in Fig. 14.7 gives rise to a first-order differential equation describing the activity changes. Given initial conditions and the rate of intake $\dot{I}(t)$ into compartment ST, the four equations can be solved for the activities in each section as functions of time. Activity entering the GI tract from the respiratory system is included in $\dot{I}(t)$. Radioactive daughters are included with the parent in the ICRP calculations. The table in Fig. 14.7 gives the assumed masses of the sections and their contents and the clearance rates. The activity transferred to the body fluids as a function of time is given by $\lambda_B q_{SI}(t)$.

When the source organ is a section of the GI tract, the committed dose equivalent is estimated for the mucosal layer of the walls of each section for penetrating and non-penetrating radiations. Other organs are also irradiated by sources in the contents of the GI tract, and the tract is irradiated by materials located in other parts of the body. The ICRP has compiled values of U_S for various ingested radionuclides and daughters in the sections of the GI tract. It also gives values of the SEE for sections of the GI tract as target organs with the contents of these sections and other organs of the body as source organs.

14.9 DOSIMETRIC MODEL FOR BONE

A great deal of effort has been devoted to studying the intake, deposition, and retention of radionuclides in the skeleton and its substructures. We shall make only a few remarks here about bone dosimetry.

ICRP Publication 30 states,

> The cells at carcinogenic risk in the skeleton have been identified as the haematopoietic stem cells of marrow, and among the osteogenic cells, particularly those on endosteal surfaces, and certain epithelial cells close to bone surfaces (ICRP Publication 11). The haematopoietic stem cells in adults are assumed to be randomly distributed predominantly throughout the hae-matopoietic marrow within trabecular bone (ICRP Publication 11). Therefore, dose equiva-lent to those cells is calculated as the average over the tissue which entirely fills the cavities within trabecular bone. For the osteogenic tissue on endosteal surfaces and epithelium on bone surfaces the Commission recommends that dose equivalent should be calculated as an average over tissue up to a distance of 10 μm from the relevant bone surfaces (para. 47, ICRP Publication 26).

The two principal target tissues for bone dosimetry are thus the active red bone marrow and cells near the bone surfaces. Except for gamma emitters, the source tissues in bone are cortical and trabecular bone. The ICRP gives specific details for estimating the number of transformations in trabecular and cortical bone and the needed absorbed fractions. Some inhaled or ingested elements are assumed to become distributed throughout the bone volume, while others are assumed to attach on bone surfaces. Generally, an alkaline-earth radionuclide with a radioactive half-life greater than 15 days belongs to the former group, while the shorter lived elements reside on the bone surfaces.

The alkaline-earth elements, which include calcium and strontium, have received much attention in bone dosimetry. A special task group of the ICRP was formed to study their metabolism in man. The extensive human data and experience with radium in the bone was mentioned in Section 11.7. Bone-seeking radionuclides, which also include those of strontium and plutonium, are considered dangerous because they irradiate the sensitive cells of the marrow. They all produce cancer in laboratory animals at sufficiently high levels of exposure.

14.10 DOSIMETRIC MODEL FOR SUBMERSION IN A RADIOACTIVE GAS CLOUD

The last important model to be described is that for submersion in a cloud of radioactive gas. In this case the organs of the body can be irradiated by gas outside the body, absorbed in the body's tissues, and contained in the lungs.

To see how these sources limit the exposure of a radiation worker, the ICRP treats a cloud of infinite extent with a constant, uniform concentration C Bq/m^3 of a gaseous radionuclide. For a person submerged in the cloud, they consider (a) the dose-equivalent rate $\dot{H}_E$ to any tissue from external radiation, (b) the rate $\dot{H}_A$ to the tissue from the gas absorbed internally in the body, and (c) the rate $\dot{H}_L$ to the lung from the gas contained in it. We discuss each of these in turn.

For irradiation from outside the body, we let s represent the dose-equivalent rate in sieverts per hour (Sv/hr) in the air per Bq/g of air. The rate in the air from C Bq/m^3 is Cs/ρ_A, where ρ_A is the density of air (~ 1300 g/m^3). The dose-equivalent rate at the body surface of a person submerged in the cloud is then given by Csk/ρ_A, where k is the ratio of the mass stopping powers of tissue and air ($k \cong 1$). Therefore, one can express the dose-equivalent rate in a small volume of tissue in the body by writing

$$\dot{H}_E = \frac{Cskg_E}{\rho_A} \text{ Sv/hr,} \tag{14.49}$$

where g_E is a geometrical factor that allows for shielding by intervening tissues. For alpha particles and low-energy beta particles, such as those of tritium, $g_E = 0$. These radiations cannot penetrate to the lens of the eye (at a depth of 3 mm) or to the basal layer of the epidermis (at a depth of 70 μm). For most other beta emitters and for low-energy photons, $g_E \cong 0.5$ near the body surfaces and approaches zero with increasing depth. For high-energy photons, $g_E \cong 1$ throughout the body.

For irradiation from gas absorbed in the body, the ICRP considers a prolonged exposure to the cloud, which results in equilibrium concentrations of the gas in the air and in tissue. The concentration C_T of gas in the tissue is then given by

$$C_T = \frac{\delta C}{\rho_T} \text{ Bq/g,} \tag{14.50}$$

where ρ_T is the density of tissue ($\sim 10^6$ g/m^3) and δ is the solubility of the gas in tissue, expressed as the volume of gas in equilibrium with a unit volume of tissue at atmospheric pressure. The solubility increases with the atomic weight of the gas, varying in water at body temperature from ~ 0.02 for hydrogen to ~ 0.1 for xenon. For adipose tissue the values may be larger by a factor of 3–20. For the dose-equivalent rate in tissue from absorbed gas, the ICRP writes

$$\dot{H}_A = \frac{s\delta Cg_A}{\rho_T} \text{ Sv/hr,} \tag{14.51}$$

where g_A is another geometric factor, depending on the size of a person and the range of the radiation. For alpha and beta particles and low-energy photons, $g_A \cong 1$ for tissues deep inside the body and $g_A \cong 0.5$ for tissues at the surface. For energetic photons $g_A \ll 1$.

The dose-equivalent rate in the lung from the gas it contains can be written

$$\dot{H}_L = \frac{sCV_L g_L}{M_L} \text{ Sv/hr,} \tag{14.52}$$

where V_L is the average volume of air in the lungs ($\sim 3 \times 10^{-3}$ m^3), M_L is the mass of the lungs (1000 g), and g_L is a geometrical factor ($\cong 1$ for alpha and beta particles and low-energy photons and decreasing with increasing photon energy).

The three rates (14.49), (14.51), and (14.52) can be applied in the following way. For tritium, $\dot{H}_E = 0$ for all relevant tissues of the body, because this nuclide emits only low-energy beta particles. The ratio of the dose-equivalent rate in any tissue from absorbed gas to the rate in the lung from the gas it contains is

$$\frac{\dot{H}_A}{\dot{H}_L} = \frac{\delta g_A M_L}{V_L g_L \rho_T}. \tag{14.53}$$

With $g_A \cong g_L \cong 1$ for tritium, substitution of values given above leads to the ratio

$$\frac{\dot{H}_A}{\dot{H}_L} \cong \frac{\delta \times 1 \times 1000 \text{ g}}{3 \times 10^{-3} \text{ m}^3 \times 1 \times 10^6 \text{ g m}^{-3}} = \frac{\delta}{3}. \tag{14.54}$$

The value of δ for tritium is ~ 0.02 for aqueous tissues and ~ 0.05 for adipose tissues. It follows that the dose-equivalent rate to the lung from the gas contained in it, which is some 60–150 times greater than the rate to any tissue from absorbed gas, is the limiting factor for submersion in a cloud of elemental tritium. (The limit for titrated water, which is much smaller than that for elemental tritium, restricts most practical cases of exposure to tritium.)

Many noble gases emit photons and energetic beta particles. Then, for tissues near the surface of the body, $g_E \cong 0.5$. From Eqs. (14.49) and (14.52) one has

$$\frac{\dot{H}_E}{\dot{H}_L} = \frac{kg_E M_L}{\rho_A V_L g_L} \cong \frac{1 \times 0.5 \times 1000}{1300 \times 3 \times 10^{-3} g_L} \gtrsim 130, \tag{14.55}$$

where the inequality reflects the fact that $g_L \leq 1$. Thus the dose-equivalent rate to tissues near the body surfaces is more than 130 times $\dot{H}_L$. From (14.49) and (14.51) we find

$$\frac{\dot{H}_E}{\dot{H}_A} = \frac{kg_E \rho_T}{\rho_A \delta g_A} \cong \frac{1 \times 0.5 \times 10^6}{1300 \delta g_A} \tag{14.56}$$

$$\cong \frac{400}{\delta g_A} \gtrsim \frac{400}{\delta}, \tag{14.57}$$

since $g_A \leq 1$. Since $\delta \leq 2$ always,

$$\frac{\dot{H}_E}{\dot{H}_A} \gtrsim 200. \tag{14.58}$$

For these noble gases, the external radiation will be the limiting factor for a person submerged in a cloud.

Conversion factors are used by the ICRP to apply the results obtained for infinite clouds to exposure in rooms of sizes from 100 m^3 to 1000 m^3.

14.11 METABOLIC DATA AND LIMITS FOR SELECTED RADIONUCLIDES

In this section we summarize some of the metabolic data for selected elements and the ALI and DAC values for some radionuclides from ICRP Publication 30. The values were calculated for occupationally exposed adults by the procedures outlined in this chapter. As stressed in Section 14.3, the annual limit on intake is a secondary limit designed to meet the basic exposure–risk criteria set forth in ICRP Publication 26. The less fundamental derived air concentration is offered as a guide for controlling exposures and should be applied with judgment in specific situations. It is based on a 40 hr work week.

As discussed in Section 14.3, the intake of a radionuclide can be limited by the risk from either stochastic or nonstochastic effects. In the data that follow, some values of ALI are determined by the nonstochastic limit in a particular organ or tissue. If this is the case, then the ALI value that results in the stochastic limit is also given (in parentheses) together with the organ or tissue to which the nonstochastic limit applies (for example, see ^{90}Sr, below).

Hydrogen

Reference-man data:

Hydrogen content of the body	7,000 g
of soft tissue	6,300 g
Daily intake of hydrogen	350 g
Water content of the body	42,000 g
Daily intake of water, including water of oxidation	3,000 g.

Water comprises about 80% of the mass of some soft tissues. As discussed in the last section, exposure to *elemental* tritium is limited by the dose equivalent from tritium in the lung. In contrast, tritiated water that is inhaled, ingested, or absorbed through the skin is assumed to become instantaneously and uniformly distributed throughout all the soft tissues of the body. While some tritium from tritiated water can become organically bound in the body, the ICRP assumes a single-exponential retention function for the body, based on tritiated water alone, with a biological half-life of 10 days. The fraction of tritium, taken in as tritiated water at time $t = 0$, retained at time t days is given by

$$R(t) = e^{-0.693t/10}. \tag{14.59}$$

If the body contains q Bq, then the concentration in soft tissue (mass 63,000 g) is $q/63,000$ Bq/g. The levels recommended in ICRP Publication 30 (ALI in Bq and DAC in Bq/m^3) follow.

		Oral	Inhalation
^{3}H (Tritiated water)	ALI	3×10^9	3×10^9
	DAC	—	8×10^5
^{3}H (Elemental tritium)	ALI	—	—
	DAC	—	2×10^{10}

Strontium

Reference-man data:

Strontium content of the body		0.32 g
	of the skeleton	0.32 g
	of soft tissues	3.3 mg
Daily intake in food and fluids		1.9 mg.

Based on human and animal data, the ICRP uses $f_1 = 0.3$ for soluble salts and 0.01 for SrTiO$_3$ as the fractional uptake of ingested strontium by the body fluids [Eq. (14.48)]. For inhalation, soluble compounds are assumed to be in class D and SrTiO$_3$ in class Y (Section 14.7). As discussed in Section 14.9, strontium isotopes ^{90}Sr, ^{85}Sr, and ^{89}Sr, having half-lives greater than 15 days, are assumed to be distributed uniformly in the volume of mineral bone. In contrast, other strontium isotopes, with shorter half-lives, are assumed to be distributed uniformly over bone surfaces. The detailed metabolic model is used to estimate the number of transformations in soft tissue, cortical bone, and trabecular bone during the 50 yr following the introduction of unit activity into the transfer compartment of the body (Fig. 14.4). The calculated values for ^{90}Sr follow (ALI in Bq and DAC in Bq/m^3).

| | | Oral | | Inhalation | |
| | | | | Class D | Class Y |
		$f_1 = 0.3$	$f_1 = 0.01$	$f_1 = 0.3$	$f_1 = 0.01$
^{90}Sr	ALI	1×10^6	2×10^7	7×10^5	1×10^5
		(1×10^6)		(8×10^5)	
		Bone surface		Bone surface	
	DAC	—	—	3×10^2	6×10^1

Iodine

Reference-man data:
Iodine content of the body	11.0 mg
of the thyroid	10.0 mg
Daily intake in food and fluids	0.2 mg

Iodine is absorbed rapidly and almost completely from the gut, and so $f_1 = 1$ is used. All data indicate that compounds of iodine belong to inhalation class D. When iodine enters the transfer compartment, the fraction 0.3 is assumed to be taken up by the thyroid and the rest excreted directly. Iodine in the thyroid has a biological half-life of 120 days and leaves the gland as organic iodine. The organic iodine becomes distributed uniformly in the other tissues of the body with a half-life of 12 days. One-tenth of this organic iodine is then assumed to be excreted while the rest returns to the transfer compartment. For ^{131}I, the recommended values of ALI (Bq) and DAC (Bq/m^3) follow.

| | | Oral | Inhalation (Class D) |
		$f_1 = 1$	$f_1 = 1$
^{131}I	ALI	1×10^6	2×10^6
		(4×10^6)	(6×10^6)
		Thyroid	Thyroid
	DAC	—	7×10^2

Cesium

Reference-man data:
Cesium content of the body	1.5 mg
of muscle	0.57 mg
of bone	0.16 mg
Daily intake in food and fluids	10.0 μg

Cesium compounds are usually rapidly and almost completely absorbed in the GI tract, and so $f_1 = 1$. They are assigned to inhalation class D. A two-component retention function is used for cesium:

$$R(t) = ae^{-0.693t/T_1} + (1 - a)e^{-0.693t/T_2}. \tag{14.60}$$

When the element enters the transfer compartment, the fraction $a = 0.1$ is transferred to one tissue compartment (Fig. 14.4) and retained there with a metabolic half-life $T_1 = 2$

days; the remainder, $1 - a = 0.9$, is transferred to another tissue compartment and kept there with a half-life T_2 of 110 days. The cesium in both of these compartments is assumed to be distributed uniformly throughout the body. The values of ALI (Bq) and DAC (Bq/m^3) for ^{137}Cs follow.

		Oral	Inhalation (Class D)
		$f_1 = 1$	$f_1 = 1$
^{137}Cs	ALI	4×10^6	6×10^6
	DAC	—	2×10^3

Radium

Reference-man data:	
Radium content of the body	31.0 pg
of the skeleton	27.0 pg
Daily intake in food and fluids	2.3 pg

Available data lead to the choices $f_1 = 0.2$ and inhalation class W for all commonly occurring compounds of radium. A comprehensive retention model for radium in adults is used to calculate the numbers of transformations in soft tissue, cortical bone, and trabecular bone for obtaining the committed dose equivalent per unit intake. Values of the ALI (Bq) and DAC (Bq/m^3) for ^{226}Ra follow.

		Oral	Inhalation (Class W)
		$f_1 = 0.2$	$f_1 = 0.2$
^{226}Ra	ALI	7×10^4 (2×10^5) Bone surface	2×10^4
	DAC	—	1×10^1

Plutonium

No data are given in the Reference Man Report (ICRP Publication 23) for the man-made element, plutonium. As a result of nuclear weapons tests, however, measurable quantities of the element are found in foods and in human tissues. Some data are available on plutonium in humans from occupationally exposed persons. A great deal of metabolic information comes from animal experiments. Plutonium is taken up poorly in the GI tract. In ICRP Publication 30, the value $f_1 = 10^{-5}$ is used for oxides and hydroxides of plutonium and $f_1 = 10^{-4}$ for all other commonly occurring compounds. (Higher uptake rates are noted for some other types of compounds, but these are unlikely to be encountered in occupational exposures.) The commonly occurring plutonium compounds are assigned to inhalation class W, except PuO$_2$, which is in class Y. (The translocation rate of ^{238}PuO$_2$ from the lung is more rapid than that of ^{239}PuO$_2$.) Plutonium in the blood is deposited primarily in the liver and skeleton, where the element has long biological half-lives ($\sim$40 yr in the liver and $\sim$100 yr in the skeleton). In ICRP Publication 30, it is assumed, for plutonium entering the transfer compartment, that the fraction 0.45 goes to the liver and 0.45 to the bone. An additional small fraction is transferred to the gonads (3.5×10^{-4} for males and 1.1×10^{-4} for females) and the rest is excreted directly. The plutonium in the gonads is assumed to be retained there indefinitely. The recommended values for the ALI (Bq) and DAC (Bq/m^3) for ^{239}Pu are listed below:

| | | | Oral | | Inhalation | |
		$f_1 = 10^{-4}$	$f_1 = 10^{-5}$	Class W $f_1 = 10^{-4}$	Class Y $f_1 = 10^{-5}$
^{239}Pu	ALI	2×10^5 (4×10^5) Bone surf.	2×10^6 (3×10^6) Bone surf.	2×10^2 (4×10^2) Bone surf.	5×10^2 (6×10^2) Bone surf.
	DAC	—	—	8×10^{-2}	2×10^{-1}

14.12 PROBLEMS

1. What is reference man and what is its role in internal dosimetry?
2. In what way does the occupational limit for nonstochastic effects differ for external radiation and for intakes of radioactive materials?
3. The annual limit on intake of ^{32}P by ingestion, 2×10^7 Bq, is determined by the stochastic limit on dose equivalent. What is the weighted sum of the committed dose equivalents to all organs and tissues of the body per unit activity of ^{32}P ingested?
4. The ALI for inhalation of ^{131}I, 2×10^6 Bq, is determined by the nonstochastic limit on dose equivalent to the thyroid. What is the committed dose equivalent to the thyroid per unit activity of ^{131}I inhaled?
5. The ALI for inhalation of ^{235}U aerosols having retention times of the order of days or less in the pulmonary region is 5×10^4 Bq. What is the corresponding derived air concentration?
6. As the result of the single intake of 6.3×10^3 Bq of a radionuclide, a certain organ of the body will receive an average dose during the next 50 yr of 10 mrad from low-energy beta rays and 15 mrad from alpha rays. These are the only radiations emitted. The organ has a weighting factor of $w_T = 0.06$ (Table 12.3).
   (a) What is the committed dose equivalent to the organ?
   (b) If this is the only organ or tissue that receives appreciable irradiation as a result of such an intake, calculate the ALI for intake by this route, assuming that it is determined by the nonstochastic limit for dose equivalent.
7. What is the dose equivalent to the kidneys per transformation from a source in the lungs that emits a single 500 keV photon per transformation, if the specific effective energy (kidney ← lung) is 5.82×10^{-9} MeV/g?
8. What is the absorbed fraction (kidney ← lung) in the last problem? (Mass of the kidneys is given in Table 14.1.)
9. The specific absorbed fraction for irradiation of the red bone marrow by 200 keV photons from a source in the liver is 4.64×10^{-6}. Calculate the specific effective energy for the liver (source organ) and red marrow (target tissue) for a gamma source in the liver that emits only a 200 keV photon in 85% of its transformations.
10. What is the committed dose equivalent to the red marrow from the source in the liver in the last problem if 2.23×10^{15} transformations occur in the liver over a 50 yr period?
11. What are the specific absorbed fractions for various target organs for the pure beta emitter ^{14}C in the liver as source organ?
12. What are the corresponding values of the specific effective energies in the last problem?
13. What is the effective half-life of a radionuclide in the body-fluid transfer compartment, if its radioactive half-life is 8 hr?
14. An activity of 5×10^6 Bq of a radioisotope, having a half-life of 2 days, enters the body fluids. What fraction of the original activity remains in this compartment at the end of 4 days?
15. Show that Eq. (14.24) follows from (14.20).
16. By letting $\lambda_a \to \infty$ in Eq. (14.24), show that the second term represents the number of transformations that would have occurred in compartment b, had the material been transferred to it instantaneously. (The first term, therefore, represents the effect on U_b of the finite residence time in compartment a.)

17. Use the two-compartment model described in Section 14.6, Eqs. (14.17)–(14.20). An activity of 10^6 Bq of a radionuclide, having a half-life of 18 hr, enters compartment a (body fluids). The fraction that goes to organ b when it leaves a is 0.30, and the metabolic half-life in b is 2 days. Calculate the number of transformations in compartments a and b during the two days after the radionuclide enters a.

18. Assume that the radionuclide in the last problem is an alpha or low-energy beta emitter with a stable daughter. What fraction of the committed dose equivalent is delivered to organ b in the 2 days after entry of the radionuclide into a?

19. (a) Repeat Problem 17 for a radionuclide that has a radioactive half-life of 90 yr and a metabolic half-life in compartment b of 40 yr. (b) How many transformations occur in compartments a and b over the 50 yr of the committed dose equivalent?

20. What fraction of inhaled aerosols with an AMAD of 1 μm is assumed to be deposited in the bronchial tree in the ICRP model of the respiratory system?

21. Write a differential equation, analogous to Eq. (14.44), that describes the rate of change $\dot{q}_b(t)$ of the activity in compartment b of the respiratory-system model (Fig. 14.5) for a rate $\dot{I}(t)$ of inhalation.

22. Given the inhalation rate $\dot{I}(t)$, write two differential equations that describe the rates of change $\dot{q}_i(t)$ and $\dot{q}_j(t)$ of activity in the lymph-node compartments of the respiratory-system model in Fig. 14.5.

23. At a certain time following inhalation of a class W aerosol, the activities in compartments b and d of the respiratory-system model are, respectively, 7.8×10^4 Bq and 1.5×10^4 Bq. What is the rate of transfer of activity to the gastrointestinal tract?

24. In the dosimetric model for the GI tract (Fig. 14.7), show that λ_B can be estimated from f_1, the fraction of a stable element that reaches the body fluids after ingestion, by writing $\lambda_B = f_1 \lambda_{SI} / (1 - f_1)$.

25. At a certain time following ingestion of a radionuclide, the activity in the contents of the small intestine is 7.57 μCi. If the fraction of the stable element that reaches the body fluids after ingestion is 0.41, what is the rate of transfer of activity from the small intestine to the body fluids?

26. Write a differential equation that describes the rate of change of the activity $\dot{q}_{SI}(t)$ in the small intestine in terms of the parameters shown in Fig. 14.7 and the activities in the other compartments.

27. Why is bone dosimetry of particular importance?

28. What are the target tissues for bone?

29. Why is the dosimetric model for submersion in a radioactive gas cloud different from the model for the respiratory system?

30. Calculate the dose-equivalent rate in air in Sv/hr due to 1 Bq of ^{14}C per gram of air at STP.

31. Estimate the dose-equivalent rate at the surface of the skin of a person immersed in air (at STP) containing 2.4×10^3 Bq/m^3 of ^{14}CO$_2$.

32. Calculate the dose-equivalent rate in a large air volume (at STP) that contains a uniform distribution of 2×10^3 Bq/m^3 of ^{137}Cs (the DAC).

33. Why is the DAC for tritiated water so much smaller than that for elemental tritium?

34. (a) What activity of tritium, distributed uniformly in the soft tissue of the body (reference man), would result in a dose-equivalent rate of 0.05 Sv/yr? (b) What would be the total mass of tritium in the soft tissue?

35. Estimate the time it takes for the body to expel by normal processes 95% of the tritium ingested in a single intake of tritiated water. Would the retention time be affected by increasing the intake or liquids?

36. What limit in what tissue determines the ALI for inhalation of ^{90}Sr when the retention is characteristic of class D? Refer to Section 14.11.

37. What fraction of the iodine in the total body of reference man is in the thyroid?

38. Is the stochastic or nonstochastic limit the determining factor for the ALI of ^{226}Ra by ingestion?

APPENDIX A
PHYSICAL CONSTANTS

Planck's constant, $h = 6.6256 \times 10^{-27}$ erg sec

$\hbar = h/2\pi = 1.0545 \times 10^{-27}$ erg sec

Electron charge, $e = -4.80298 \times 10^{-10}$ esu

$= -1.60210 \times 10^{-19}$ C

Velocity of light in vacuum, $c = 2.997925 \times 10^{10}$ cm/sec

Avogadro's number, $N_0 = 6.02252 \times 10^{23}$ mole^{-1}

Molar volume at STP (0°C, 760 torr) = 22.4136 L

Density of air at STP (0°C, 760 torr) = 1.293×10^{-3} g/cm^3

$= 1.293$ kg/m^3

Rydberg constant for hydrogen, $R_H = 109{,}677.58$ cm^{-1}

First Bohr orbit radius in hydrogen, $a_0 = 5.29167 \times 10^{-9}$ cm

Ratio proton and electron masses = 1836.10

Neutron mass = 1.008665 AMU = 939.550 MeV = 1.67482×10^{-24} g

Proton mass = 1.007277 AMU = 938.256 MeV = 1.67252×10^{-24} g

Electron mass = 0.000549 AMU = 0.511006 MeV = 9.1091×10^{-28} g

H atom mass = 1.007825 AMU = 938.766 MeV = 1.67343×10^{-24} g

Boltzmann's constant, $k = 1.38054 \times 10^{-16}$ erg K^{-1}

APPENDIX B
UNITS AND CONVERSION FACTORS

$1 \text{ cm} = 10^4 \ \mu\text{m} = 10^8 \ \text{Å}$

1 in. = 2.5400 cm (exactly)

$1 \text{ barn} = 10^{-24} \text{ cm}^2$

$1 \text{ cm}^3 = 0.99997 \text{ mL} = 9.9997 \times 10^{-4} \text{ L}$

1 gal = 3.785297 L

$1 \text{ dyne} = 1 \text{ g cm/sec}^2 = 10^{-5} \text{ kg m/sec}^2 = 10^{-5} \text{ N}$

1 kg = 2.205 lb

$1 \text{ erg} = 1 \text{ dyne cm} = 1 \text{ g cm}^2/\text{sec}^2 = 1 \text{ esu}^2/\text{cm}$

$1 \text{ J} = 1 \text{ N m} = 1 \text{ kg m}^2/\text{sec}^2 = 8.987590 \times 10^9 \text{ C}^2/\text{m}$

$10^7 \text{ erg} = 1 \text{ J}$

$1 \text{ eV} = 1.6021 \times 10^{-12} \text{ erg} = 1.6021 \times 10^{-19} \text{ J}$

$1 \text{ AMU} = 931.48 \text{ MeV} = 1.66043 \times 10^{-24} \text{ g}$

1 gram calorie = 4.186 J

1 W = 1 J/sec = 1 V A

1 statvolt = 299.8 V

$1 \text{ esu} = 3.336 \times 10^{-10} \text{ C}$

1 A = 1 C/sec

1 C = 1 V F

$1 \text{ Ci} = 3.700 \times 10^{10} \text{ sec}^{-1} = 3.700 \times 10^{10} \text{ Bq}$

$1 \text{ R} = 2.58 \times 10^{-4} \text{ C/kg air} (= 1 \text{ esu/cm}^3 \text{ air at STP})$

1 rad = 100 erg/g = 0.01 Gy

1 Gy = 1 J/kg = 100 rad

1 Sv = 100 rem

0°C = 273 K

1 atmosphere = 760 mm Hg = 760 torr

1 day = 86,400 sec

$1 \text{ yr} = 365 \text{ days} = 3.1536 \times 10^7 \text{ sec}$

1 radian = 57.30°

APPENDIX C
SOME BASIC FORMULAS OF PHYSICS
(CGS and MKS units)

Classical Mechanics

Momentum = mass × velocity, $p = mv$

  units: g cm/sec; kg m/sec

Kinetic energy, $T = \frac{1}{2} mv^2 = p^2/2m$

  units: 1 erg = 1 g cm^2/sec^2; 1 J = 1 kg m^2/sec^2

Force = mass × acceleration, $F = ma$

  units: 1 dyne = 1 g cm/sec^2; 1 N = 1 kg m/sec^2

Work = force × distance = change in energy

  units: 1 erg = 1 dyne cm = 1 g cm^2/sec^2;

        1 J = 1 N m = 1 kg m^2/sec^2

Impulse = force × time = change in momentum, $I = Ft = \Delta p$

  units: 1 dyne sec = 1 g cm/sec; 1 N sec = 1 kg m/sec

Angular momentum, uniform circular motion, $L = mvr$

  units: 1 g cm^2/sec = 1 erg sec; 1 kg m^2/sec = 1 J sec

Centripetal acceleration, uniform circular motion, $a = v^2/r$

  units: cm/sec^2, m/sec^2

Relativistic Mechanics (units same as in classical mechanics)

Relativistic quantities:

  v = speed of object

  c = speed of light in vacuum

  $\beta = v/c$, dimensionless, $0 \leq \beta < 1$

  $\gamma = 1/\sqrt{1 - \beta^2}$, dimensionless, $1 \leq \gamma < \infty$

Rest energy, $E_0 = mc^2$, m = rest mass

Relativistic mass, $m/\sqrt{1 - \beta^2} = \gamma m$

Total energy, $E_T = mc^2/\sqrt{1 - \beta^2} = \gamma mc^2$

Kinetic energy = total energy − rest energy,

$$T = E_T - E_0 = mc^2(\gamma - 1) = mc^2\left(\frac{1}{\sqrt{1 - \beta^2}} - 1\right)$$

Momentum, $p = \gamma mv = mv/\sqrt{1 - \beta^2}$

Relationship between energy and momentum,
$$E_T^2 = p^2c^2 + m^2c^4 = (mc^2 + T)^2$$

Electromagnetic Theory

Force F between two point charges, q_1 and q_2, at separation r (Coulomb's law) in vacuum,

CGS: $F = q_1q_2/r^2$, q_1, q_2 in esu (statcoulombs), r in cm, and F in dynes

MKS: $F = k_0q_1q_2/r^2$, q_1, q_2 in coulombs, r in m, F in newtons, and

  $k_0 = 8.987590 \times 10^9$ N m^2/C^2

  ($= 1/(4\pi\epsilon_0)$ in terms of permittivity constant ϵ_0)

Potential energy of two point charges at separation r in vacuum,

CGS: PE $= q_1q_2/r$, 1 erg $= 1$ esu^2/cm

MKS: PE $= k_0q_1q_2/r$, 1 J $= 8.987590 \times 10^9$ C^2/m

Electric field strength (force per unit charge), $E = F/q$

  units: 1 dyne/esu $= 1$ statvolt/cm;

   1 N/C $= 1$ V/m

Electric field strength between parallel plates at separation d and potential difference V, $E = V/d$

  units: 1 dyne/esu $= 1$ statvolt/cm;

   1 N/C $= 1$ V/m

Capacitance, Q/V

  units: 1 F $= 1$ C/V

Current, $I = Q/t$ (charge per unit time)

  units: 1 A $= 1$ C/sec

Power, $P = VI$ (potential difference $\times$ current)

  units: 1 W $= 1$ V A $= 1$ J/sec

Relationship between wavelength λ and frequency ν of light in vacuum (speed of light $= c$), $\lambda\nu = c$

Quantum Mechanics

de Broglie wavelength, $\lambda = h/p = h/mv = h/\gamma mv$

Photon energy, $E = h\nu$

Photon momentum, $p = E/c = h\nu/c$

Bohr quantization condition for angular momentum, $L = n\hbar$

Bohr energy levels, $E = -13.6Z^2/n^2$ eV

Uncertainty relations, $\Delta p_x \Delta x \gtrsim \hbar$, $\Delta E \Delta t \gtrsim \hbar$.

APPENDIX D
SELECTED DATA ON NUCLIDES†

Nuclide	Natural abundance (%)	Mass difference $\Delta = M - A$ (MeV) (at. mass − at. mass No.)	Type of decay	Half-life	Major radiations, Energies (MeV), and Frequency per disintegration (%)
^1_0n	—	8.0714	β^-	12 min	β^-: 0.78 max
^1_1H	99.985	7.2890	—	—	—
^2_1H	0.015	13.1359	—	—	—
^3_1H	—	14.9500	β^-	12.3 yr	β^-: 0.0186 max. No γ
^3_2He	0.00013	14.9313	—	—	—
^4_2He	99.99+	2.4248	—	—	—
^6_3Li	7.42	14.088	—	—	—
$^{11}_6C$	—	10.648	β^+ 99+% EC 0.19%	20.3 min	β^+: 0.97 max γ: 0.511 (200%, $\gamma^\pm$)
$^{12}_6C$	98.892	0	—	—	—
$^{14}_6C$	—	3.0198	β^-	5730 yr	β^-: 0.156 max (avg. 0.045) No γ
$^{22}_{10}Ne$	8.82	−8.025	—	—	—
$^{22}_{11}Na$	—	−5.182	β^+ 90.6% EC 9.4%	2.60 yr	β^+: 1.820 max (0.05%) 0.545 max γ: 1.275 (100%), 0.511 (180%, $\gamma^\pm$), Ne X-rays
$^{24}_{11}Na$	—	−8.418	β^-	15.0 hr	β^-: 1.389 max γ: 1.369 (100%), 2.754 (100%)
$^{24}_{12}Mg$	78.60	−13.933	—	—	—
$^{26}_{12}Mg$	11.3	−16.214	—	—	—
$^{26}_{13}Al$	—	−12.211	β^+ 85% EC 15%	7.4×10^5 yr	β^+: 1.17 max γ: 1.12 (4%), 1.81 (100%), 0.511 (170%, $\gamma^\pm$), Mg X-rays
$^{26m}_{13}Al$	—	−11.982	β^+	6.4 sec	β^+: 3.21 max γ: 0.511 (200%, $\gamma^\pm$)
$^{32}_{15}P$	—	−24.303	β^-	14.3 day	β^-: 1.170 max. No γ
$^{32}_{16}S$	95.0	−26.013	—	—	—
$^{35}_{16}S$	—	−28.847	β^-	87.9 day	β^-: 0.167 max. No γ
$^{35}_{17}Cl$	75.53	−29.015	—	—	—
$^{37}_{17}Cl$	24.47	−31.765	—	—	—
$^{35}_{18}Ar$	—	−23.05	β^+	1.8 sec	β^+: 4.94 max γ: 1.22 (5%), 1.76 (2%), 0.511 (200%, $\gamma^\pm$)
$^{37}_{18}Ar$	—	−30.951	EC	35.1 day	γ: Cl X-rays

continued

Appendix D (Continued)

Nuclide	Natural abundance (%)	Mass-difference $\Delta = M - A$ (MeV) (at. mass − at. mass No.)	Type of decay	Half-life	Major radiations, Energies (MeV), and Frequency per disintegration (%)
$^{40}_{19}$K	0.0117	−33.533	β^- 89% EC 11% β^+ 0.001%	1.26×10^9 yr	β^-: 1.314 max β^+: 0.483 max γ: 1.460 (11%), Ar X-rays
$^{42}_{19}$K	—	−35.02	β^-	12.4 hr	β^-: 3.52 max (82%), 2.00 (18%) γ: 0.31 (0.2%), 1.524 (18%)
$^{55}_{25}$Mn	100	−57.705	—	—	—
$^{55}_{26}$Fe	—	−57.474	EC	2.60 yr	γ: Mn X-rays
$^{59}_{26}$Fe	—	−60.660	β^-	45.6 day	β^-: 1.573 max (0.3%), 0.475 max (53.5%), 0.283 max (45.4%), 0.140 max (0.8%) γ: 0.143 (0.8%), 0.192 (2.8%), 1.098 (56.3%), 1.290 (43.4%)
$^{60}_{27}$Co	—	−61.651	β^-	5.26 yr	β^-: 1.48 max (0.12%), 0.314 max (99+%) γ: 1.173 (100%), 1.332 (100%)
$^{60}_{28}$Ni	26.16	−64.471	—	—	—
$^{65}_{29}$Cu	30.9	−67.27	—	—	—
$^{65}_{30}$Zn	—	−65.92	EC 98.3% β^+ 1.7%	245 day	β^+: 0.327 max γ: 1.115 (49%), 0.511 (3.4%, $\gamma^\pm$), Cu X-rays e^-: 1.106
$^{85}_{36}$Kr	—	−81.48	β^-	10.8 yr	β^-: 0.67 max γ: 0.514 (0.41%)
$^{90}_{38}$Sr	—	−85.95	β^-	27.7 yr	β^-: 0.546 max. No γ Daughter radiations from ^{90}Y
$^{85}_{39}$Y	—	−77.79	β^+ 70% EC 30%	5.0 hr	β^+: 2.24 max γ: 0.231 (13%), 0.77 (8%), 2.16 (9%), 0.511 (140%, $\gamma^\pm$), Sr X-rays e^-: 0.215
$^{90}_{39}$Y	—	−86.50	β^-	64.0 hr	β^-: 2.27 max. No γ
$^{99}_{43}$Tc	—	−87.33	β^-	2.12×10^5 yr	β^-: 0.292 max. No. γ
$^{99m}_{43}$Tc	—	−87.18	IT	6.0 hr	γ: 0.140 (90%), Tc X-rays e^-: 0.001, 0.119
$^{103}_{45}$Rh	100	−88.014	—	—	—
$^{103m}_{45}$Rh	—	−87.974	IT	57.5 min	γ: 0.040 (0.4%), Rh X-rays e^-: 0.017, 0.037
$^{103}_{46}$Pd	—	−87.46	EC	17.0 day	γ: 0.297 (0.011%), 0.362 (0.06%), 0.498 (0.011%), Rh X-rays Daughter radiations from ^{103m}Rh

Appendix D (Continued)

Nuclide	Natural abundance (%)	Mass difference $\Delta = M - A$ (MeV) (at. mass – at. mass No.)	Type of decay	Half-life	Major radiations, Energies (MeV), and Frequency per disintegration (%)
$^{126}_{52}$Te	18.71	−90.05	—	—	—
$^{126}_{53}$I	—	−87.90	EC 55% β^- 44% β^+ 1.3%	12.8 day	β^-: 1.25 max β^+: 1.13 max γ: 0.386 (34%), 0.667 (33%), 0.511 (2.6%, $\gamma^\pm$), Te X-rays
$^{131}_{53}$I	—	−87.441	β^-	8.05 day	β^-: 0.806 max (0.6%), 0.606 max γ: 0.080 (2.6%), 0.284 (5.4%), 0.364 (82%), 0.637 (6.8%), 0.723 (1.6%) e^-: 0.046, 0.330 Daughter radiations from ^{131m}Xe
$^{126}_{54}$Xe	0.090	−89.15	—	—	—
$^{137}_{55}$Cs	—	−86.9	β^-	30.0 yr	β^-: 1.176 max (7%), 0.514 max γ: 0.662 (85%), Ba X-rays e^-: 0.624, 0.656
$^{137}_{56}$Ba	11.3	−88.0	—	—	—
$^{191}_{76}$Os	—	−36.4	β^-	15.0 day	β^-: 0.143 max γ: 0.129 (25%), Ir X-rays e^-: 0.030, 0.042, 0.053, 0.116, 0.127 Daughter radiations from ^{191m}Ir included above
$^{191m}_{77}$Ir	—	−36.5	IT	4.9 sec	γ: 0.129 (25%), Ir X-rays e^-: 0.30, 0.042, 0.053, 0.116, 0.127
$^{198}_{79}$Au	—	−29.59	β^-	2.70 day	β^-: 0.962 max (99%), 0.286 (1%) γ: 0.412 (95%), 0.676 (1%), 1.088 (0.2%) e^-: 0.329, 0.398
$^{198}_{80}$Hg	10.02	−30.97	—	—	—
$^{203}_{80}$Hg	—	−25.26	β^-	46.9 day	β^-: 0.214 max γ: 0.279 (77%) e^-: 0.194, 0.264, 0.275
$^{210}_{84}$Po	—	−15.95	α	138 day	α: 5.305 (100%) γ: 0.803 (0.0011%)
$^{218}_{84}$Po	—	8.38	α	3.05 min	α: 6.00 Daughter radiations from ^{214}Pb, ^{214}Bi, ^{214}Po
$^{222}_{86}$Rn	—	16.39	α	3.82 day	α: 5.49 γ: 0.510 (0.07%) Daughter radiations from ^{218}Po, ^{214}Pb, ^{214}Bi, ^{214}Po

Appendix D (Continued)

Nuclide	Natural abundance (%)	Mass difference $\Delta = M - A$ (MeV) (at. mass − at. mass No.)	Type of decay	Half-life	Major radiations, Energies (MeV), and Frequency per disintegration (%)
$^{226}_{88}$Ra	—	23.69	α	1602 yr	α: 4.78 (95%), 4.60 (5%) γ: 0.186 (4%), 0.26 (0.007%), Rn X-rays e^-: 0.087, 0.170 Daughter radiations from ^{222}Rn, ^{218}Po, ^{214}Pb, ^{214}Bi, ^{214}Po
$^{235}_{92}$U	0.720	40.93	α	7.1×10^8 yr	α: 4.58 (8%), 4.40 (57%), 4.37 (18%) γ: 0.143 (11%), 0.185 (54%), 0.204 (5%), Th X-rays Daughter radiations from ^{231}Th, etc.
$^{238}_{92}$U	99.276	47.33	α	4.91×10^9 yr	α: 4.20 (75%), 4.15 (25%) γ: Th X-rays e^-: 0.030, 0.043 Daughter radiations from ^{234}Th, ^{234m}Pa
$^{239}_{94}$Pu	—	48.60	α	2.44×10^4 yr	α: 5.16 (88%), 5.11 (11%) γ: 0.039 (0.007%), 0.052 (0.020%), 0.129 (0.007%), plus others, U X-rays e^-: 0.008, 0.019, 0.033, 0.047

†Adapted from *Radiological Health Handbook*, U.S. Public Health Service Publ. No. 2016, Bureau of Radiological Health, Rockville, MD (1970).

APPENDIX E
STATISTICS

Radioactive decay and particle detection are two examples of common statistical phenomena in health physics. Some concepts and formulas are summarized here.

Let p be the probability that a certain random event occurs when a single trial is made and let q be the probability that the event does not occur. Then $p + q = 1$. The terms in the binomial expansion of $(p + q)^k$ give the probabilities that the event will occur exactly a given number of times, n, when k random trials are made:

$$(p + q)^k = p^k + kp^{k-1}q + \frac{k(k-1)}{2!} p^{k-2}q^2$$

$$+ \frac{k(k-1)(k-2)}{3!} p^{k-3}q^3 + \ldots + q^k. \qquad (E.1)$$

The first term, p^k, is the probability that the event will happen $n = k$ times in k trials; the second term, $kp^{k-1}q$, is the probability that the event will happen $n = k - 1$ times in k trials; and so on. The last term, q^k, is the probability that the event will not occur at all ($n = 0$) in k trials.

Example
A die is thrown four times. What are the probabilities that a 6 will appear $n = 4, 3, 2, 1,$ or 0 times?

Solution
The probability of throwing a 6 in a single roll with a nonloaded die is $p = \frac{1}{6}$. The probability of not throwing a 6 is $q = \frac{5}{6}$. Equation (E.1) gives, for $k = 4$,

$$\left(\frac{1}{6} + \frac{5}{6}\right)^4 = \left(\frac{1}{6}\right)^4 + 4\left(\frac{1}{6}\right)^3\left(\frac{5}{6}\right) + \frac{4\cdot3}{2!}\left(\frac{1}{6}\right)^2\left(\frac{5}{6}\right)^2 + \frac{4\cdot3\cdot2}{3!}\left(\frac{1}{6}\right)\left(\frac{5}{6}\right)^3 + \frac{4\cdot3\cdot2\cdot1}{4!}\left(\frac{5}{6}\right)^4 \qquad (E.2)$$

$$= \frac{1}{1296} + \frac{20}{1296} + \frac{150}{1296} + \frac{500}{1296} + \frac{625}{1296} = 1. \qquad (E.3)$$

The five fractions give, respectively, the probabilities of there being exactly $n = 4, 3, 2, 1,$ or 0 6s in the four rolls. Note that the terms sum to unity, as they must.

When k becomes very large, the probabilities for different values of n approach a normal, or Gaussian, distribution about the mean $\bar{n}$. The probability that the event will occur exactly n times is then given by

$$p(n) = \frac{1}{\sigma\sqrt{2\pi}} e^{-(n-\bar{n})^2/2\sigma^2}, \qquad (E.4)$$

where σ is the standard deviation. For practical purposes, the normal distribution can be used in place of the binomial when $n \gtrsim 30$. The standard deviation gives a measure of the spread of the distribution. The area under the symmetric curve (E.4) is unity. Table E.1

Table E.1. Areas Under Various Intervals About the Mean $\bar{n}$ of Normal
Distribution with Standard Deviation σ

Interval	Area
$\bar{n} \pm 0.674\sigma$	0.500
$\bar{n} \pm 1.00\sigma$	0.683
$\bar{n} \pm 1.65\sigma$	0.900
$\bar{n} \pm 1.96\sigma$	0.950
$\bar{n} \pm 2.00\sigma$	0.954
$\bar{n} \pm 2.58\sigma$	0.990
$\bar{n} \pm 3.00\sigma$	0.997

gives the area contained under different intervals about the mean $\bar{n}$. The areas give cumulative probabilities; e.g., there is a 5% chance that the random occurrence will fall outside the interval $\bar{n} \pm 1.96\sigma$.

If $p(n) \ll 1$, the binomial distribution approaches the Poisson,

$$p(n) = \frac{\bar{n}^n e^{-\bar{n}}}{n!}. \tag{E.5}$$

Whereas the normal distribution (E.4) depends upon the two parameters $\bar{n}$ and σ, the Poisson formula (E.5) contains only the single parameter $\bar{n}$, which must therefore also govern the spread. The standard deviation of the Poisson distribution is given by the square root of the mean,

$$\sigma = \sqrt{\bar{n}}. \tag{E.6}$$

When $n \gtrsim 20$, the Poisson and normal distributions very nearly coincide, and Table E.1 applies with the simplification (E.6). The relative error is

$$R = \frac{\sigma}{\bar{n}} = \frac{1}{\sqrt{\bar{n}}}. \tag{E.7}$$

Example
A sample of a long-lived radionuclide gives 1018 counts in 5 min. (a) What is the mean count rate? (b) What is the relative error? (c) What is the relative error if the sample is counted for 10 min? (d) How long should the sample be counted to be 99% certain that the measured count rate is within 5% of the true count rate? Background is negligible.

Solution
(a) The mean count rate is 1018 counts/$(5 \times 60 \text{ sec}) = 3.39$ cps. (b) We use the original number of counts as an estimate of the mean, $\bar{n} = 1018$, that one would find in a large number of 5 min trials. The relative error is then $R = 1/\sqrt{1018} = 0.0313 = 3.13\%$. (c) If the sample is counted for 10 min (i.e., twice as long), then $\bar{n} = 2 \times 1018 = 2036$ and $R = 1/\sqrt{2036} = 0.0222 = 2.22\%$. The relative error thus depends on the total number of counts taken and can be made smaller by counting for longer times. (d) According to Table E.1 to attain 99% confidence, the uncertainty must be written $\pm 2.58\ \sigma = \pm 2.58\sqrt{\bar{n}}$. If the uncertainty is 5%, then the number $\bar{n}$ of counts needed is given by

$$0.05\ \bar{n} = 2.58\sqrt{\bar{n}}. \tag{E.8}$$

It follows that

$$\bar{n} = \left(\frac{2.58}{0.05}\right)^2 = 2663, \tag{E.9}$$

giving $t = 2663/3.39 = 786$ sec $= 13.1$ min.

The use of Poisson statistics requires that $p(n)$ in Eq. (E.5) be very small compared with unity. This is usually the case in counting a radioactive sample, where a large number of atoms are present, or in making pulse-height measurements, when a large number of ions are collected. In this example, if the sample is counted for only 10 sec, then $\bar{n} = 33.9 \cong 34$. The probability of observing exactly this number of counts is, by Eq. (E.5),

$$p(34) = \frac{34^{34}e^{-34}}{34!} = \frac{1.18 \times 10^{52} \times 1.71 \times 10^{-15}}{2.95 \times 10^{38}} = 0.0683. \tag{E.10}$$

Alternatively, since the Poisson and Gaussian distributions are nearly the same when $n > 20$, we can apply Eq. (E.4) with $n = \bar{n} = 34$ and $\sigma = \sqrt{34}$. Then, simply, $p(34) = 1/\sqrt{68\pi} = 0.0684$.

Usually, in determining the activity of a radioactive sample, allowance must be made for the counts registered from natural background. Typically, a gross count rate r_g is observed by counting for a time t_g and a background count rate r_{bg} is measured, in the absence of the sample, over a longer time t_{bg}. The net count rate, due to the sample alone, is then $r_n = r_g - r_{bg}$. The standard deviation in the net count rate is given by

$$\sigma_n = \sqrt{\sigma_g^2 + \sigma_{bg}^2}, \tag{E.11}$$

where σ_g and σ_{bg} are the standard deviations in the gross and background count rates. The standard deviation in the *number* n_g of gross counts registered is $\sqrt{n_g}$, and so the standard deviation in the gross *count rate* is $\sigma_g = \sqrt{n_g}/t_g$. Since $n_g = r_g t_g$, it follows that

$$\sigma_g = \sqrt{\frac{r_g}{t_g}}. \tag{E.12}$$

A similar equation holds for σ_{bg}. Therefore, in place of Eq. (E.11) we may write

$$\sigma_n = \sqrt{\frac{r_g}{t_g} + \frac{r_{bg}}{t_{bg}}}. \tag{E.13}$$

Example
A sample is placed in a counter for 10 min and 1426 counts are registered. The sample is then removed and 2561 background counts are observed in 90 min. (a) What is the net count rate of the sample and its standard deviation? (b) How long would the sample have to be counted in order to be 95% certain that the measured count rate is within ±5% of its true value?

Solution
(a) With $t_g = 10$ min and $t_{bg} = 90$ min, the gross and background count rates are $r_g = 1426/10 = 143$ cpm and $r_{bg} = 2561/90 = 28.5$ cpm. The net count rate is $r_n = 143 - 28.5 = 115$ cpm. The standard deviation in the net count rate is, by Eq. (E.13),

$$\sigma_n = \sqrt{\frac{143}{10} + \frac{28.5}{90}} = 3.82 \text{ cpm.} \tag{E.14}$$

(b) A 5% uncertainty in the net count rate is $0.05 \, r_n = 0.05(115) = 5.75$ cpm. For the true net count rate to be within this range of the mean at the 95% confidence level (1.96σ) requires that $1.96 \, \sigma_n = 5.75$, or $\sigma_n = 2.93$ cpm. We see from Eq. (E.14) that σ_n can be reduced to the desired value by counting the sample for a time t_g such that

$$\sigma_n = \sqrt{\frac{143}{t_g} + \frac{28.5}{90}} = 2.93 \text{ cpm.} \tag{E.15}$$

Solving gives $t_g = 17.3$ min. Note that σ_n can also be reduced by taking a longer background count.

In pulse-height measurements, energy resolution at a peak is commonly expressed as the full width at half-maximum (FWHM). An example is shown in Fig. 9.25. If the peak has Gaussian shape, then FWHM = 2.35σ. In the Poisson limit, the fractional resolution is given by

$$\frac{\text{FWHM}}{\bar{n}} = \frac{2.35}{\sqrt{\bar{n}}}. \tag{E.16}$$

Multiplication by 100 then gives the percentage resolution, as shown in Fig. 9.25.

Measurements show that considerably better energy resolution can be achieved than that predicted by Eq. (E.16), which assumes that individual events occur randomly. The departure of ionization events from complete randomness is not surprising when we consider that a particle has only a finite amount of energy to spend. Energy conservation alone limits the number of ionizations that it can produce. The departure of observed statistical fluctuations from a Poisson distribution can be described quantitatively. In pulse-height measurements, the Fano factor F is defined as the ratio of the observed variance σ^2 and the variance $\bar{n}$ predicted by Poisson statistics. If k pulses are observed with the distribution $n_1, n_2, \ldots, n_k$ in the number of ions, then the variance is

$$\sigma^2 = \frac{1}{k-1} \sum_{i=1}^{k} (n_i - \bar{n})^2, \tag{E.17}$$

and the Fano factor is given by

$$F = \frac{\sigma^2}{\bar{n}} = \frac{1}{\bar{n}(k-1)} \sum_{i=1}^{k} (n_i - \bar{n})^2. \tag{E.18}$$

In terms of F, the resolution (E.16) becomes

$$\frac{\text{FWHM}}{\bar{n}} = \frac{2.35\sigma}{\bar{n}} = 2.35 \sqrt{\frac{F}{\bar{n}}}. \tag{E.19}$$

The Fano factor can be considerably less than unity in proportional counters and semiconductor detectors. It tends to be about unity in scintillation detectors.

GENERAL SUBJECT INDEX

ABOUT THE AUTHOR

James E. Turner is a senior research scientist at Oak Ridge National Laboratory and a part-time professor of physics at the University of Tennessee. He received a B.A. degree from Emory University, an M.S. from Harvard University, and a Ph.D. in physics from Vanderbilt University. He was also a Fulbright scholar at the Georg-August Universität in Göttingen, Federal Republic of Germany. Before joining Oak Ridge National Laboratory in 1962, he was an instructor of physics at Yale and a radiological physicist with the Atomic Energy Commission. Dr. Turner has served as a member of the National Council on Radiation Protection and Measurements and on the Board of Directors of the Health Physics Society. He was an editor of *Health Physics* and an associate editor of *Radiation Research*. He has been a certified health physicist since 1966.